# Photonic Crystals - Introduction, Theory and Applications

# Photonic Crystals - Introduction, Theory and Applications

**Editor:**

Klaus Petritsch, Ph.D.

ArclerPress

www.arclerpress.com

**Photonic Crystals - Introduction, Theory and Applications**

*Editor: Klaus Petritsch, Ph.D.*



**708 3rd Avenue, 6th Floor**
**New York**
**NY 10017**
**United States of America**
**www.arclerpress.com**

ISBN: 978-1-68094-365-8

Library of Congress Control Number: 2016950945

# About the Editor

## Klaus Petritsch, Ph.D.

Klaus obtained his PhD in science after working as research assistant at the University of Cambridge, UK. He published over 20 papers, gave presentations to large international science audiences, taught physics, and trained technical staff in a Silicon Valley solar energy start up. In Finland, a global leader in education, Klaus complemented his degrees with certificates in vocational teaching, instructional design, and eLearning. He regularly conducts workshops on learning at one of Finland's leading universities in teacher education and freelances as learning consultant, editor and instructional designer.

# Contents

# List of Contributors

**Claudia Pacholski**
Department of New Materials and Biosystems, Max Planck Institute for Intelligent Systems, Heisenbergstr. 3, Stuttgart 70569, Germany

**Mark G. Scullion**
SUPA, School of Physics and Astronomy, University of St Andrews, St Andrews, Scotland KY16 9SS, UK

**Thomas F. Krauss**
SUPA, School of Physics and Astronomy, University of St Andrews, St Andrews, Scotland KY16 9SS, UK

**Andrea Di Falco**
SUPA, School of Physics and Astronomy, University of St Andrews, St Andrews, Scotland KY16 9SS, UK

**Longju Liu**
Department of Electrical and Computer Engineering, Iowa State University, Ames, Iowa 50011, USA

**JingxiangZhang**
Department of Electrical and Computer Engineering, Iowa State University, Ames, Iowa 50011, USA

**MohsinAli Badshah**
School of Mechanical Engineering, Chung-Ang University, Seoul 156-756, Republic of Korea

**Liang Dong[1]**
Department of Mechanical Engineering, University of Hawaii, Honolulu, Hawaii 96822, USA

**Jingjing Li**
Department of Mechanical Engineering, University of Hawaii, Honolulu, Hawaii 96822, USA

**Seok-min Kim**
School of Mechanical Engineering, Chung-Ang University, Seoul 156-756, Republic of Korea

**Meng Lu**
Department of Electrical and Computer Engineering, Iowa State University, Ames, Iowa 50011, USA
Department of Mechanical Engineering, Iowa State University, Ames, Iowa 50011, USA

**Je´re´mie Teyssier**
Department of Quantum Matter Physics, University of Geneva, Geneva 1211, Switzerland

**Suzanne V. Saenko**
Laboratory of Artificial and Natural Evolution (LANE), Department of Genetics and Evolution, University of Geneva, Sciences III, 30, Quai Ernest-Ansermet, Geneva 1211, Switzerland

**Dirk van der Marel**
Department of Quantum Matter Physics, University of Geneva, Geneva 1211, Switzerland

**Michel C. Milinkovitch**
Laboratory of Artificial and Natural Evolution (LANE), Department of Genetics and Evolution, University of Geneva, Sciences III, 30, Quai Ernest-Ansermet, Geneva 1211, Switzerland

**Zhi-Hui Chen**
Key Lab of Advanced Transducers and Intelligent Control System, Ministry of Education and Shanxi Province, College of Physics and Optoelectronics, Taiyuan University of Technology, Taiyuan 030024, China
State Key Laboratory of Information Photonics and Optical Communications, Beijing University of Posts and Telecommunications, Beijing 100876, China.

**Na Qiao**
Key Lab of Advanced Transducers and Intelligent Control System, Ministry of Education and Shanxi Province, College of Physics and Optoelectronics, Taiyuan University of Technology, Taiyuan 030024, China

**YibiaoYang**
Key Lab of Advanced Transducers and Intelligent Control System, Ministry of Education and Shanxi Province, College of Physics and Optoelectronics, Taiyuan University of Technology, Taiyuan 030024, China

**HanYe**
State Key Laboratory of Information Photonics and Optical Communications, Beijing University of Posts and Telecommunications, Beijing 100876, China.

**Shaoding Liu**
Key Lab of Advanced Transducers and Intelligent Control System, Ministry of Education and Shanxi Province, College of Physics and Optoelectronics, Taiyuan University of Technology, Taiyuan 030024, China

**Wenjie Wang**
Key Lab of Advanced Transducers and Intelligent Control System, Ministry of Education and Shanxi Province, College of Physics and Optoelectronics, Taiyuan University of Technology, Taiyuan 030024, China

**Yuncai Wang**
Key Lab of Advanced Transducers and Intelligent Control System, Ministry of Education and Shanxi Province, College of Physics and Optoelectronics, Taiyuan University of Technology, Taiyuan 030024, China

**J S Perez-Huerta**
Instituto de Ciencias F´ısicas, Universidad Nacional Autonoma de M ´ exico, ´ Apartado Postal 48-3, 62251 Cuernavaca, Morelos, Mexico

**Guillermo P Ortiz**
Departamento de F´ısica, Facultad de Ciencias Exactas, Naturales y Agrimensura, Universidad Nacional del Nordeste, Avenida Libertad 5400 Campus-UNNE, W3404AAS Corrientes, Argentina

**Bernardo S Mendoza**
Division of Photonics, Centro de Investigaciones en Optica, Leon, Guanajuato, ´ Mexico

**W Luis Mochan**
Instituto de Ciencias F´ısicas, Universidad Nacional Autonoma de M ´ exico, ´ Apartado Postal 48-3, 62251 Cuernavaca, Morelos, Mexico

**Kang Xie**
School of Instrument Science and Opto-electronic Engineering, Hefei University of Technology, Hefei, Anhui 230009, China

**Wei Zhang**
School of Instrument Science and Opto-electronic Engineering, Hefei University of Technology, Hefei, Anhui 230009, China

**Allan D Boardman**
Department of Physics, University of Salford, Salford, Manchester M5 4WT, UK

**Haiming Jiang**
School of Instrument Science and Opto-electronic Engineering, Hefei University of Technology, Hefei, Anhui 230009, China

**Zhijia Hu**
School of Instrument Science and Opto-electronic Engineering, Hefei University of Technology, Hefei, Anhui 230009, China
Department of Polymer Science and Engineering, University of Science and Technology of China, Hefei, Anhui 230026, China

**Yong Liu**
School of Optoelectronic Information, University of Electronic Science and Technology of China, Chengdu, Sichuan 610054, China

**Ming Xie**
Department of Physics, University of Salford, Salford, Manchester M5 4WT, UK

**Qiuping Mao**
School of Instrument Science and Opto-electronic Engineering, Hefei University of Technology, Hefei, Anhui 230009, China

**Lei Hu**
School of Instrument Science and Opto-electronic Engineering, Hefei University of Technology, Hefei, Anhui 230009, China

**Qian Li**
School of Instrument Science and Opto-electronic Engineering, Hefei University of Technology, Hefei, Anhui 230009, China

**Tianyu Yang**
School of Instrument Science and Opto-electronic Engineering, Hefei University of Technology, Hefei, Anhui 230009, China

**Fei Wen**
School of Instrument Science and Opto-electronic Engineering, Hefei University of Technology, Hefei, Anhui 230009, China

**Erlei Wang**
School of Instrument Science and Opto-electronic Engineering, Hefei University of Technology, Hefei, Anhui 230009, China

**Ping-Yuan Lo**
Department of Physics and Center for Quantum Information Science, National Cheng Kung University, Tainan 70101, Taiwan, Republic of China.

**Heng-Na Xiong**
Department of Physics and Center for Quantum Information Science, National Cheng Kung University, Tainan 70101, Taiwan, Republic of China.

**Wei-Min Zhang**
Department of Physics and Center for Quantum Information Science, National Cheng Kung University, Tainan 70101, Taiwan, Republic of China.

**Mikhail V. Rybin**
Ioffe Institute, St Petersburg 194021, Russia
Department of Nanophotonics and Metamaterials, ITMO University, St Petersburg 197101, Russia

**Dmitry S. Filonov**
Department of Nanophotonics and Metamaterials, ITMO University, St Petersburg 197101, Russia

**Kirill B. Samusev**
Ioffe Institute, St Petersburg 194021, Russia
Department of Nanophotonics and Metamaterials, ITMO University, St Petersburg 197101, Russia

**Pavel A. Belov**
Department of Nanophotonics and Metamaterials, ITMO University, St Petersburg 197101, Russia

**Yuri S. Kivshar**
Department of Nanophotonics and Metamaterials, ITMO University, St Petersburg 197101, Russia
Nonlinear Physics Center and the ARC Center of Excellence CUDOS, Australian National University, Canberra Australian Capital Territory 0200, Australia

**Mikhail F. Limonov**
Ioffe Institute, St Petersburg 194021, Russia
Department of Nanophotonics and Metamaterials, ITMO University, St Petersburg 197101, Russia

**Francesco Pagliano**
Applied Physics Department, COBRA Research Institute, Eindhoven University of Technology, 5600 MB Eindhoven, The Netherlands

**YongJin Cho**
Applied Physics Department, COBRA Research Institute, Eindhoven University of Technology, 5600 MB Eindhoven, The Netherlands

**Tian Xia**
Applied Physics Department, COBRA Research Institute, Eindhoven University of Technology, 5600 MB Eindhoven, The Netherlands

**Frank van Otten**
Applied Physics Department, COBRA Research Institute, Eindhoven University of Technology, 5600 MB Eindhoven, The Netherlands

**Robert Johne**
Applied Physics Department, COBRA Research Institute, Eindhoven University of Technology, 5600 MB Eindhoven, The Netherlands
Max-Planck-Institute for the Physics of Complex Systems, No¨thnitzer Strae 38, 01187 Dresden, Germany

**Andrea Fiore**
Applied Physics Department, COBRA Research Institute, Eindhoven University of Technology, 5600 MB Eindhoven, The Netherlands

**V. Peano**
Institute for Theoretical Physics, Friedrich-Alexander-Universität Erlangen-Nürnberg, Staudtstraße 7, 91058 Erlangen, Germany

**C. Brendel**
Institute for Theoretical Physics, Friedrich-Alexander-Universität Erlangen-Nürnberg, Staudtstraße 7, 91058 Erlangen, Germany

**M. Schmidt**
Institute for Theoretical Physics, Friedrich-Alexander-Universität Erlangen-Nürnberg, Staudtstraße 7, 91058 Erlangen, Germany

**F. Marquardt**
Institute for Theoretical Physics, Friedrich-Alexander-Universität Erlangen-Nürnberg, Staudtstraße 7, 91058 Erlangen, Germany
Max Planck Institute for the Science of Light, Günther-Scharowsky-Straße 1, 91058 Erlangen, Germany

**C-L Hung**
Norman Bridge Laboratory of Physics 12-33, California Institute of Technology, Pasadena, CA 91125, USA
Institute for Quantum Information and Matter, California Institute of Technology, Pasadena, CA 91125, USA

**S M Meenehan**
Institute for Quantum Information and Matter, California Institute of Technology, Pasadena, CA 91125, USA
Thomas J Watson, Sr, Laboratory of Applied Physics, California Institute of Technology, Pasadena, CA 91125, USA

**D E Chang**
ICFO—Institut de Ciencies Fotoniques, Mediterranean Technology Park, E-08860 Castelldefels, Barcelona, Spain

**O Painter**
Institute for Quantum Information and Matter, California Institute of Technology, Pasadena, CA 91125, USA
Thomas J Watson, Sr, Laboratory of Applied Physics, California Institute of Technology, Pasadena, CA 91125, USA

**H J Kimble**
Norman Bridge Laboratory of Physics 12-33, California Institute of Technology, Pasadena, CA 91125, USA
Institute for Quantum Information and Matter, California Institute of Technology, Pasadena, CA 91125, USA

**Qian Liu**
Department of System Engineering, University of Arkansas at Little Rock, 2801 S University Ave, Little Rock, AR 72204, USA

**Eric Sandgren**
Department of System Engineering, University of Arkansas at Little Rock, 2801 S University Ave, Little Rock, AR 72204, USA

**Miles Barnhart**
Department of Mechanical & Aerospace Engineering, University of Missouri, E3422, Lafferre Hall, Columbia, MO 65211, USA

**Rui Zhu**
Department of Mechanical & Aerospace Engineering, University of Missouri, E3422, Lafferre Hall, Columbia, MO 65211, USA

**Guoliang Huang**
Department of Mechanical & Aerospace Engineering, University of Missouri, E3422, Lafferre Hall, Columbia, MO 65211, USA

**Cihui Liu**
State Key Laboratory of Bioelectronics, School of Biological Science and Medical Engineering, Southeast University, Nanjing 210096, China

**Xing Liu**
State Key Laboratory of Bioelectronics, School of Biological Science and Medical Engineering, Southeast University, Nanjing 210096, China

**Hongyun Xuan**
State Key Laboratory of Bioelectronics, School of Biological Science and Medical Engineering, Southeast University, Nanjing 210096, China

**Jiaoyu Ren**
State Key Laboratory of Bioelectronics, School of Biological Science and Medical Engineering, Southeast University, Nanjing 210096, China

**Liqin Ge**
State Key Laboratory of Bioelectronics, School of Biological Science and Medical Engineering, Southeast University, Nanjing 210096, China

**Marcelo Wu**
Institute for Quantum Science and Technology, University of Calgary, Calgary, Alberta T2N 1N4, Canada
National Institute for Nanotechnology, Edmonton, Alberta T6G 2M9, Canada

**Aaron C. Hryciw**
National Institute for Nanotechnology, Edmonton, Alberta T6G 2M9, Canada

**Chris Healey**
Institute for Quantum Science and Technology, University of Calgary, Calgary, Alberta T2N 1N4, Canada
National Institute for Nanotechnology, Edmonton, Alberta T6G 2M9, Canada

**David P. Lake**
Institute for Quantum Science and Technology, University of Calgary, Calgary, Alberta T2N 1N4, Canada

**Harishankar Jayakumar**
Institute for Quantum Science and Technology, University of Calgary, Calgary, Alberta T2N 1N4, Canada

**Mark R. Freeman**
National Institute for Nanotechnology, Edmonton, Alberta T6G 2M9, Canada
Department of Physics, University of Alberta, Edmonton, Alberta T6G 2E9, Canada

**John P. Davis**
Department of Physics, University of Alberta, Edmonton, Alberta T6G 2E9, Canada

**Paul E. Barclay**
Institute for Quantum Science and Technology, University of Calgary, Calgary, Alberta T2N 1N4, Canada
National Institute for Nanotechnology, Edmonton, Alberta T6G 2M9, Canada

**Peifeng Jing**
Department of Electrical Engineering, University of Washington, Seattle WA, 98195, USA

**JingdaWu**
Department of Electrical Engineering, University of Washington, Seattle WA, 98195, USA

**GaryW. Liu**
Department of Bioengineering, University of Washington

**EthanG. Keeler**
Department of Electrical Engineering, University of Washington, Seattle WA, 98195, USA

**Suzie H. Pun**
Department of Bioengineering, University of Washington

**LihY. Lin**
Department of Electrical Engineering, University of Washington, Seattle WA, 98195, USA

**Yin Fang**
Department of Chemical Engineering, University of Florida, Museum Road, Gainesville, Florida 32611, USA

**Yongliang Ni**
Department of Mechanical and Aerospace Engineering, University of Florida, Gainesville, Florida 32611, USA

**Sin-Yen Leo**
Department of Chemical Engineering, University of Florida, Museum Road, Gainesville, Florida 32611, USA

**Curtis Taylor**
Department of Mechanical and Aerospace Engineering, University of Florida, Gainesville, Florida 32611, USA

**Vito Basile**
ITIA-CNR, Industrial Technologies and Automation Institute, National Council of Research, Via Bassini, 15, 20133 Milano, Italy

**Peng Jiang**
Department of Chemical Engineering, University of Florida, Museum Road, Gainesville, Florida 32611, USA

**Ming-Xing Luo**
Information Security and National Computing Grid Laboratory, Southwest Jiaotong University, Chengdu 610031, China

**Hui-Ran Li**
Information Security and National Computing Grid Laboratory, Southwest Jiaotong University, Chengdu 610031, China

**Hong Lai**
School of Computer and Information Science, Southwest University, Chongqing 400715, China

**XiaojunWang**
School of Electronic Engineering, Dublin City University, Dublin 9, Ireland

# Preface

Photonics, the science and technology of generating, controlling, and detecting photons, has brought mankind laser surgery and manufacturing, global fiber telecommunication, and more recently, optical sensing, measuring, controlling, and the beginning of optical computing. The 21st century could well become the age of photonics and metamaterials, like the 20th century was the age of electronics and semiconductors. The United Nations declared 2015 the "International Year of Light and Light-based Technologies".

Photonic crystals are periodic optical structures (in one, two or three dimensions) that can control the flow of light in the same way that the periodic potential in a semiconductor crystal affects the flow of electrons. Since Eli Yablonovitch and Sajeev John have proposed both the term "photonic crystal" and the light-photon analogy in 1987, scientists have used band theory to describe many properties of photonic crystals. A simple and important example is the analogous use of allowed and forbidden energy bands: if light is forbidden to propagate inside a lattice, i.e. its frequency falls within the photonic bandgap, the lattice acts like a mirror throughout, which can be used to create effective waveguides that are already used commercially. These waveguides use interference instead of the more conventional internal reflection. Photonic crystal devices can also do many other interesting things such as guiding light around sharp bends or even trap it entirely. Other conventional concepts like reciprocal space, Brillouin zones, Bloch wave functions, and dispersion relations are also used to describe optical waves.

The first part of this book reviews basic concepts of photonic crystals in the context of sensor applications. This is followed by a discussion how absorption in photonic crystals can be tuned or created through an artificial molding process or naturally, through emotionally triggered skin modifications like those performed by chameleons. Effectively enhanced broadband absorption is then discussed, followed by an efficient method to calculate macroscopic optical responses of periodic composite systems. The latter marks the beginning of the more theoretical part of the book: light guiding beyond photonic bandgaps, a breakdown of the Bose-Einstein distribution, a phase diagram to distinguish photonic crystals form dielectric metamaterials, dynamically controlled emission of single excitons in photonic crystal cavities, topological phases of sound and light, and trapped atoms in one-dimensional photonic crystals. The last part of this collection aims to provide insight into the extraordinary wide range of powerful applications of photonic crystals such as optimized solar cells, "smart" supercapacitors, a torque sensor, photonic tweezers, pressure-responsive shape memory, and controlled NOT gates for the use in future quantum computers.

**Editor**
**Klaus Petritsch, Ph.D.**

# CHAPTER 1 PHOTONIC CRYSTAL SENSORS BASED ON POROUS SILICON

Claudia Pacholski

Department of New Materials and Biosystems, Max Planck Institute for Intelligent Systems, Heisenbergstr. 3, Stuttgart 70569, Germany

## ABSTRACT

**Porous silicon has been established as an excellent sensing platform for the optical detection of hazardous chemicals and biomolecular interactions such as DNA hybridization, antigen/antibody binding, and enzymatic reactions. Its porous nature provides a high surface area within a small volume, which can be easily controlled by changing the pore sizes. As the porosity and consequently the refractive index of an etched porous silicon layer depends on the electrochemial etching conditions photonic crystals composed of multilayered porous silicon films with well-resolved and narrow optical reflectivity features can easily be obtained. The prominent optical response of the photonic crystal decreases the detection limit and therefore increases the sensitivity of porous silicon sensors in comparison to sensors utilizing Fabry-Pérot based optical transduction. Development of porous silicon photonic crystal sensors which allow for the detection of analytes by the naked eye using a simple color change or the fabrication of stacked porous silicon photonic crystals showing two distinct optical features which can be utilized for the discrimination of analytes emphasize its high application potential.**

## INTRODUCTION

Optical sensors for the detection and quantification of hazardous chemicals, investigation of biomolecular interactions or studies on cellular systems have been developed for decades and are still a field of extensive research (reviewed for example in Reference [1]). Label-free optical sensors use mainly surface plasmon resonance and interferometry as transduction methods whose performance complements each other. A recent study on the sensitivity of localized surface plasmon resonance (LSPR) transducers in comparison to interferometric sensors identified the superiority of LSPR based devices for the analysis of thin (several nm) analyte and recognition interfaces and emphasized the advantage of interferometric sensors for the investigation of thicker layers [2]. The high potential of porous silicon for fabrication of interferometric sensors originates from its easily controllable fabrication process resulting in layers with defined porosity (refractive index), its high surface area, simple surface chemistry, and full compatibility with microprocessing techniques. In the early stages porous silicon based sensors were composed of a single porous layer on the silicon substrate leading to Fabry-Pérot interference. Here, the reflectivity spectrum shows interference fringes which correspond to constructive and destructive interference from light reflected at the air/porous silicon and porous silicon/crystalline silicon interfaces. Changes in the average refractive index of the porous silicon layer caused by infiltration or adsorption of analytes are detected by spectral shifts in the reflectivity spectrum [3,4].

**Citation:** Claudia Pacholski, Photonic Crystal Sensors Based on Porous Silicon, doi:10.3390/s130404694.

Photonic crystals are composed of alternating regions of high and low dielectric constants and can be obtained in 1, 2, or 3 dimensional periodic array arrangements (Figure 1). By choosing appropriate dielectric constants and geometry these materials can exhibit a photonic band gap (stop band) which is characterized by the prevention of light propagation at a range of frequencies defined by the internal structure of the photonic crystal. Hence, photonic crystals can be employed as optical filters and allow for the isolation of narrow reflection bands. Since the conceptual introduction of photonic crystals by Yablonovitch and John in 1987 [5,6] photonic crystals have been fabricated from diverse materials including semiconductors, polymers, oxides, and porous silicon.

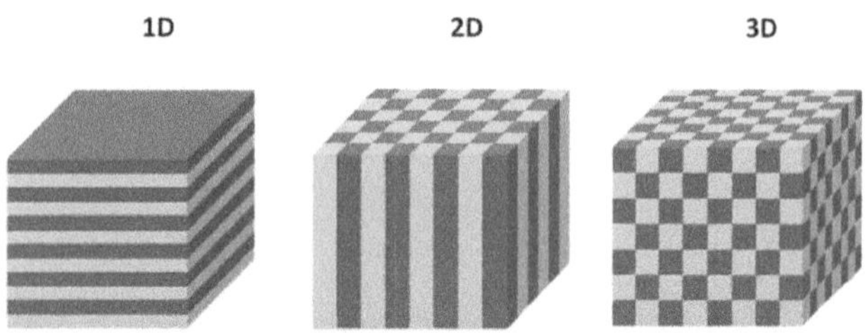


**Figure 1:** Photonic crystals (adapted from Reference [7]).

The implementation of more complicated optical structures, e.g., photonic crystals, into porous silicon based sensors improved their sensing capabilities in two ways. On the one hand the sensitivity and specificity provided by the porous silicon sensor was considerable enhanced. The sharp resonant optical response of the photonic crystal makes it much easier to detect small shifts in the reflectivity spectrum leading to detection limits on the femtomolar level. The incorporation of a lateral porosity gradient provides a size exclusion filter resulting in improved specificity of a porous sensor [8]. On the other hand photonic crystal sensors allow for the detection of analytes by the naked eye. Based on their internal structure photonic crystal solely reflect light at distinct frequencies and therefore appear as a pure color to the eye. Penetration of analytes into the pores consequently cause easily noticeable color changes in the photonic crystal sensors.

## FABRICATION OF POROUS SILICON PHOTONIC CRYSTALS

Porous silicon was accidently discovered in the mid-1950s by Uhlir and Uhlir, who tried to find a convenient method for electropolishing silicon wafers [9]. They found that upon electrochemical etching of silicon wafers in fluoride containing solutions small holes can propagate in the <100> direction in the Si wafer. The overall electrochemical reaction for Si etching is given by Equation (1):

$$Si + 6\,F^- + 2\,H^+ + 2\,h^+ \rightarrow SiF_6^{2-} + H_2 \tag{1}$$

in which $h^+$ is a hole injected into the valence band of the semiconductor. The simplicity of this reaction equation belies the complexity of porous silicon formation which involves electronic as well as chemical factors. Numerous parameters such as the applied voltage, the chosen silicon substrate (dopant type and concentration), the electrolyte composition, temperature and light intensity have a considerable influence on the resulting silicon nanostructure. A detailed discussion of porous silicon formation is beyond the scope of this review and can be found in reference [10]. However, in general pores nucleate randomly but homogenously on the silicon surface upon electrochemical etching leading to pores with a narrow pore diameter distribution. The pore diameters can be easily controlled and varied between a few and several thousands of nanometers. Figure 2(a) shows a schematic of the porous silicon formation process. Etching occurs mainly at the pore tips as holes are directed to the tips by the electric field and etching of the pore walls is prevented by passivation upon etching. Hence, dissolution of silicon is primarily obtained at the porous silicon/crystalline silicon interface. An example for an applied current density versus time waveform for electrochemical etching and a corresponding SEM image of an etched porous silicon layer are displayed in Figure 2(b,c), respectively.

## 1D Photonic Crystals

The first porous silicon based photonic crystal was electrochemically etched into a silicon wafer by Vincent in 1994 [12]. The key to its successful fabrication was the observation that the amount of silicon dissolved at the interface of the pores and the crystalline silicon is directly related to the current passed at that moment. Thus, changes in the applied current result in corresponding changes in the porosity of the formed porous silicon leading to porous silicon layers possessing different refractive indices. Vincent varied the applied current between two values leading to alternating discrete porous silicon layers of two different refractive indices. By carefully choosing appropriate porosities and thicknesses of the porous silicon layers he was able to create a Bragg reflector which provides a photonic bandgap rejection of a wide range of wavelengths of light. This band is called the stop band of the structure and its position is defined by Equation (2):

$$\lambda_{SB} = 4n_iL_i \tag{2}$$

in which the product nL represents the optical thickness which is composed of the refractive index (n) and the thickness (L) of the porous silicon layer. A microcavity is composed of a Fabry Pérot cavity sandwiched between two Bragg reflectors. Thereby the periodicity of the refractive index profile is disturbed and a sharp resonant dip in the stop band is obtained. In contrast to the abrupt changes between two refractive indices in Bragg stacks the refractive index profil in rugate filters is characterized by a smooth, sinusoidal variation in the refractive index. Figure 3 summarizes the presented 1D photonic crystal structures prepared with porous silicon including the approximate current density waveforms and representative spectra.

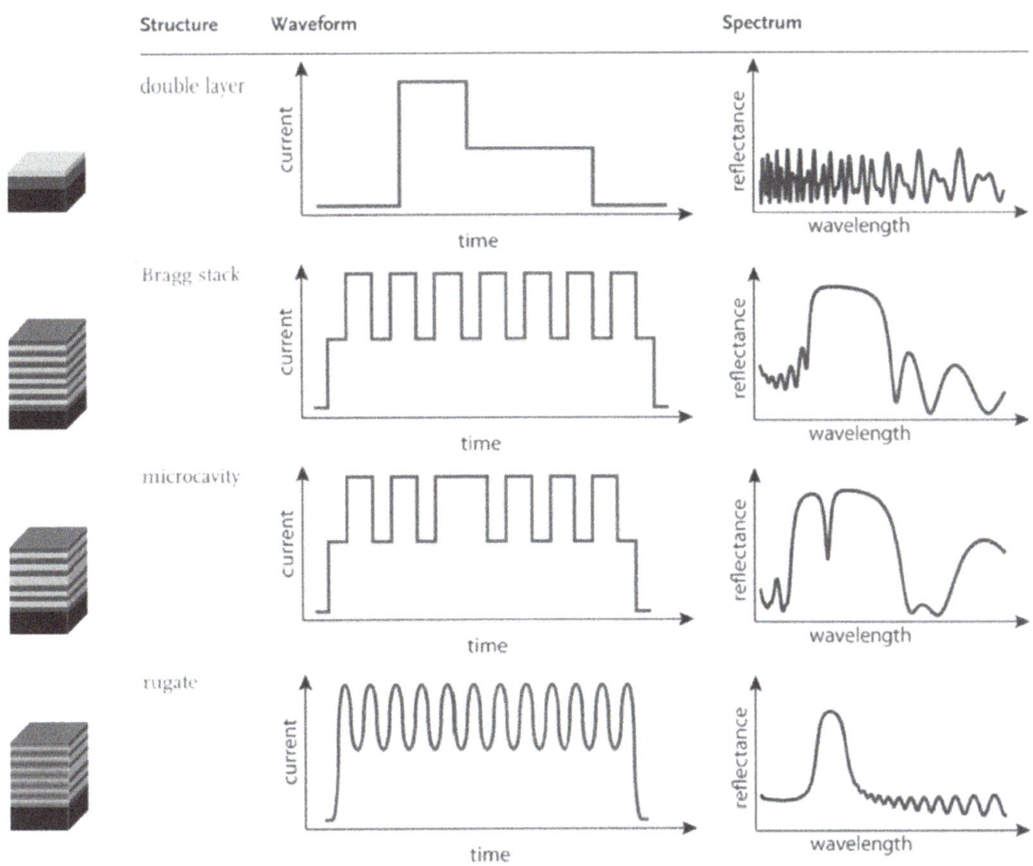


**Figure 3:** Porous silicon based 1D photonic crystals. Reprinted from Reference [13] with permission. Copyright Wiley-VCH 2012.

## 2D and 3D Photonic Crystals

2D photonic crystals are often realized either by a square lattice of rods with high dielectric constant which are embedded in a medium with a low dielectric constant or by a hexagonal lattice of holes with a low dielectric constant surrounded by a material with a high dielectric constant [14]. Lehmann *et al.*electrochemically etched ordered macropores into silicon and thereby fabricated structures which showed a complete 2D bandgap in the near-infrared for the first time [15]. Macroporous silicon is in general obtained by pre-patterning an n-type silicon wafer using standard lithographic methods followed by electrochemical etching in hydrofluoric acid containing solutions under backside illumination. The illumination generates electronic holes which promote the dissolution of silicon at the pore tips leading to the formation of straight pores with very high aspect ratio [16,17]. The pore arrangement in x, y direction is defined by the lithographic mask. The diameter of the pores in depth (z direction) can be varied by changing the illumination intensity, the applied current density and by post-treatment of 2D structures with basic solutions. Thereby 3D photonic crystals based on macroporous silicon were fabricated [18–21]. In Figure 4 schematics of 2D and 3D porous silicon photonic crystals are displayed. Other approaches to 2D and 3D photonic crystals include the utilization of porous silicon as low dielectric constant material in hexagonal arrays or as sacrificial layer for the generation of more complicated structures in silicon [22,23].

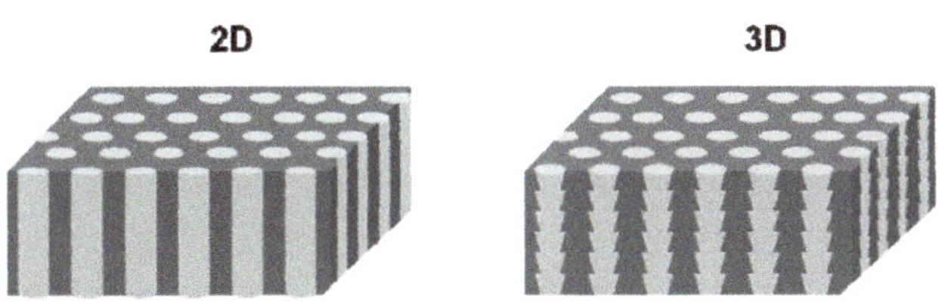


**Figure 4:** Exemplary schematics of 2D and 3D porous silicon photonic crystals.

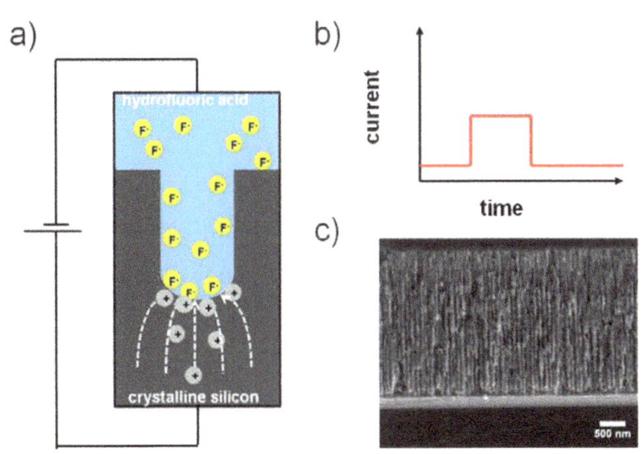


**Figure 2:** Fabrication of porous silicon. (**a**) Schematic of porous silicon formation by electrochemical etching. Adapted from Reference [11]. (**b**) Applied current density versus time waveform and corresponding SEM image of etched a porous silicon layer (**c**).

## OPTICAL SENSORS

Porous silicon films on crystalline silicon show a so-called Fabry-Pérot fringe pattern in their reflectivity spectrum which is caused by constructive and destructive interference of reflected light rays at the interfaces bordering the porous layer. The wavelengths of the fringe maxima $\lambda_{max}$ in the reflectivity spectra can be calculated by using Equation (3):

$$m\lambda_{max} = 2nL \qquad (3)$$

where m is an integer corresponding to the spectral order of the fringe, n is the average refractive index of the porous layer and L is its thickness [24]. Optical sensors based on porous silicon mainly detect analytes by shifts of the fringe pattern on the wavelength scale resulting from changes in the average refractive index of the porous layer. For this purpose the effective optical thickness (EOT, 2 nL) is often estimated by applying a Fourier transform to a plot of reflected light intensity versus frequency. The position of the resulting single peak in the Fourier transform corresponds to the effective optical thickness of the porous layer. Here, the refractive index n is a composite index which is composed of the refractive index of the medium filling the pores and the refractive index of the material from which the porous structure has been created (Si).

Reflectometric sensing modes which have been reported for porous silicon based sensors are summarized in Figure 5. If the refractive index of the medium filling the pores increases, the fringe pattern in the reflectivity spectrum shifts to the red end of the spectrum (longer wavelengths) and consequently to higher EOT values (Figure 5(A)). This sensing mode has extensively been employed for the detection of hazardous chemicals and biomolecular recognition events. A blue shift (shorter wavelengths) of the interference pattern is obtained if the average refractive index of the porous layer decreases (Figure 5(B)). The decrease in refractive index was correlated to the corrosion of the porous silicon caused by analytes such as negatively charged DNA and transition metal complexes. Finally, changes in the amount of reflected light resulting from changes in the refractive index contrast at the porous silicon layer interfaces can be used for the detection of analytes which do not penetrate the porous layer (Figure 5(C)). All of the described reflectometric sensing modes for single-layered porous silicon can be transferred to multi-layer porous structures whose more elaborate optical properties increase the sensitivity of the porous silicon sensor and allow for background correction [13].

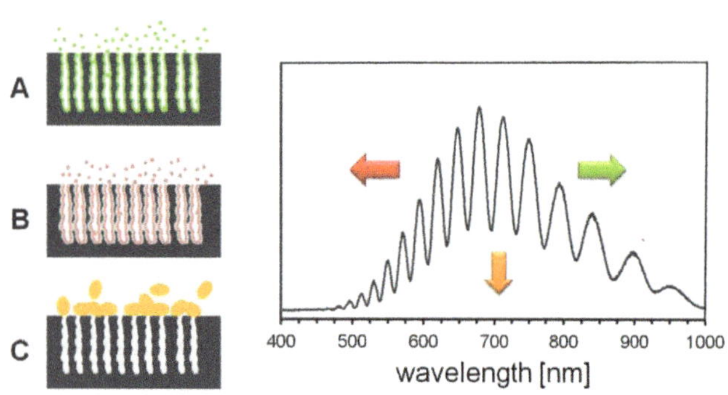


**Figure 5:** Optical sensors based on porous silicon. (**A**) Increase of the average refractive index of the porous layer leads to a shift of the interference pattern to longer wavelengths. (**B**) If the average refractive index of the porous layer decreases the fringe pattern in the reflectivity spectrum shifts to shorter wavelengths. (**C**) Adsorption or binding of the analyte to the medium/porous silicon interphase changes the refractive index contrast and decreases the amount of reflected light.

## Stabilization and Functionalization of Porous Silicon

The surface chemistry of porous silicon is governed by the reactivity of silicon-silicon and silicon-hydrogen bonds which are formed upon electrochemical etching. Both species can react with atmospheric oxygen and water at ambient conditions leading to the oxidation of the porous silicon structure. The resulting $SiO_2$ nanostructure is unstable in water-based media and will especially be dissolved in aqueous solutions containing nucleophilic molecules and tensides. Hence, the porous silicon surface has to be stabilized for (bio)sensing applications. For this purpose two strategies have been developed resulting in the formation of either silicon-oxygen of silicon-carbon bonds (reviewed in [25,26]). Silicon-oxygen bonds are often generated by thermal oxidation in air, ozone oxidation, and oxidation by organic molecules. The oxidized silicon surface presents silanol groups which can be utilized for further functionalization reactions including the introduction of amines and subsequent coupling of biomolecules to the porous silicon surface via the formation of amide bonds. However, the porous $SiO_2$ structure remains prone to degradation in aqueous media despite its functionalization with silanes as all layers containing silicon-oxygen bonds are susceptible to hydrolysis at high pH (higher than pH 7) in general. A more stable surface functionalization was achieved by the formation of silicon carbon bonds. Based on the low electronegativity of carbon the nucleophilic attack of silicon carbon bonds by water or hydroxide is hampered. Silicon-carbon bonds are formed by hydrosilylation, chemical or electrochemical grafting techniques as well as carbonization reactions. Notwithstanding the obvious advantages of porous silicon stabilized by silicon-carbon bonds its application was neglected several years due to its tedious fabrication in oxygen excluding conditions (inert atmosphere).

## 1D Photonic Crystal Sensors

### *Gas Sensing*

Nanoporous material is able to concentrate vapor by microcapillary condensation. This phenomenon can be described using the Kelvin Equation (4) which relates the pore radius r to the partial pressure at which condensation of vapor will be observed at a defined temperature [27]:

$$r = \frac{\gamma V}{RT\ln(P/P^0)} \qquad (4)$$

Here, γ represents the surface tension at the gas/liquid interface, V represents the molar volume of the liquid, R represents the gas constant, P is the observed vapor pressure and $P^0$ the pressure of the vapor at the given temperature T. This relation emphasizes the suitability of porous silicon sensors for the detection of vapors such as volatile organic compounds (VOCs), organophosphate nerve agents, ammonia, and hydrofluoric acid [28]. Zangooie *et al.*, stimulated the following intensive research on optical porous silicon gas sensors in 1997 by studying the optical changes in thin porous layers caused by vapor exposure using spectroscopic ellipsometry [29]. Two years later Snow *et al.* [30] reported on drastic changes in the reflectivity spectra of porous silicon 1D photonic crystals resulting from exposure to volatile VOCs. A Bragg mirror was fabricated by electrochemical etching of silicon and the shift of the reflectivity peak on the wavelength scale over time was monitored. The experiments showed that different organic vapors, e.g., chlorobenzene and acetone, lead to distinct shifts of the reflectivity peak which can be correlated to their refractive index. The results confirmed the condensation of vapor in the pores of the employed porous silicon sensor. A challenging task of until then developed porous silicon gas sensors was the discrimination between vapors of substances with similar refractive indices. Mulloni and Pavesi addressed this issue by the introduction of an optical sensor which detects two different optical properties at the same time [31]. For this purpose a microcavity structure was etched into lightly doped silicon which additionally shows high luminescence efficiency. Exposure of this sensor to organic solvents led to large redshifts of the resonance peak of the microcavity which were attributed to changes in the composite refractive index of the porous layers. At the same time luminescence intensities were measured and correlated with differences in the dielectric functions of the analytes. Thereby an increased specificity and sensitivity of porous silicon based gas sensors for VOCs was achieved. This concept was extended to the fabrication of a porous silicon microcavity sensor which monitored three different parameters, *i.e.*, the resonance wavelength of the microcavity, the luminescence, and the electrical conductance [32]. However, the more complicated sensor lay-out of the multi-parametric sensors and the encouraging performance of porous silicon microcavity sensors detecting only changes in the resonance wavelength pushed the research on the latter in the following years. De Stefano *et al.*, investigated systematically the response of porous silicon microcavity structures on exposure to flammable substances [33]. A deeper insight into capillary condensation in porous silicon optical microcavities was gained by time-resolved sensing of iso-propanol [34]. The experimental results as well as the corresponding numerical simulations of changes in the position of the cavity peak on the wavelength scale demonstrated an immediate and homogenous filling of the pores. Similar behavior was reported for porous silicon rugate structures [35,36]. By removing the porous silicon film from the silicon substrate using ultrasonication microparticles referred to as "smart dust particles" were obtained which were successfully employed for remote sensing of chemicals [37]. An even more practical application was suggested by King *et al.*, who glued a porous silion rugate film on top of an optical fiber and used this sensor for monitoring VOCs breakthrough in an activated carbon filtration bed [38].

Pretty soon the importance of the surface chemistry of the porous silicon based sensors for its gas sensing performance was noticed and investigated. Chapron *et al.*, fabricated a porous silicon rugate filter whose reflectivity peak was located in the infra-red spectral range and studied the influence of surface modifications such as chemical oxidation and hydrosilylation with octadecene on the optical response of the porous silicon sensor [39]. Only small differences in the sensitivity of the two rugate sensors were observed. However, a thorough analysis of the reflectivity spectra allowed for a quantitative determination of the porosity as well as the thickness of the $SiO_2$ and alkyl layer formed on the porous silicon surface. A systematic study on this topic was later published by Ruminski *et al.*, who demonstrated a strong impact of the surface chemistry on the sensitivity and stability of the porous silicon gas sensors [40]. Exposure of the sensors to either more hydrophilic (isopropanol) or hydrophobic VOCs (heptane) led to pronounced redshifts of the rugate peaks in the reflectivity spectrum which were in all cases stronger for hydrophilic chemicals expect for carbonized samples resulting from decomposition of acetylene on freshly etched porous silicon. The carbonized porous silicon showed higher sensitivity for heptane vapors. At the same time the stability of the porous silicon sensors was investigated and it turned out that especially carbonized, thermally oxidized, and dimethylsiloxyl-terminated porous silicon structures are stable for a long time qualifying them for environmental monitoring applications. Furthermore composites of porous silicon and carbon have a one order of magnitude lower detection limit for toluene in comparison to freshly etched porous silicon sensors [41]. Another strategy for increasing the sensitivity of porous silicon 1D photonic crystal sensors was reported by Jang *et al.* [42]. They incorporated copper(II) sulfate in a porous silicon Bragg mirror and used a LED as light source. By replacing the commonly used tungsten halogen lamp by an LED and monitoring changes in the reflectivity intensity instead of shifts in the Bragg peak position the detection limit of the sensor for triethyl phosphate was lowered from 1.4 ppm to 150 ppb.

Another challenge for porous silicon based optical sensors is zero-point drift of the reflectivity spectrum caused by fluctuations in light source intensity, temperature, and humidity. In order to correct for the inevitable background noise the introduction of a reference channel directly on the porous silicon sensor platform was suggested. King *et al.* realized this reference channel by etching a porous silicon 1D photonic crystal showing two reflectivity maxima in the spectrum and infiltration of a pH sensitive probe (bromothymol blue) into its pores [43]. The absorbance maximum of the indicator dye overlapped with one of the reflectance peaks of the photonic crystal upon exposure to $NH_3$ vapor resulting in a considerable intensity decrease of this reflectance peak. At the same time the second reflectance peak located in a different spectral range was unaffected. Hence, a ratio of the intensity of the two peaks can correct for experimental fluctuations and thereby increase the sensitivity of the sensor for $NH_3$ vapor. The concept of an internal reference channel was also used for the detection of HF and $Cl_2$[44]. An even more intriguing exploitation of this strategy was the fabrication of double stacks of rugate filters whose surfaces were functionalized with different molecules [45]. By giving the top stack hydrophobic and the bottom stack hydrophilic properties and calculating the weighted differences between the two peak frequencies of the two rugate filters a humidity-compensating sensor for VOCs was obtained. Figure 6 shows a reflectivity spectrum and a backscatter SEM image of two separate rugate structures etched on top of each other. It has to be emphasized that the stacking order of the rugate filters is important for the appearance of both rugate peaks in the reflectivity spectrum [46].

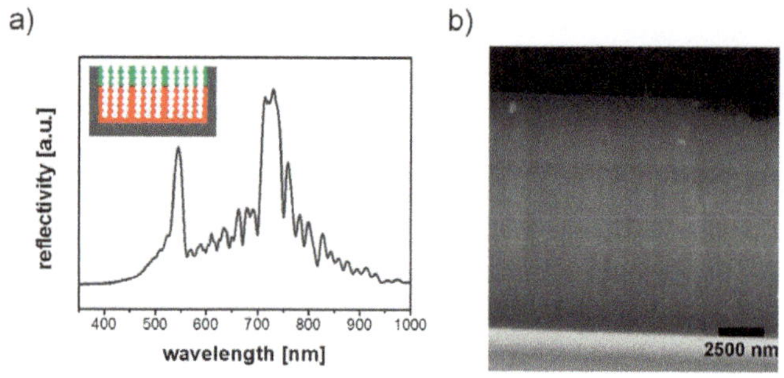


**Figure 6:** Example for a porous silicon gas sensor. (**a**) Reflectivity spectrum of a double layer rugate structure which has been utilized for the elimination of humidity based changes in the reflectivity spectrum interfering with the detection of organic vapors. (**b**) Backscatter SEM image of a double layer rugate structure.

Recently, another approach for improving the specificity of porous silicon gas sensors was introduced by King *et al.*, They mounted a 1D phtonic crystal on a thermoelectric Peltier device and monitored shifts in the spectral position of the photonic stop band upon exposure to VOCs as a function of temperature [47]. The observed hysteresis loops were explained by differences in the diffusion and adsorption properties of the vapors within the porous structure and turned out to be characteristic for each investigated analyte. Kelly *et al.*, also used the parameter diffusion for the identification and quantification of VOCs with porous silicon photonic crystal sensor [48]. Here, the porous silicon sensor consisted of three stacked mesoporous rugate filters displaying three distinct peaks in the reflectivity spectrum. The layer in the middle serves as "drift tube" and delays the diffusion of the vapor into the bottom layer. The optical response of the three layers was monitored over time and the sequential response of the three layers was successfully correlated to the rate of analyte diffusion through the porous silicon structure. This method allowed for the identification of VOCs in parts per million concentrations.

## *Biosensing*

The first porous silicon based biosensor was fabricated in 1997 and consisted of a single porous silicon layer whose Fabry-Pérot fringes in the reflectivity spectrum shifted on the wavelength scale upon binding of biomolecules (DNA and proteins) to the functionalized porous silicon surface [3]. The performance of these single layer porous silicon sensors was compared to the sensitivity of photonic crystal structures etched into silicon, e.g., Bragg stack and microcavity, by Anderson *et al.* [49]. Changes in the reflectivity spectrum upon infiltration of sucrose solutions were theoretically calculated and experimentally measured. As expected the 1D photonic crystal structures showed higher sensitivities for refractive index changes in the pores than the single layer sensor. However, in contradiction to the theoretical predictions the Bragg stack displayed a larger redshift of the resonance peak on the wavelength scale in comparison to the microcavity structure emphasizing the importance of a well-thought-out design for the multilayer sensors. Ouyang *et al.*, systematically optimized the etching parameter for macroporous silicon microcavities in order to allow for the detection of macromolecule infiltration into the porous structure [11,50]. The sensitivity of the prepared macroporous mirocavity structures depended on several parameters such as the pore size, the porosity contrast between the porous layers, and the thickness of the layers. Thereby it was crucial to adjust the macroporous structure to the biosensing task. The excellent optical sensing properties of the developed macroporous sensor were demonstrated by infiltration of rabbit IgG and detection of biotin/streptavidin interaction with a detection limit of 0.3 ng/mm$^2$. An in-depth study on this topic was published by DeLouise *et al.*, who investigated a series of microcavity biosensor devices for immobilization of glutathione-S-transferase within the porous matrix [51]. The amount of immobilized enzyme was indirectly quantified by measuring the enzyme activity and successfully cross-correlated to the magnitude of the optical response of the microcavity. Detection limits as low as 50–250 pg/mm$^2$were achieved for the optimized macroporous silicon sensors—values which could compete with standard SPR devices. Two years later, in 2007, macroporous silicon microcavity photonic bandgap sensors were used for selective and quantitative detection of protein interactions [52].

Almost all of the previously described porous silicon biosensors focused on an increase in the composite refractive index of the porous photonic crystal structure by infiltration of substances into the pores which possess higher refractive indices and consequently monitored shifts of the resonance peaks to higher wavelengths (redshift). In contrast, Kilian *et al.*, reported on peptide-modified porous silicon rugate filters for the detection of protease activity by pronounced blue shifts of the reflectivity peak [53]. Here, the high refractive index peptides were removed from the porous matrix by enzyme cleavage leading to a decrease in the composite refractive index of the porous layer. A tailor-made surface functionalization strategy allowed them to detect protease concentrations of 37 nM.

Pacholski and Sailor monitored intensity changes in the reflectivity spectrum of a porous silicon rugate structure upon exposure to sucrose and bovine serum albumin (BSA) solutions [54]. The reflectivity spectrum of a 1D porous silicon photonic crystal consists of a strong peak corresponding to the optical response of the photonic crystal (rugate filter) and Etalon fringes (Fabry-Pérot interference) originating from reflections at the interfaces of the porous structure (Figure 7(a)). By following the intensity changes of these two optical features instead of shifts in the position of the "rugate" peak on the wavelength scale the refractive index contrast at the interface medium/porous layer was investigated rather than refractive index changes in the porous layer. In Figure 7(b) the intensity changes of the rugate peak and the intensity changes of the Fabry-Pérot (here represented by the effective optical thickness) upon analyte exposure are displayed. Both signals were not suitable for a reliable detection of BSA and sucrose. However, calculating the ratio between the intensity of the EOT and the "rugate" peak intensities led to a sensitive detection of BSA. It was concluded that the intensity of the "rugate" peak can serve as reference channel whereas the EOT intensity detects binding events to the sensor surface.

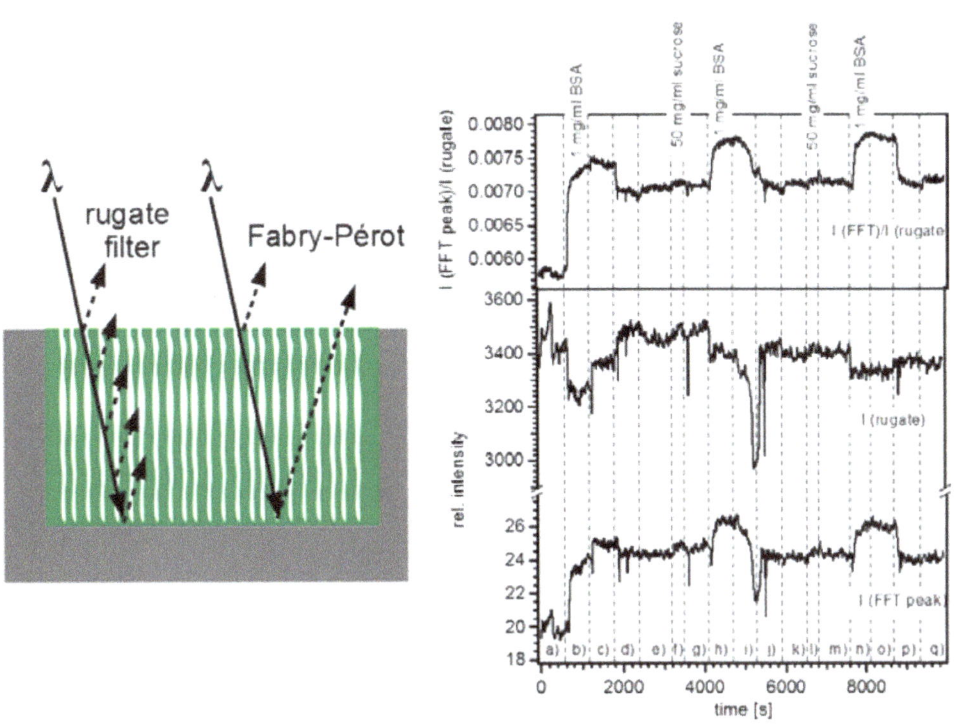


**Figure 7:** Porous silicon photonic crystal considered as optical double layers. (**a**) Schematic of a 1D porous silicon photonic crystal. Interference of light beams occurs from all layers of the multilayered structure (rugate) as well as from reflections at the interfaces bordering the porous layer (Fabry-Pérot interference). (**b**) Sensogram of the structure upon exposure to sucrose and BSA solutions. Reprinted from Reference [54] with permission. Copyright Wiley-VCH 2007.

A new and exciting concept for the detection of biomolecules by porous silicon 1D photonic crystal sensors was introduced by Orosco *et al.*, in 2006 [55]. They determined protease activity by color changes that were observable by the naked eye making the use of a spectrometer for optical signal transduction dispensable. A schematic of the sensor principle is depicted in Figure 8(a). A porous silicon rugate filter was electrochemically etched. Methyl groups were grafted onto its surface in order to provide hydrophobicity and thereby prevent water penetration into the pores. After coating the sensor surface with a hydrophobic protein, small droplets of water containing different amounts of proteases were placed on the sensor surface. The proteases digested the hydrophobic protein and the proteolytic cleavage products penetrated the hydrophobic pores leading to an increase in the composite refractive index of the porous layer and consequently to a color change. Protease concentrations as low as 7.2 pmol were detectable by the naked eye (Figure 8(b)). By taking reflectivity spectra the detection limit could not be improved (Figure 8(c)). Gao *et al.*, transferred this concept to hydrophilic proteins [56]. Further development of porous silicon 1D photonic crystal sensors for visual colorimetric readout was reported by Bonanno and De Louise in 2010 [57]. Here, the integration of a stimuli-responsive hydrogel into the porous structure whose cross-linking densities is controlled by interactions with the analyte is the key to achieve direct optical detection.

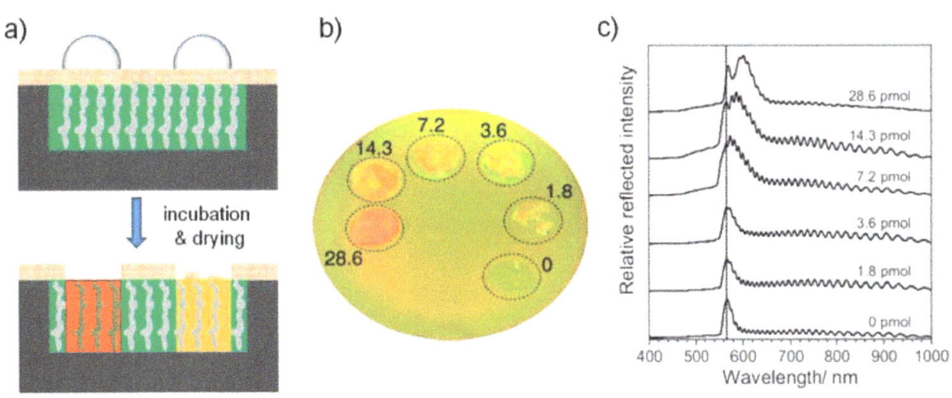


**Figure 8:** Optical detection of protease activity using a protein-coated porous silicon sensor. (**a**) Schematic of the sensor principle. A hydrophobic porous silicon rugate structure is coated with a hydrophobic protein. Addition of small amounts of proteases leads to digestion of the protein layer. The proteolytic cleavage products infiltrate into the porous structure resulting in a visible color change. (**b**) Photo of a protein-coated 1D photonic crystal which has been treated with different amounts of proteases (given in picomoles). (**c**) Reflectivity spectra taken from the

spots on the protein-coated porous silicon sensor shown in (b). Parts (a) & (b) reprinted from Reference [55] with permission. Copyright Wiley-VCH 2006.

A push towards commercial application of porous silicon sensors was evoked by Bonanno and DeLouise in 2010. They developed a label-free porous silicon competitive inhibition immunosensor for the rapid and reliable detection of opiates in urine [58]. Competitive binding assays enable the detection of small molecules like opiates by optical sensors which transduce refractive index changes into an optical signal. Tiny refractive index changes caused by infiltration of small molecules into the pores lead to a very small change in the composite refractive index of the porous layer. Hence, the corresponding optical response of the sensor is not sufficient for reliable sensing. By specific binding of appropriate antibodies to the analyte the change in the refractive index is considerably increased resulting in a sensitive and reproducible optical response of the sensor to the analyte. This principle has been exploited for the porous silicon competitive binding immunosensor. Here, a porous silicon 1D photonic crystal was functionalized with an opiate analogue. If this sensor was exposed to a solution of antimorphine IgG the antibody was solely bound to the immobilized opiate analogue leading to a maximum wavelength shift of the stop band peak (reference). For a mixture of a test solution containing opiates and the same amount of antibody a smaller red-shift in the biosensor signal was obtained. This observation is based on the fact that less antibody is available for binding to the sensor surface as the opiates in the test solution compete with the opiate analogue bound to the sensor surface. The performance of such a label-free porous silicon immunosensor assay was positively evaluated for the detection of opiates in a blind clinical study [59].

Not only biomolecules have been sensed by porous silicon optical sensors but also biological organisms such as bacteria, eukaryotes, and prokaryotes. First studies on this topic were performed using changes in the photoluminescence of a porous silicon microcavity as optical transduction. Tailor-made ligands for Gram-(−) bacteria were bound to the pore walls of these sensors and allowed for discrimination of Gram-(−) and Gram-(+) bacteria in lysed bacteria solutions [60,61]. In 2006 Schwartz *et al.*, investigated changes in the intensity of light scattered from a porous silicon 1D photonic crystal for monitoring morphology changes in an assembly of hepatocyte cells upon exposure to toxins [62]. The same method was employed for the detection of virus infection in bacteria [63]. To the best of my knowledge there has been no report on porous silicon photonic crystal sensors which detect whole cells by changes in their reflectivity spectrum. Indirect observation of live cells was achieved by seeding cells which secreted proteases on a biopolymer loaded porous silicon rugate filter. These enzymes digested the biopolymer incorporated in the porous layer and thereby provoked a blue-shift of the reflectivity peak [64].

## New Concepts

Recently, fractal photonic crystals have been fabricated by electrochemical etching of silicon. They possess a deterministic nonperiodic modulation of the refractive index. Examples for 1D nonperiodic structure are Thue-Morse, Cantor, and Fibonacci [65–67]. A direct comparison of the optical sensor performance of periodic and deterministic nonperiodic porous silicon structures has been published by Moretti *et al.*, in 2007 [68]. They exposed porous silicon Bragg mirrors and Thue-Morse multilayers to VOCs and determined a higher sensitivity for the nonperiodic porous silicon structure. These results were explained by the narrower resonance peaks and the smaller amount of interfaces in the Thus-Morse structure in comparison to the Bragg stack. Appropriately designed porous silicon Cantor structures also showed higher sensitivity for the detection of silanization and further functionalization of the pore walls than porous silicon microcavities [69]. In Figure 9(a) a schematic and a SEM image of a Cantor structure etched into silicon are displayed. Another type of photonic structures so-called polybasic multilayers were suggested for the fabrication of highly sensitive porous silicon biosensors. Lv *et al.*, demonstrated their application to the detection of antigen-antibody interactions and DNA hybridization [70,71].

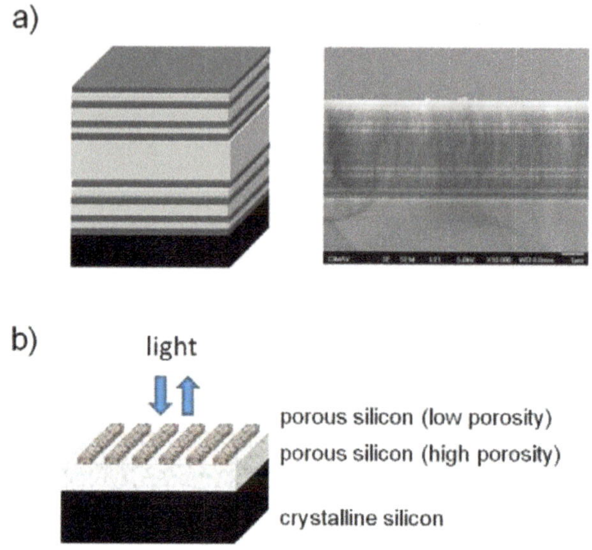


**Figure 9:** New porous silicon sensor lay-outs. (**a**) Schematic and SEM image of a Cantor structure. Image courtesy: V. Agarwal. (**b**) Schematic of a planar photonic crystal sensor.

Besides these more exotic photonic structures classical 2D photonic crystal sensors based on porous silicon have been investigated for several years. The photonic band gap of 2D photonic crystal is observed for light propagation within the plane of periodicity. The fabrication of porous silicon 2D photonic crystals is more challenging than the production of a 1D photonic structure. Most porous silicon 2D photonic crystals were prepared from silicon-on-insulator substrates using standard lithography for the generation of an etching mask and reactive ion etching. Special care has to be taken for measuring the optical properties of these devices as the photons have to encounter the porous structure perpendicular to the pore axis. The demanding fabrication process for 2D photonic porous silicon structures has hindered its widespread application to sensor research. However, their high sensitivity makes them ideal candidates for the detection of very small amounts of analytes or even single molecule detection. The latter can be achieved by introduction of point defects in the lattice leading to a strong localization of light. Chow *et al.*, tested the sensitivity of a 2D photonic crystal microcavity to refractive index changes and determined a resolution of better than $\Delta = 0.001$ [72]. Lee and Fauchet reported on the detection of a single latex sphere with a size which is comparable to viruses [73]. In addition they used a similar 2D photonic crystal sensor for monitoring glutaraldehyde and BSA binding to the internal sensor surface [73]. The extreme high sensitivity of 2D photonic crystal microcavity sensors was confirmed by studies performed in 2009. An amount of only $21 \times 10^{-18}$ g of anti-biotin was detected overtrumping SPR based sensors [74]. The next generation of 2D photonic crystal sensors based on porous silicon possessed linear defects instead of point defects in the crystal lattice—2D photonic crystal waveguides [75]. By coupling micro- or nanocavities to silicon photonic crystal waveguides further advancements in the development of ultrasensitive optical sensors were made [76,77].

As already mentioned before the previously described 2D photonic crystal sensors suffer from time-consuming and expensive fabrication methods as well as complicated optical set-ups. In order to circumvent these high demands on the sensor device production and application, new concepts were developed focusing on the design of surface wave and planar 1D photonic crystal devices. Surface wave porous silicon sensors are obtained by etching a periodic porous silicon multilayer with a defect layer on top. The sensor is operated in Otto configuration which uses a single prism for coupling light into the structure. An evanescent wave is generated when light hits the sensor surface comparable to the well-known sensing principle of surface plasmon resonance sensors. Sensitive detection of ethanol vapor and amine grafting to the porous silicon surface wave sensors was demonstrated [78,79]. An example for a planar 1D photonic crystal is shown in Figure 9(b). The suggested planar photonic crystal sensor is composed of a periodic array of porous silicon bars on top of a porous silicon layer with a higher porosity and consequently a lower refractive index. This structure is characterized by the appearance of extreme narrow optical features—so-called Fano resonances—and can be used at normal incidence omitting complicated optical set-ups. Jamois *et al.*, have demonstrated the superior sensor performance for the detection of DNA hybridization [80]. Theoretical simulations predict that the sensitivity of these sensors could even be improved by periodic patterning of porous silicon multilayers [81].

## CONCLUSIONS/OUTLOOK

Porous silicon is a very attractive material for the design of tailor-made photonic crystal sensors. The ability to control the porosity and consequently the refractive index of the porous structure at will opens up countless possibilities for the construction of optical devices with the desired photonic properties. The fabrication of porous silicon sensors is relatively simple and cost-efficient allowing for mass-production. In combination with its very convenient and controllable surface chemistry porous silicon has to be taken seriously as competitor for established optical sensor devices like surface plasmon resonance based systems. The future will show if porous silicon will find its way from research labs into commercial applications.

## ACKNOWLEDGMENTS

This work was supported by the German Federal Ministry of Education and Research (BMBF, project PhoNa, contract No. 03IS2101E) and the Max Planck Society.

## REFERENCES

1. Qazi, H.H.; bin Mohammad, A.; Akram, M. Recent progress in optical chemical sensors. *Sensors*2012, *12*, 16522–16556.
2. Kedem, O.; Vaskevich, A.; Rubinstein, I. Comparative assessment of the sensitivity of localized surface plasmon resonance transducers and interference-based Fabry-Perot transducers. *Ann. Phys.* 2012, *524*, 713–722.
3. Lin, V.S.Y.; Motesharei, K.; Dancil, K.P.S.; Sailor, M.J.; Ghadiri, M.R. A porous silicon-based optical interferometric biosensor. *Science* 1997, *278*, 840–843.
4. Dancil, K.P.S.; Greiner, D.P.; Sailor, M.J. A porous silicon optical biosensor: Detection of reversible binding of IgG to a protein A-modified surface. *J. Am. Chem. Soc.* 1999, *121*, 7925–7930.
5. Yablonovitch, E. Inhibited spontaneous emission in solid-state physics and electronics. *Phys. Rev. Lett.* 1987, *58*, 2059–2062.
6. John, S. Strong localization of photons in certain disordered dielectric superlattices. *Phys. Rev. Lett.* 1987, *58*, 2486–2489.
7. Joannopoulos, J.D.; Johnson, S.G.; Winn, J.N.; Meade, R.D. *Photonic Crystals—Molding the Flow of Light*, 2nd ed.; Princeton University Press: Princeton, NJ, USA, 2008.
8. Bonanno, L.M.; DeLouise, L.A. Whole blood optical biosensor. *Biosens. Bioelectron* 2007, *23*, 444–448.
9. Uhlir, A. Electrolytic shaping of germanium and silicon. *Bell Syst. Technol. J.* 1956, *35*, 333–347.
10. Kolasinski, K.W. Etching of silicon in fluoride solutions. *Surf. Sci.* 2009, *603*, 1904–1911.
11. Huimin, O.; Fauchet, P.M. Biosensing using porous silicon photonic bandgap structures. *Proc. SPIE* 2005, *6005*.
12. Vincent, G. Optical-properties of porous silicon superlattices. *Appl. Phys. Lett.* 1994, *64*, 2367–2369.
13. Sailor, M.J. *Porous Silicon in Practice*; Wiley-VCH Verlag & Co. KGaA: Weinheim, Germany, 2012; p. 249.
14. Birner, A.; Wehrspohn, R.B.; Gosele, U.M.; Busch, K. Silicon-based photonic crystals. *Adv. Mater*2001, *13*, 377–388.
15. Gruning, U.; Lehmann, V.; Engelhardt, C.M. 2-Dimensional infrared photonic band-gap structure-based on porous silicon. *Appl. Phys. Lett.* 1995, *66*, 3254–3256.
16. Lehmann, V. The physics of macropore formation in low doped n-type silicon. *J. Electrochem. Soc.* 1993, *140*, 2836–2843.
17. Lehmann, V.; Gruning, U. The limits of macropore array fabrication. *Thin Solid Films* 1997, *297*, 13–17.
18. Garin, M.; Trifonov, T.; Rodriguez, A.; Marsal, L.F.; Alcubilla, R. Optical properties of 3D macroporous silicon structures. *Mater. Sci. Eng. B Adv. Funct. Solid State Mater* 2008, *149*, 275–280.
19. Trifonov, T.; Rodriguez, A.; Marsal, L.F.; Pallares, J.; Alcubilla, R. Macroporous silicon: A versatile material for 3D structure fabrication. *Sens. Actuators A Phys.* 2008, *141*, 662–669.

20. Matthias, S.; Muller, F.; Gosele, U. Simple cubic three-dimensional photonic crystals based on macroporous silicon and anisotropic posttreatment. *J. Appl. Phys.* 2005, *98*.

21. Chelnokov, A.; Wang, K.; Rowson, S.; Garoche, P.; Lourtioz, J.M. Near-infrared Yablonovite-like photonic crystals by focused-ion-beam etching of macroporous silicon. *Appl. Phys. Lett.* 2000,*77*, 2943–2945.

22. Dang, Z.; Breese, M.B.H.; Recio-Sanchez, G.; Azimi, S.; Song, J.; Liang, H.; Banas, A.; Torres-Costa, V.; Jose, M.-P.R. Silicon-based photonic crystals fabricated using proton beam writing combined with electrochemical etching method. *Nanoscale Res. Lett.* 2012, *7*.

23. Martin-Palma, R.J.; Manso, M.; Arroyo-Hernandez, M.; Torres-Costa, V.; Martinez-Duart, J.M. Nanostructured-porous-silicon-based two-dimensional photonic crystals. *Appl. Phys. Lett.* 2006,*89*.

24. Hecht, E. *Optik*, 5th ed.; Oldenbourg Wissenschaftsverlag GmbH: München, Germany, 2009; pp. 627–711.

25. Kilian, K.A.; Boecking, T.; Gooding, J.J. The importance of surface chemistry in mesoporous materials: Lessons from porous silicon biosensors. *Chem. Commun.* 2009, 630–640.

26. Buriak, J.M.; Stewart, M.P.; Geders, T.W.; Allen, M.J.; Choi, H.C.; Smith, J.; Raftery, D.; Canham, L.T. Lewis acid mediated hydrosilylation on porous silicon surfaces. *J. Am. Chem. Soc.*1999, *121*, 11491–11502.

27. Galvin, K.P. A conceptually simple derivation of the Kelvin equation. *Chem. Eng. Sci.* 2005, *60*, 4659–4660.

28. Sailor, M.J.; Link, J.R. "Smart dust": Nanostructured devices in a grain of sand. *Chem. Commun.*2005, 1375–1383.

29. Zangooie, S.; Bjorklund, R.; Arwin, H. Vapor sensitivity of thin porous silicon layers. *Sens. Actuators B Chem.* 1997, *43*, 168–174.

30. Snow, P.A.; Squire, E.K.; Russell, P.S.J.; Canham, L.T. Vapor sensing using the optical properties of porous silicon Bragg mirrors. *J. Appl. Phys.* 1999, *86*, 1781–1784.

31. Mulloni, V.; Pavesi, L. Porous silicon microcavities as optical chemical sensors. *Appl. Phys. Lett.*2000, *76*, 2523–2525.

32. Baratto, C.; Faglia, G.; Sberveglieri, G.; Gaburro, Z.; Pancheri, L.; Oton, C.; Pavesi, L. Multiparametric porous silicon sensors. *Sensors* 2002, *2*, 121–126.

33. De Stefano, L.; Rendina, I.; Moretti, L.; Rossi, A.M. Optical sensing of flammable substances using porous silicon microcavities. *Mater. Sci. Eng. B Solid State Mater. Adv. Technol.* 2003,*100*, 271–274.

34. De Stefano, L.; Moretti, L.; Rendina, I.; Rossi, A.M. Time-resolved sensing of chemical species in porous silicon optical microcavity. *Sens. Actuators B Chem.* 2004, *100*, 168–172.

35. Salem, M.S.; Sailor, M.J.; Fukami, K.; Sakka, T.; Ogata, Y.H. Sensitivity of porous silicon rugate filters for chemical vapor detection. *J. Appl. Phys.* 2008, *103*.

36. Dorvee, J.; Sailor, M.J. A low-power sensor for volatile organic compounds based on porous silicon photonic crystals. *Phys. Status Solidi A Appl. Mater. Sci.* 2005, *202*, 1619–1623.

37. Schmedake, T.A.; Cunin, F.; Link, J.R.; Sailor, M.J. Standoff detection of chemicals using porous silicon "smart dust" particles. *Adv. Mater* 2002, *14*, 1270–1272.

38. King, B.H.; Ruminski, A.M.; Snyder, J.L.; Sailor, M.J. Optical-fiber-mounted porous silicon photonic crystals for sensing organic vapor breakthrough in activated carbon. *Adv. Mater* 2007,*19*, 4530–4534.

39. Chapron, J.; Alekseev, S.A.; Lysenko, V.; Zaitsev, V.N.; Barbier, D. Analysis of interaction between chemical agents and porous Si nanostructures using optical sensing properties of infra-red Rugate filters. *Sens. Actuators B Chem.* 2007, *120*, 706–711.

40. Ruminski, A.M.; King, B.H.; Salonen, J.; Snyder, J.L.; Sailor, M.J. Porous silicon-based optical microsensors for volatile organic analytes: Effect of surface chemistry on stability and specificity.*Adv. Funct. Mater* 2010, *20*, 2874–2883.

41. Kelly, T.L.; Gao, T.; Sailor, M.J. Carbon and carbon/silicon composites templated in rugate filters for the adsorption and detection of organic vapors. *Adv. Mater* 2011, *23*, 1776–1781.

42. Seunghyun, J.; Youngdae, K.; Jihoon, K.; Jaehyun, P.; Cheolyoung, P.; Sungjin, K.; Sungdong, C.; Youngchun, K.; Honglae, S. Detection of organophosphates based on surface-modified DBR porous silicon using LED light. *Mater. Lett.* 2008, *62*, 552–555.

43. King, B.H.; Gramada, A.; Link, J.R.; Sailor, M.J. Internally referenced ammonia sensor based on an electrochemically prepared porous $SiO_2$ photonic crystal. *Adv. Mater* 2007, *19*, 4044–4048.

44. Ruminski, A.M.; Barillaro, G.; Chaffin, C.; Sailor, M.J. Internally referenced remote sensors for HF and $Cl_2$ using reactive porous silicon photonic crystals. *Adv. Funct. Mater* 2011, *21*, 1511–1525.

45. Ruminski, A.M.; Moore, M.M.; Sailor, M.J. Humidity-compensating sensor for volatile organic compounds using stacked porous silicon photonic crystals. *Adv. Funct. Mater* 2008, *18*, 3418–3426.

46. Jalkanen, T.; Salonen, J.; Torres-Costa, V.; Fukami, K.; Sakka, T.; Ogata, Y.H. Structural considerations on multistopband mesoporous silicon rugate filters prepared for gas sensing purposes. *Opt. Express* 2011, *19*, 13291–13305.

47. King, B.H.; Wong, T.; Sailor, M.J. Detection of pure chemical vapors in a thermally cycled porous silica photonic crystal. *Langmuir* 2011, *27*, 8576–8585.

48. Kelly, T.L.; Sega, A.G.; Sailor, M.J. Identification and quantification of organic vapors by time-resolved diffusion in stacked mesoporous photonic crystals. *Nano Lett.* 2011, *11*, 3169–3173.

49. Anderson, M.A.; Tinsley-Bown, A.; Allcock, P.; Perkins, E.A.; Snow, P.; Hollings, M.; Smith, R.G.; Reeves, C.; Squirrell, D.J.; Nicklin, S.; *et al.* Sensitivity of the optical properties of porous silicon layers to the refractive index of liquid in the pores. *Phys. Status Solid A Appl. Res.* 2003,*197*, 528–533.

50. Ouyang, H.; Christophersen, M.; Viard, R.; Miller, B.L.; Fauchet, P.M. Macroporous silicon microcavities for macromolecule detection. *Adv. Funct. Mater* 2005, *15*, 1851–1859.

51. DeLouise, L.A.; Kou, P.M.; Miller, B.L. Cross-correlation of optical microcavity biosensor response with immobilized enzyme activity. Insights into biosensor sensitivity. *Anal. Chem.*2005, *77*, 3222–3230.

52. Ouyang, H.; de Louise, L.A.; Miller, B.L.; Fauchet, P.M. Label-free quantitative detection of protein using macroporous silicon photonic bandgap biosensors. *Anal. Chem.* 2007, *79*, 1502–1506.

53. Kilian, K.A.; Boecking, T.; Gaus, K.; Gal, M.; Gooding, J.J. Peptide-modified optical filters for detecting protease activity. *ACS Nano.* 2007, *1*, 355–361.

54. Pacholski, C.; Sailor, M.J. Sensing with Porous Silicon Double Layers: A General Approach for Background Suppression. Physica Status Solidi C—Current Topics in Solid State Physics. Proceeding of 14th Applied Surface Analysis Workshop, Kaiserslautern, Germany, 17– 20 September 2006; Stutzmann, M., Ed.; Wiley-VCH: Weinheim, Germany, 2007; Volume 4, pp. 2088–2092.

55. Orosco, M.M.; Pacholski, C.; Miskelly, G.M.; Sailor, M.J. Protein-coated porous-silicon photonic crystals for amplified optical detection of protease activity. *Adv. Mater* 2006, *18*, 1393–1396.

56. Gao, L.; Mbonu, N.; Cao, L.; Gao, D. Label-free colorimetric detection of Gelatinases on nanoporous silicon photonic films. *Anal. Chem.* 2008, *80*, 1468–1473.

57. Bonanno, L.M.; de Louise, L.A. Integration of a chemical-responsive hydrogel into a porous silicon photonic sensor for visual colorimetric readout. *Adv. Funct. Mater* 2010, *20*, 573–578.

58. Bonanno, L.M.; de Louise, L.A. Tunable detection sensitivity of opiates in urine via a label-free porous silicon competitive inhibition immunosensor. *Anal. Chem.* 2010, *82*, 714–722.

59. Bonanno, L.M.; Kwong, T.C.; de Louise, L.A. Label-free porous silicon immunosensor for broad detection of opiates in a blind clinical study and results comparison to commercial analytical chemistry techniques. *Anal. Chem.* 2010, *82*, 9711–9718.

60. Hubbard, R.D.; Horner, S.R.; Miller, B.L. Highly substituted ter-cyclopentanes as receptors for lipid A. *J. Am. Chem. Soc.* 2001, *123*, 5810–5811.

61. Chan, S.; Horner, S.R.; Fauchet, P.M.; Miller, B.L. Identification of gram negative bacteria using nanoscale silicon microcavities. *J. Am. Chem. Soc.* 2001, *123*, 11797–11798.

62. Schwartz, M.P.; Derfus, A.M.; Alvarez, S.D.; Bhatia, S.N.; Sailor, M.J. The smart petri dish: A nanostructured photonic crystal for real-time monitoring of living cells. *Langmuir* 2006, *22*, 7084–7090.

63. Alvarez, S.D.; Schwartz, M.P.; Migliori, B.; Rang, C.U.; Chao, L.; Sailor, M.J. Using a porous silicon photonic crystal for bacterial cell-based biosensing. *Phys. Status Solidi A Appl. Mater. Sci.*2007, *204*, 1439–1443.

64. Kilian, K.A.; Lai, L.M.H.; Magenau, A.; Cartland, S.; Boecking, T.; di Girolamo, N.; Gal, M.; Gaus, K.; Gooding, J.J. Smart tissue culture: in situ monitoring of the activity of protease enzymes secreted from live cells using nanostructured photonic crystals. *Nano Lett.* 2009, *9*, 2021–2025.

65. Agarwal, V.; Mora-Ramos, M.E. Optical characterization of polytype Fibonacci and Thue-Morse quasiregular dielectric structures made of porous silicon multilayers. *J. Phys. D Appl. Phys.*2007, *40*, 3203–3211.

66. Agarwal, V.; Soto-Urueta, J.A.; Becerra, D.; Mora-Ramos, M.E. Light propagation in polytype Thue-Morse structures made of porous silicon. *Photonics Nanostruct. Fundam. Appl.* 2005, *3*, 155–161.

67. Moretti, L.; Rea, I.; Rotiroti, L.; Rendina, I.; Abbate, G.; Marino, A.; de Stefano, L. Photonic band gaps analysis of Thue-Morse multilayers made of porous silicon. *Opt. Express* 2006, *14*, 6264–6272.

68. Moretti, L.; Rea, I.; de Stefano, L.; Rendina, I. Periodic versus aperiodic: Enhancing the sensitivity of porous silicon based optical sensors. *Appl. Phys. Lett.* 2007, *90*.

69. Escorcia-Garcia, J.; Gaggero-Sager, L.M.; Palestino-Escobedo, A.G.; Agarwal, V. Optical properties of Cantor nanostructures made from porous silicon: A sensing application. *Photonics Nanostruct. Fundam. Appl.* 2012, *10*, 452–458.

70. Lv, X.; Mo, J.; Jiang, T.; Zhong, F.; Jia, Z.; Li, J.; Zhang, F. Novel multilayered porous silicon-based immunosensor for determining Hydroxysafflor yellow A. *Appl. Surf. Sci.* 2011, *257*, 1906–1910.

71. Lv, X.; Chen, L.; Zhang, H.; Mo, J.; Zhong, F.; Lv, C.; Ma, J.; Jia, Z. Hybridization assay of insect antifreezing protein gene by novel multilayered porous silicon nucleic acid biosensor.*Biosens. Bioelectron* 2013, *39*, 329–333.

72. Chow, E.; Grot, A.; Mirkarimi, L.W.; Sigalas, M.; Girolami, G. Ultracompact biochemical sensor built with two-dimensional photonic crystal microcavity. *Opt. Lett.* 2004, *29*, 1093–1095.

73. Lee, M.R.; Fauchet, P.M. Nanoscale microcavity sensor for single particle detection. *Opt. Lett.*2007, *32*, 3284–3286.

74. Zlatanovic, S.; Mirkarimi, L.W.; Sigalas, M.M.; Bynum, M.A.; Chow, E.; Robotti, K.M.; Burr, G.W.; Esener, S.; Grot, A. Photonic crystal microcavity sensor for ultracompact monitoring of reaction kinetics and protein concentration. *Sens. Actuators B Chem.* 2009, *141*, 13–19.

75. Martin-Palma, R.J.; Torres-Costa, V.; Manso, M.; Martinez-Duart, J.M. Finite-thickness photonic crystals based on nanostructured porous silicon for optical sensing. *J. Nanophotonics* 2009, *3*.

76. Pal, S.; Guillermain, E.; Sriram, R.; Miller, B.; Fauchet, P.M. Microcavities in Photonic Crystal Waveguides for Biosensor Applications. In Frontiers in Pathogen Detection: From Nanosensors to Systems, Proc. SPIE 7553, San Francisco, CA, USA, 23 January 2010; Fauchet, P.M., Miller, B.L., Eds.; Volume 7553.

77. Pal, S.; Guillermain, E.; Sriram, R.; Miller, B.L.;

Fauchet, P.M. Silicon photonic crystal nanocavity-coupled waveguides for error-corrected optical biosensing. *Biosens. Bioelectron*2011, *26*, 4024–4031.

78. Descrovi, E.; Frascella, F.; Sciacca, B.; Geobaldo, F.; Dominici, L.; Michelotti, F. Coupling of surface waves in highly defined one-dimensional porous silicon photonic crystals for gas sensing applications. *Appl. Phys. Lett.* 2007, *91*.

79. Guillermain, E.; Lysenko, V.; Orobtchouk, R.; Benyattou, T.; Roux, S.; Pillonnet, A.; Perriat, P. Bragg surface wave device based on porous silicon and its application for sensing. *Appl. Phys. Lett.* 2007, *90*.

80. Jamois, C.; Li, C.; Gerelli, E.; Chevolot, Y.; Monnier, V.; Skryshevskyi, R.; Orobtchouk, R.; Souteyrand, E.; Benyattou, T. Porous-silicon based planar photonic crystals for sensing applications. *Proc. SPIE* 2010, *7713*.

81. Jamois, C.; Li, C.; Orobtchouk, R.; Benyattou, T. Slow Bloch surface wave devices on porous silicon for sensing applications. *Photonics Nanostruct. Fundam. Appl.* 2010, *8*, 72–77.

# CHAPTER 2 SLOTTED PHOTONIC CRYSTAL SENSORS

Mark G. Scullion , Thomas F. Krauss and Andrea Di Falco

SUPA, School of Physics and Astronomy, University of St Andrews, St Andrews, Scotland KY16 9SS, UK

## ABSTRACT

**Optical biosensors are increasingly being considered for lab-on-a-chip applications due to their benefits such as small size, biocompatibility, passive behavior and lack of the need for fluorescent labels. The light guiding mechanisms used by many of them results in poor overlap of the optical field with the target molecules, reducing the maximum sensitivity achievable. This review article presents a new platform for optical biosensors, namely slotted photonic crystals, which provide higher sensitivities due to their ability to confine, spatially and temporally, the optical mode peak within the analytic itself. Loss measurements showed values comparable to standard photonic crystals, confirming their ability to be used in real devices. A novel resonant coupler was designed, simulated, and experimentally tested, and was found to perform better than other solutions within the literature. Combining with cavities, microfluidics and biological functionalization allowed proof-of-principle demonstrations of protein binding to be carried out. Higher sensitivities were observed in smaller structures than possible with most competing devices reported in the literature. This body of work presents slotted photonic crystals as a realistic platform for complete on-chip bio sensing; addressing key design, performance and application issues, whilst also opening up exciting new ideas for future study.**

## INTRODUCTION

Small optical biosensors form a substantial part of the growing 'lab-on-a-chip' (LOC) paradigm. LOC epitomises the main goal of much biosensor research, that being the ability to shrink down many of the analytical capabilities of a biomedical research lab into a small disposable chip. In one scenario, we can imagine a small chip into which a single drop of blood is placed. This drop is then rapidly screened by a multitude of different sensor elements on the device for many different diseases or other important factors. The full breakdown of the sample is then read-out to the user (for example, a doctor), who can react accordingly. If the device could be made cheap, simple to use and disposable, then it could be kept sterile and would be suitable for rapid, high throughput testing rather than labour intensive laboratory tests, which require much experience, time and training. In essence, the goal is to make devices for detecting biomedical material that are similar to the home pregnancy tests available today: cheap and simple to use, and providing a clear result that can be obtained and read by a non-specialist.

Several challenges are evident to make this vision a reality, but many technologies offer potential solutions to the problem. A growing presence in this area is that of integrated optics,

**Citation:** Mark G. Scullion, Thomas F. Krauss and Andrea Di Falco, Slotted Photonic Crystal Sensors, doi:10.3390/s130303675.

which combines several micro- or nano-photonic components on the same chip, using fabrication techniques borrowed from the electronics industry. The small size of each component means that hundreds of devices be fabricated on the same chip, offering multi-functionality and multiplexing capabilities. Many micro-photonic devices have been proposed as optical biosensors, such as ring resonators [1–3], surface plasmons [4–11], interferometers [12–15], slot waveguides [16–19] and photonic crystals [20–34]. Whilst micro-photonic devices, by definition, are very small, many of them require bulky external support equipment, such as sources and spectrometers, to function fully as sensors. A key challenge is to integrate all of these functions on-chip (lab-on-a-chip), rather than the current reality of chip-in-a-lab.

Some desktop-sized devices containing photonic chips and microfluidic circuits have been developed commercially, such as the surface plasmon based Biacore [35] and the ring resonator based Genalyte [36]. Whilst these devices are very sensitive, they are housed with bulky and expensive instrumentation and are thus not truly lab-on-a-chip. Performing all functions on a chip engenders stability, good alignment and compactness, thus the environmental requirements may not be as challenging as for the above machines. Another issue is getting the target molecules to the sensing region of the device. Andreas Manz and co-workers pioneered the field of Micro Total Analysis Systems (MicroTAS) [37], combining electrical and optical components with small capillaries that can be used to control different liquid samples. Groups such as his, Stephen Quake's [38], George Whitesides' [39] and others have further developed these microfluidic circuits, allowing a variety of micro-scale chemistry and biology experiments on-chip.

A new player in the field of optical biosensors is the slotted photonic crystal [40–51]. Slotted photonic crystals are typically fabricated in silicon-on-insulator substrates using lithographic techniques, and combine two types of photonic structures, namely the slot waveguide and the photonic crystal to squeeze light down into very small volumes of air. This strong confinement of light in air is the key advantage of the slotted photonic crystal, as it promotes strong light-matter interactions with a substance of interest. Many of the other photonic sensors mentioned above utilize only the evanescent tail of the optical mode for detecting biomolecules, which limits their sensitivity. As shall be seen, the small footprint of the slotted photonic crystal device, as small as a few microns, not only presents opportunities for dense arrays of sensing elements in lab-on-a-chip applications, but also provides potential for some unique and interesting experiments with biological material. The principles behind these devices, and demonstrations of biosensing, are reviewed below.

The review is structured as follows: in Section 2 the principle of operation of optical biosensors, slot waveguides, photonic crystals and slotted photonic crystals are explored and compared. Section 3 presents the general fabrication and characterization protocols. Coupling and propagation losses are investigated in Section 4, whilst Section 5 looks at the integration of slotted photonic crystals with microfluidics. Detection of proteins using slotted photonic crystals is shown in Section 6; before the key challenges facing photonic biosensor technologies are discussed in Section 7 and a summary and outlook at the end.

## PRINCIPLE OF OPERATION

### Optical Biosensors

Optical sensors function by observing changes in some property of light as it passes through the substance of interest. Many optical sensors rely on a spectral feature, usually a sharp peak or dip, which is formed by a resonance (e.g., a cavity). Sensitive refractive index measurements can be made by tracking the peak wavelength of the resonance. In order to make the sensor more specific to a target biomolecule, the surface is coated with antibodies via chemically activated functional groups. These surface coated antibodies act as specific capture agents: protein binding to these surface receptors induces a response from the sensor. When a sample is flown across such a functionalized sensor, in addition to a bulk refractive index change, there will also be a surface refractive index change due to the antigen-antibody binding. In this case, the response curve of the device is shaped by the presence of target molecules within the sample. As they diffuse in solution, some of the target molecules will be captured by the functionalized surface receptors, until some saturation level is reached. Higher concentrations of target molecules induce sharper diffusion curves, with higher plateaus. The concentration of binding antigens can therefore be estimated from the shape of this curve. Other antigens cannot bind to a particular antibody as they lack the correct geometry, and thus do not induce the surface refractive index change. This way the sensor response can be attributed to one antigen type alone and is not confused by the presence of multiple antigens in a single solution. Optical biosensors are distinguished from simple optical refractive index sensors by this mechanism, and it is the binding dynamics and the density of surface bound material we measure rather than just a bulk refractive index change.

### Slot Waveguides

First proposed in 2004 by Almeida and Barrios at Cornell University, slot waveguides [52] have seen much interest in a number of applications due to their unusual optical properties. They allow strong spatial confinement of light within a narrow air slot inside a material of high refractive index. Standard optical waveguide designs rely on total internal reflection as a means of guiding and confining light. This phenomenon requires for the waveguide cladding to be of lower refractive index than the waveguide itself. The case of the slot waveguide, where light is confined inside a lower refractive index medium, therefore seems a bit strange. This can be explained, however, by the imperfect spatial confinement of light by total internal reflection.

Light incident on a boundary between two non-absorbing dielectrics of permittivity $\varepsilon_1$ and $\varepsilon_2$ experiences a combination of transmission and reflection effects. The Maxwell boundary conditions require that the tangential components of the electric field **E** and the magnetic field **H**, and the normal components of the electric flux density **D** and the magnetic flux density **B**, be continuous across the boundary in order to satisfy the conservation of energy. As $\mathbf{D} = \varepsilon\mathbf{E}$, the boundary condition for the electric flux density can be expressed in the form [53]:

$$\left[\varepsilon_1\mathbf{E}_{i0} + \varepsilon_1\mathbf{E}_{r0} - \varepsilon_2\mathbf{E}_{t0}\right]\cdot\hat{\mathbf{n}} = 0 \tag{1}$$

where $\hat{\mathbf{n}}$ is a unit vector normal to the boundary and i, r and t subscripts refer to the incident, reflected and transmitted fields respectively, and the 0 subscript refers to the maximum value. In the case of the electric field being parallel to the plane of incidence the normal components can be shown to be [52]:

$$\frac{E_{in}}{E_{in} - E_{rn}} = \frac{n_1^2}{n_2^2} \qquad (2)$$

where $n = \sqrt{\varepsilon}$ (for a dielectric). The normal component of the electric field just inside $n_2$ (the low index) is therefore larger than that just inside $n_1$ (the high index) by a factor of $n_1^2/n_2^2$.

For total internal reflection, the transmitted angle in Snell's Law becomes complex in nature, with a purely imaginary cosine angle, and the transmitted wave propagates only along the interface [53]. The purely imaginary nature of the normal component (i.e., the cosine angle) of the transmitted field means that it decays exponentially from the maximum value in Equation (2) into the cladding from the boundary. This field is said to be evanescent in nature. High refractive index contrast boundaries can result in quite a large discontinuity in the normal component of the electric field inside the low index medium, for example at an air/silicon interface, the discontinuity immediately inside the air would be almost a factor of 12 greater [52].

If we now consider two optical waveguides in close proximity such that these field discontinuities can interact, which necessitates their separation to be within the decay length of the evanescent field, then these two discontinuities can combine and be enhanced within the narrow air region separating them. Two closely spaced waveguides, or equivalently a narrow slot in a single waveguide, can therefore form a mode of propagation in which most of the light intensity is contained within the low refractive index region, even though total internal reflection is used for guiding. The sharp evanescently decaying discontinuities at each slot wall combine to produce high levels of field inside the air slot as shown in Figure 1(b). This is the basis behind the operation of slot waveguides.

a) b)

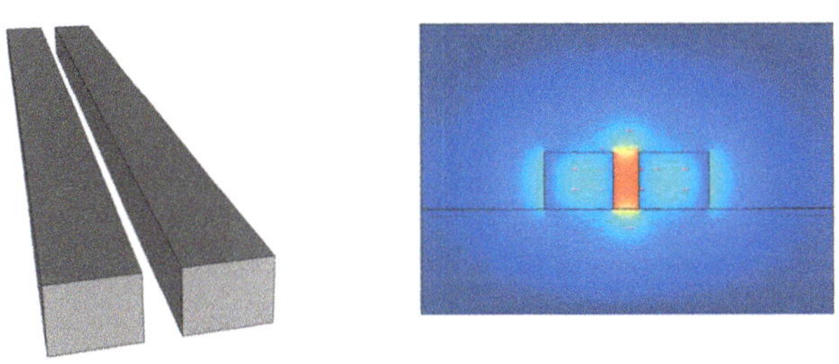

**Figure 1:** (**a**) Slot waveguide geometry. A narrow air slot is introduced into a ridge waveguide. Alternatively, this can be viewed as two separate waveguides separated by a small region of air. Light is guided mostly in air; (**b**) Field distribution of a slot mode.

The slot architecture allows for increased light interactions with biomolecules, as the majority of the modal field interacts with the analyte, rather than just the tail as in the standard waveguide geometry. Currently, many slot waveguide biosensors are based on ring resonators. In one such case tested by Barrios and co-workers [16,17] a slot ring resonator is fabricated in silicon nitride and functionalized to detect BSA, the lowest limit being 0.042 µg/mL. A similar device featuring an array of slot ring resonators has been combined with microfluidics, splitters, a photodiode array and grating couplers, and has shown detection of anti-BSA binding (0.125 µg/mL limit) [18]. A smaller biotin coated silicon-on-insulator slot ring [19] that was used to detect avidin at 100 µg/mL concentration has also been reported.

Slot waveguides have also been used to demonstrate optical guiding of sub-100 nm particles along the slot [54]. Whilst this is not strictly 'trapping', trapping of sub-100 nm dielectric particles may be possible using slot cavities (e.g., slotted photonic crystals). Trapping of such small dielectric particles is difficult to achieve with more conventional optical traps due to the low polarisability of dielectrics; though slots provide a strong intensity gradient that could compensate for this. In addition, slot waveguides have been used as sensitive refractive index sensors. In one case a refractive index changes as small as $10^{-4}$ RIU caused by the presence of acetylene gas was demonstrated by Lipson's group at Cornell by incorporating a gas flow cell onto the chip [55].

## Photonic Crystals

First proposed by Yablonivitch [56] and John [57], but with origins stretching back as far as Bragg and Rayleigh, photonic crystals have been the centre of much research in integrated optics due to the control of light they offer. In such structures, light can be confined to high quality factor cavities, directed, split or even slowed down. Photonic crystals consist of periodic arrangements of material of different dielectric constant. The physics describing the behaviour of light within them is in many ways analogous to the behaviour of electrons in semiconductors. In both cases, it is the periodicity of the structure that is most important: a periodic structure can only support certain wavelengths. In the case of semiconductors it is a periodic arrangement of atoms, and hence electric potential, that results in the formation of allowed electronic bands. Similarly, for photonic crystals the periodic arrangement of dielectric material results in the formation of allowed photonic bands. In its simplest form, the one-dimensional case, the photonic crystal is recognizable as a Bragg mirror.

Two dimensional photonic crystals [58] are typically created using a triangular or square lattice of air or silica filled holes in a semiconductor medium such as silicon or gallium arsenide. Like the electrons in a semiconductor, photons in a photonic crystal can be thought to occupy certain energy bands corresponding to allowed modes of propagation in the crystal lattice. Two continuums of modes, one corresponding to air (or low index) and the other to dielectric (or high index) guided modes, are separated by a bandgap region. Those frequencies which satisfy the Bragg condition lie within this photonic bandgap region, and are forbidden to propagate through the lattice. Like the 1-D Bragg mirror, 2D photonic crystals can be thought of as very efficient wavelength dependent mirrors, with frequencies within the bandgap being strongly reflected. The strong wavelength dependence of these 'mirrors' give the photonic crystal its key functionality. Carving defects into the periodic lattice, for example by removing holes, allows defect states to be created within the photonic bandgap. The lattice surrounding the defect still acts as a mirror for these frequencies, thus frequencies which would normally be rejected can be confined to and guided by defect states within the photonic crystal.

One of the most common defects is realized by removing a single row of holes from the crystal, thus creating a **W1** waveguide. This creates the defect modes shown in the photonic bandstructure depicted in Figure 2. Forbidden frequencies can be guided by this defect, while light confinement out-of-plane is provided by total internal reflection.

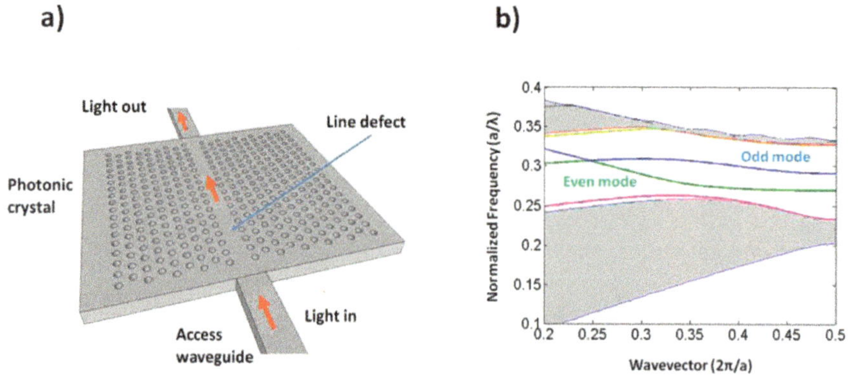


**Figure 2:** (**a**) Schematic of a 2-D slab photonic crystal, with a W1 defect. Light is delivered to and from a photonic crystal via access waveguides. The defect consists of a single row of holes being removed, which allows normally forbidden frequencies to exist within the crystal. Holes on either side of the line defect behave as very efficient mirrors for forbidden frequencies, and thus can be used to guide light; (**b**) Corresponding photonic band diagram of 2D photonic crystal with W1 defect. Even and odd defect guided modes shown.

Whilst slot waveguides specialize in the strong spatial confinement of light, photonic crystal waveguides can employ temporal light confinement mechanisms such as slow light [59,60]. Physically, slow light behaviour results from coherent backscattering at each unit cell of the photonic crystal forming slow moving interference patterns within the waveguide [59]. The main advantage of slow light is that it results in pulse length compression, which in turn leads to higher intensity and increased light-matter interactions. When more localized defects are utilized (e.g., removal of a few holes), cavities with high quality factor can also be created. In this case a single allowed state is created within the bandgap rather than the extended defect of the W1. Additionally, with careful design, a region of constant, high group index can be realized over a limited bandwidth by engineering the gradient of the photonic crystal›s dispersion curve to be relatively flat. The spectral and dispersive properties of photonic crystals can therefore be tailored through structural, rather than material, changes.

The most characteristic spectral feature created by the photonic crystal waveguide is a transmission cut-off near the bandgap. This feature is sensitive to the refractive index of the cladding, and can therefore be used as a point of reference for measuring wavelength shifts resulting from the binding of biomolecules to its surface. For example, detection of single-stranded DNA as low as 19.8 nM has been reported by tracking spectral features near the band edge of a photonic crystal waveguide [20], whilst photonic crystal waveguides have also been used to demonstrate the detection of 10 μg/mL (0.15 μM) BSA by measuring shifts in the cut-off wavelength [21].

By introducing defects into the lattice, high quality factor cavities can be created of a size comparable to a single wavelength of light. High quality factor cavities have sharper spectral features than the photonic crystal waveguide mode cut-off, allowing smaller wavelength shifts to be resolved more easily. Several photonic crystal cavities have been used to demonstrate detection of biomolecules such as BSA (2% [22,23], 1 μM [25]), anti-biotin (20 pM [24]), biotin (3 μg/mL [28], 1 nM [29]) and 3-APTES monolayers [26]. The small size of the photonic crystal cavity is a big advantage over the larger microring cavities, as it gives greater free spectral range (typically ~100–200 nm for the crystal versus ~1–10 nm for the ring), allowing many more devices of different resonance to be multiplexed. In one example of this, arrays of 1-D photonic crystals with cavities were used to demonstrate detection of anti-streptavidin (1 μg/mL limit) [27]. In this demonstration, each cavity of unique resonance is excited by a single bus waveguide, allowing the response of multiple cavities, each optimized for a different target biomolecule, to be obtained from a single spectrum. 1-D photonic crystals can also be constructed from porous silicon, which is used in many different biosensor platforms [61] due to its large surface area. In one such device [30], protease concentrations as low as 14 μM could be detected by the human eye (7 μM with a sensor) due to colour changes in the photonic crystal as a result of infiltration into the pores.

Photonic crystal structures can also support guided resonances [62], in which light is coupled to a short lived guided mode. Such modes are leaky as they exist above the light line and can therefore couple out of the slab; so guided mode resonance devices can be used at normal incidence. The resonances can result in sharp spectral features which can be used for sensing. Lin et al. [31] exploit such a resonance in a sub-wavelength grating to measure the surface attachment of cells. As the local refractive index changes on the grating due to cell attachment, the position of the spectral feature shifts accordingly. This concept has been used to image the attachment of cells to a surface in detail [32,33], which is useful in ascertaining the viability of a cell. Light is scanned over a range of angles through the grating (and an objective) and onto a CCD camera. From these images, the spectral profile at each point on the cell can be determined with a resolution of 0.61 $\mu m^2$. The device has been incorporated into a 96-well plate and is being developed commercially by SRU Biosystems [63].

## Slotted Photonic Crystals

Slotted photonic crystals combine the advantages of spatial confinement of light in air of slot waveguides with the temporal confinement of light provided by photonic crystals in a single structure. This is advantageous because most of the light interacts with the contents of the slot. Simply, a slotted photonic crystal can be viewed as an air slot defect within the guiding region of a standard W1 photonic crystal as shown inFigure 3. Early work in this area [40] has shown, however, that the optical properties of slotted photonic crystal waveguides are very different to that of their non-slotted counterparts. Slotted photonic crystals guide light primarily in air, which is in stark contrast to the conventional non-slotted photonic crystals where most of the light is confined within the high index dielectric. A typical band diagram for a slotted photonic crystal is shown in Figure 4. Comparing Figure 4 with Figure 2, it is evident that the fundamental even modes of the non-slotted and slotted photonic crystals have gradients of opposite sign. This introduces problems associated with efficiently coupling light into the structure, requiring a novel coupler as will be seen later.

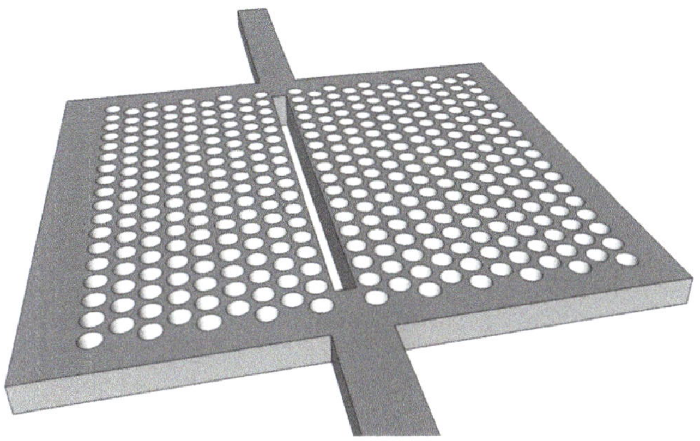

**Figure 3:** Sketch of slotted photonic crystal geometry. Slot waveguide defect introduced to guiding region of W1 photonic crystal.

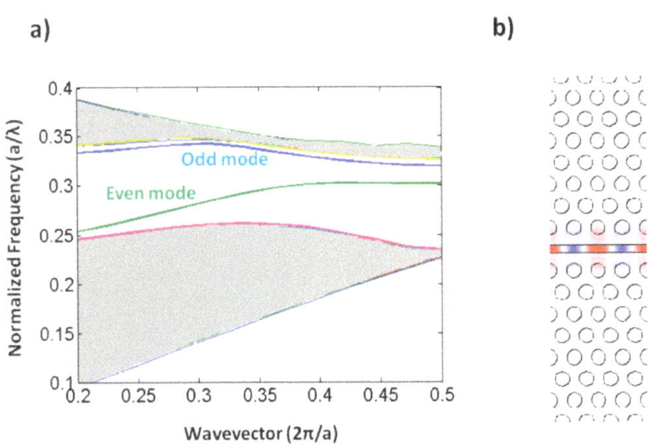


**Figure 4:** (**a**) Photonic band diagram of 2-D slot photonic crystal; (**b**) Mode profile.

Initial work with slotted photonic crystals has shown that they make excellent chemical sensors. Sensitivities of up to 1,500 nm/RIU have been reported, with detection limit $7.8 \times 10^{-6}$ RIU, by our group [41]. We note that that this sensitivity is even higher than the predicted 500 nm/RIU due to the wetting properties of the slot. Solutions containing dissolved material have better wettability than pure water. Other groups have shown detection of gases using these devices, in one case using cavities to detect a variety of gases (510 nm/RIU) [43], or by using slow light effects to enhance absorption of light by methane (100 ppm limit) [44]. The same slow light device has also been used to measure xylene in water (86 μg/L limit) [45]. A group at COBRA (Eindhoven, The Netherlands) have used a cavity created by two deflected photonic crystal nanobeams fabricated in InGaAsP substrates to demonstrate detection of sugar/water solutions with up to 900 nm/RIU sensitivity [46]. The device is covered with a single layer of quantum dots, and is excited by a CW laser. Refractive index shifts are detected by measuring changes in the photoluminescence spectrum of the quantum dots.

A few functionalized slotted photonic crystals have also been demonstrated, for example a glutaraldehyde coated slotted photonic crystal nanolaser in GaInASP/InP has been demonstrated to detect BSA with very high sensitivity (410 nm/RIU, 255 fM limit), by observing changes in the laser spectrum [50]. Whilst BSA has a high affinity for glutaraldehyde, this reaction is non-specific, and in this particular case the affinity was calculated to be four orders of magnitude higher than normal. Such an enhancement, the authors postulate, could be because of optical forces aiding the trapping of the biomolecules within the laser cavity. In another device from the University of Rochester, single latex spheres, 100 nm in diameter, are detected using a slotted microcavity photonic crystal [51]. Our group at St Andrews has developed functionalized slotted photonic crystal cavities, along with integrated microfluidic channels, and has shown detection of the protein avidin at concentrations as low as 15 nM [42]. The main advantages over ring-resonators are the compactness and larger free spectral range which allows potential for 100s of sensors to multiplexed on the same chip with different frequencies.

## Comparison of Small Optical Biosensors

Many different figures-of-merit are used to assess the performance of biosensors. These include, the minimum detected (or interpolated) concentration (in terms of parts per billion, μM, particles/L or g/L), the bound mass (in terms of pg), the number of molecules or cells, or the surface density (in terms of pg/mm$^2$) which makes it difficult to compare different designs. We suggest quoting the minimum concentration detected, as this can be readily checked and controlled, unlike other quantities which rely on assumptions such as the bound mass. This concentration performance will, however, vary between different antibody-antigen pairs, and is not what the sensor physically measures (it measures surface density of bound material). Expressing the concentration in μM gives an idea of the number of detected molecules per litre, but makes molecules of different mass hard to compare; whilst on the other hand using g/L lets you compare different mass concentrations, but not different number of molecules without additional information. Comparing biosensors is therefore not trivial, and the more figures-of-merit can be supplied the better.

The table below shows a comparison between photonic crystal, slot waveguide and slotted photonic crystal biosensors highlighting their size, sensing area, target substance, detection limit, and to aid comparison, sensor response for a given concentration (where the information is available or can be easily derived).

As can be seen from Table 1, slotted photonic crystals offer the most compact solution of the optical biosensors shown. Wavelength shifts comparable to much larger structures can also be achieved. These wavelength shifts are an order of magnitude greater than those of standard photonic crystals. These observations highlight the strengths of combining slot waveguides and photonic crystals in a single structure. The smaller size offered by the temporal confinement within the photonic crystal allows greater free spectral range and potential for dense multiplexing than standard slot waveguides. The increased wavelength shifts results from the strong spatial confinement in air offered by the slot. Bigger wavelength shifts for a given concentration reduce the performance requirements of any potential on-chip spectrometer; a necessary component for true lab-on-a-chip devices. Also evident is that many photonic crystal and slot waveguide based biosensors have demonstrated proof-of-principle by detecting substances such as streptavidin or BSA. Whilst these substances can be conjugated to a vast library of different antibodies [64], in general, most of these devices have yet to find application in a real world problem of importance. This is to be expected, however, due to the relative infancy of many of these devices when compared to the Biacore surface plasmon devices [35] sold commercially. One of most popular of these devices, the Biacore 3,000 is capable of detecting small biomolecules in concentrations below 1 nM, using samples as small as 5 µL, inside four different flow cells which can be optimised for different antibodies. Although the sensor area is on the order of square millimetres, the surrounding equipment result in a large machine that weighs 50 kg and could only be of real use within a lab. Although the recent Biacore 4,000 machine has the ability to run 4,800 interactions in a period of 24 hours, only 20 different target spots (i.e., 20 different assays) can be addressed, which is an improvement over the four of the Biacore 3,000, but a long way off the desired hundreds or thousands in protein microarrays [65]. The goal of much research in biosensor technologies is to produce a cheaper, more compact and higher throughput device that can compete with the performance of the Biacore without being restricted to high-end labs. Making the devices smaller is therefore of great benefit for multiplexing and higher throughput.

**Table 1:** Comparison of selected small optical biosensors

| Reference | Type of Device | Target Substance | Estimated Sensing Surface Area ($\mu m^2$) | Minimum Conc. Measured | $\Delta\lambda$ for 10 µg/mL Conc. (nm) |
|---|---|---|---|---|---|
| | | | **Slot Waveguide** | | |
| Barrios *et al.* [16]/ Banuls *et al.* [17] | Slot ring resonator | BSA | >250 | 0.042 µg/mL | 2.5 |
| Carlborg *et al.* [18] | Slot ring resonator | Anti-BSA | >250 | 0.125 µg/mL | 1.2 |
| Claes *et al.* [19] | Slot ring resonator | Avidin | >13 | 100 µg/mL (1.5 µM) | - |
| | | | **Photonic Crystal** | | |
| Zlatanovic *et al.* [24] | 2-D Photonic crystal cavity | Anti-biotin | 0.272 | 20 pM | - |
| Toccafondo *et al.* [20] | 2-D Photonic crystal waveguide | s-DNA | >100 | 19.8 nM | < 0.2 |
| Skivesen *et al.* [21] | 2-D Photonic crystal waveguide | BSA | >100 | 10 µg/mL (150 nM) | 0.2 |
| Dorfner *et al.* [25] | 2-D Photonic crystal cavity | BSA | 8.15 | <1 µM | - |
| Lee and Fauchet [22,23] | 2-D Photonic crystal cavity | BSA | 50 | 2% | - |
| Kang *et al.* [26] | 2-D Photonic crystal cavity | 3-APTES | - | - | - |
| Guo *et al.* [29] | 1-D Photonic crystal | Biotin-20T | - | 1 nM | 0.5 |
| Orosco *et al.* [30] | 1-D Porous silicon photonic crystal | Protease | - | 7 µM | - |
| Mandal *et al.* [27] | 1-D Photonic crystal cavity | Anti-streptavidin | 8.36 | 1 µg/mL | 0.25 |
| Konopsky and Alieva [28] | 1-D Photonic crystal | Biotin | - | 3 µg/mL | - |

| | | | | | |
|---|---|---|---|---|---|
| Scullion *et al.* [42] | Slotted photonic crystal cavity | Avidin | <2.2 | 1 μg/mL (15 nM) | 1.5 |
| Lee *et al.* [51] | Slotted microcavity photonic crystal | Latex spheres | <2 | - | - |
| Kita *et al.* [50] | Slotted photonic crystal nanolaser | BSA | <1 | 255 fM | - |
| Lai *et al.* [45] | Slow light slotted photonic crystal | Xylene in water | - | 86 μg/mL | N/A |

There may also be some room for improving experimental performance through better environmental control. Much of the cost for Biacore comes from the great lengths they go to stabilising the sample environment, minimising noise (such as temperature fluctuations) from unwanted sources. Whilst 'lab-on-a-chip' devices will also have to take environmental conditions into consideration, the integrated optics approach is perhaps more stable and requires less engineering as all the elements are on the same chip, thus variables can be cancelled out more easily using a reference.

## Fabrication and Characterization

Slotted photonic crystals are typically fabricated in Silicon-on-Insulator (SOI), which consists of a 220 nm top layer of silicon, on top of a 2 μm layer of silica on bulk silicon substrate. The silica layer is typically removed by a hydrofluoric acid (HF) wet etch in order to create a suspended membrane, which preserves vertical symmetry and takes advantage of the high refractive index of silicon for the slot waveguide effect. The fabrication procedure is outlined below in Figure 5, more details of which can be found in [66]. In brief, ZEP resist coated squares of SOI are exposed in an ebeam writer, before etching in a custom built reactive ion etching (RIE) device with a 1:1 blend of CHF3 and SF6 gases. The final structure is then cleaved with a diamond tipped scribe. An SEM image of one such structure is shown in Figure 6. The transmission line optical characterization setup is shown in Figure 7.

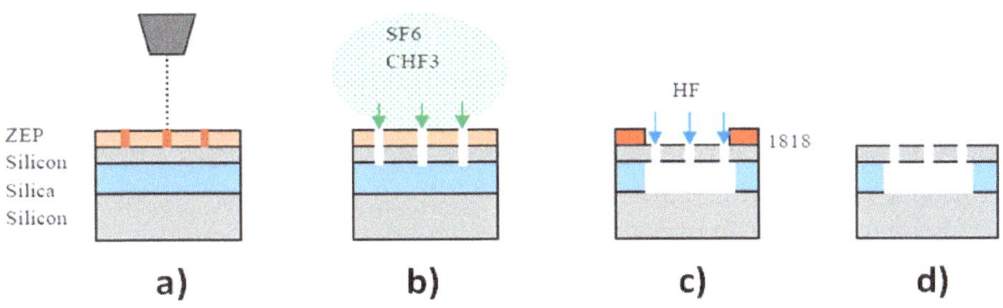


**Figure 5:** Photonic crystal membrane fabrication steps. (**a**) ZEP e-beam resist coated SOI substrate exposed to electrons in electron beam writer; (**b**) Sample developed exposing surface in pattern of design. Silicon etched using plasma of reactive ions; (**c**) Silica under photonic crystals removed using hydrofluoric acid through an S1818 photoresist patterned window. d) S1818 removed and final membrane sample cleaved.

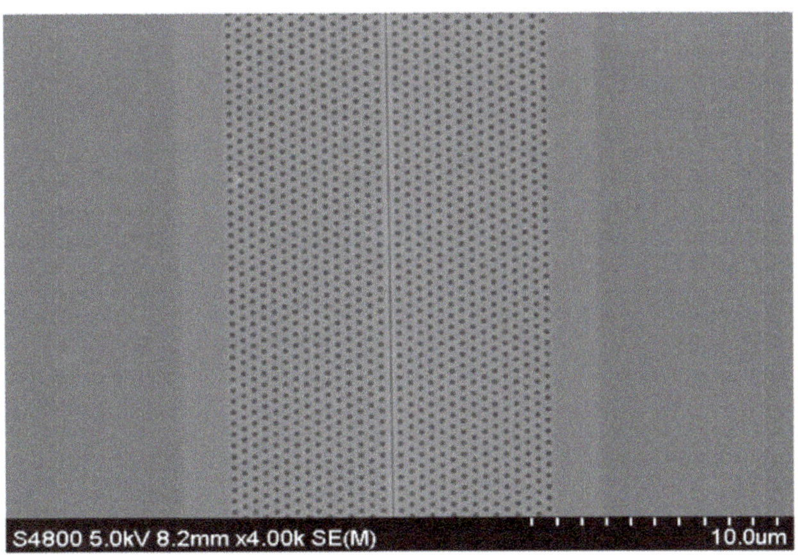


**Figure 6:** SEM image of slotted photonic crystal fabricated in SOI.

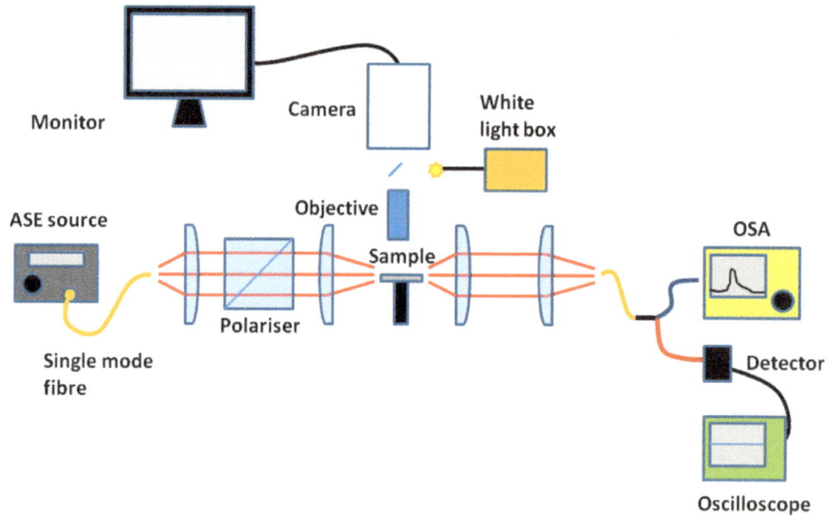


**Figure 7:** Optical endfire setup for transmission measurements. TE polarised light from broadband ASE source focused onto sample edge facet using aspheric lenses. Light collected from back facet split between OSA and photodetector. Sample using photodetector signal and camera as reference.

## LOSSES IN SLOTTED PHOTONIC CRYSTALS

### Coupling Loss

Whilst the unique optical properties of slotted photonic crystals are advantageous in many applications, this uniqueness also creates additional challenges, for example relating to the coupling of light into these structures. As seen earlier, the even modes of standard and slotted photonic crystals have very different dispersive properties; most notably, the gradient of the dispersion curve is of opposite sign. When these differing structures are coupled together, the group velocity mismatch (from the difference in sign) results in coupling into a backwards-propagating mode, resulting in strong reflection at the interface. In addition, there is also a mismatch between the shape of the slotted photonic crystal mode and that of the ridge waveguides used to deliver light to them. To take full advantage of the slotted architecture, a suitable coupler must be found.

For regular slot waveguides, one solution is to reduce the mode shape mismatch by using a sharply tapered ridge waveguide [67]. The same approach has been used for slotted photonic crystals, by firstly coupling the mode from ridge waveguide to slot waveguide via a taper, and then subsequently from slot waveguide to slotted photonic crystal [68]. Although this approach is relatively simple and goes some way to reducing the spatial mismatch of the modes, it introduces multiple interfaces (taper:slot and slot:slotted photonic crystals at both entry and exit) which all need to be optimized. For example, to optimize the slot: slotted photonic crystal interface, one needs to use a tapering of the hole position over several periods. More importantly, however, this design does not take into account the differences in dispersive properties introduced by the presence of the photonic crystal.

Due to the periodicity of a photonic crystal structure, all of its allowed optical modes can be represented using a region in its bandstructure known as the Brillouin zone. Like in solid state physics, the Brillouin zone lies below wavevector values of $\pi/a$ in the dispersion diagram (see Figure 2 for example bandstructure), this value itself corresponding to the bandgap. Above $\pi/a$, the modes are constructed from multiples of the first Brillouin zone wavevectors, thus secondary Brillouin zones and bandgaps result from the periodicity. When waveguides are used to couple light into standard W1 photonic crystals, group velocity matching is ensured through coupling in the second Brillouin zone, rather than the fundamental first, because here the two modes have the same sign of group index [69]. We have found that this is not the case for coupling between slot waveguides and slotted photonic crystals [70].

Figure 8 shows the guided modes of slot waveguides and slotted photonic crystals, along with simulated transmission spectra for coupling between them, for two different cases. In the first case shown in Figure 8(a,c) for a narrow rail slot waveguide, the guided modes of the slot waveguide (shown in blue) and the slotted photonic crystal (shown in red) intersect within the first Brillouin zone (i.e., below $k = \pi/a$). As can be clearly seen, the two modes have the same sign of gradient, and relatively good transmission between the two structures is observed. When the rails of the slot waveguide are made thicker, as in the second case shown in Figure 8(b,d), then the two modes intersect in the second Brillouin zone (i.e., above $k = \pi/a$) where they have gradients of opposite sign, and relatively poor transmission between the two structures is observed. In order to ensure favourable coupling to the slotted photonic crystal mode, the rails of the input slot waveguide must be made sufficiently narrow such that intersection occurs within the first Brillouin zone. This introduces limitations on the slots that can be used, and can lead to very fragile practical devices. Another coupling method suggested in the literature is to use multimode interference based structures [71]. In this case a standard input ridge waveguide is locally increased in width at the interface with the slotted photonic crystal. By adjusting the length of this region, the phase difference between the fundamental and second-order modes can be adjusted to give the best mode profile for coupling into the slotted photonic crystal. An enhancement of 20 dB in coupling was achieved for a 35 nm bandwidth [71].

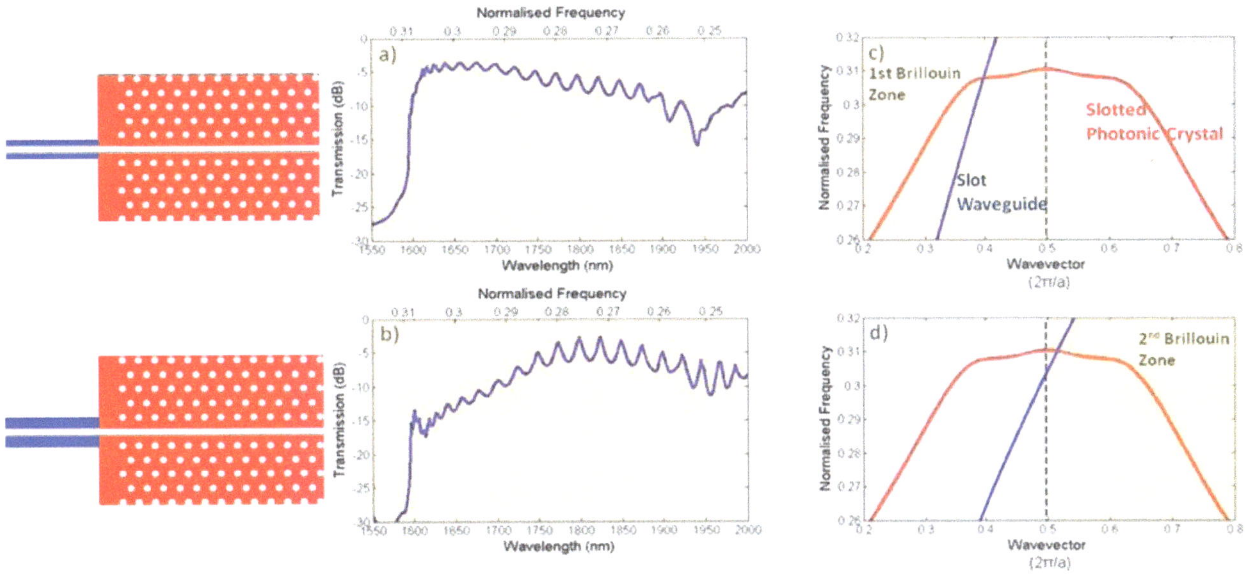


**Figure 8:** Coupling between slot waveguide and slotted photonic crystal. (**a**,**c**) Coupling in 1st Brillouin zone using narrow slot rails (196 nm wide) results in good transmission due to group velocity and phase matching. (**b**,**d**) Coupling in 2nd Brillouin zone using wide slot rails (294 nm wide) results in poor transmission due to group velocity mismatch. (Copyright © IEEE. All rights reserved. Adapted with the permission from [70]).

It has been shown by several authors [72–76] that lattice defects near the edge of a standard photonic crystal can be used to improve their directional emission. One design presented by Bauer and John [77] uses enlarged defect rods at the entrance and exit of a square lattice. Such resonant defects act as secondary sources of light that can cancel out the higher diffraction orders via destructive interference [77]. The result is a self-collimated beam that can be used to achieve very strong coupling with other structures, and operates over a large bandwidth. Recently, in [70] it has been shown that the same scheme can be used for slotted photonic crystals. Using the structure of Figure 9, coupling losses as low as 1.5 dB per interface over a bandwidth of 78 nm were realised experimentally [70]. The performance can be varied depending on the parameters used for the slot termination, the defect holes and the access waveguide width. This solution is robust in terms of fabrication, has good bandwidth and is therefore very suitable for taking full advantage of the slotted photonic crystal design.

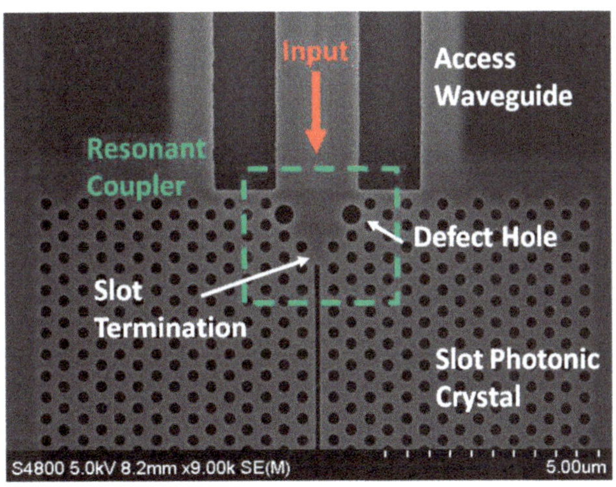


**Figure 9:** SEM image of slotted photonic crystal with resonant defect couplers and access waveguides, fabricated in SOI.

## Propagation Loss

Recently it has also been shown that the propagation losses of slotted photonic crystals waveguides can be as low as 10 dB/cm [78], and thus approaching similar levels as that of standard photonic crystals [79] and regular slot waveguides [80–83], as shown in Table 2 below. This is somewhat surprising as the surface roughness of the slot introduces another scattering surface that standard photonic crystals do not possess, thus one would expect the losses to be much higher. Broadly speaking, however, the increased scattering on the sidewalls of the slot is balanced by the reduced scattering at the sidewalls of the holes. This balance clearly depends on the slot width, and higher losses occur in narrower slots, as one would expect for the lower proportion of the field on the slot walls.

**Table 2:** Propagation loss comparison

| Reference | Device | Material | Slot Width (nm) | Propagation Loss (dB/cm) |
|---|---|---|---|---|
| Di Falco *et al.* [78] | Slotted photonic crystal | Silicon | 115 | 29 ± 10 |
| | | | 125 | 9 ± 5 |
| | | | 135 | 4 ± 7 |
| | | | 145 | 8 ± 6 |
| Baer-Jones *et al.* [80] | Slot waveguide ring resonator | Silicon | 50 | 11.6 ± 3.5 |
| | | | 60 | 7.7 ± 2.3 |
| | | | 70 | 8.1 ± 1.1 |
| Spott *et al.* [81] | Asymmetric slot waveguides | Silicon/PMMA | 130 | 1.7 ± 1.1 |
| Zhang *et al.* [82] | Slot waveguide | Silicon | 80 | 8.6 ± 0.61 |
| | | | 100 | 11.1 ± 1.15 |
| Alasaarela *et al.* [83] | Slot waveguide | Silicon/TiO2 | Uncoated | 71 |
| | | | Coated | 7 ± 2 |
| Notomi *et al.* [79] | Standard photonic crystal waveguide | Silicon | N/A | 2 |

The propagation losses of slotted photonic crystal waveguides have been shown to be comparable to regular slot waveguides, but the question remains: how low do the losses have to be to be for a useful application? The main advantage of using slotted photonic crystals is that they allow the peak of the optical mode to interact with the contents of the slot. Reducing the losses maximises the amount of light interacting with said contents, which is useful for sensing and other applications. The coupling losses can have a large impact (3 dB vs. 15 dB), though this has been rectified using the resonant coupler design. Whilst there is a strong variation in the propagation losses with slot width and wavelength, the structures used are typically less than 100 μm in length. Take for example (a) a slotted photonic crystal with loss of 39 dB/cm, and (b) another with loss of 11 dB/cm. If the structure is only 100 microns long, then (a) results in an absolute loss of 0.39 dB, whereas (b) results in an absolute loss of 0.11 dB. Both of these numbers are quite small when compared to the coupling loss, so do we care? If we use a cavity for sensing, using waveguide (b) allows more passes to be made, increasing the interaction time with the sample and producing a sharper (higher quality factor) resonance. For slow light structures, the losses also become more important due to enhanced scattering resulting from the higher intensities involved. Another point to consider is when several devices are used in series, waveguide (b) allows more to be multiplexed together than (a). It therefore depends on the application. To summarise, for short waveguide structures the propagation losses are not so important, whereas if long optical path lengths are to be used through cavities, slow light or multiplexing, then they can start to become significant. The results also show that the losses are not so much higher than standard photonic crystals to impose restrictions on their use in applications.

## MICROFLUIDIC INTEGRATION

### Background

The previous sections have shown that light can be squeezed down into nanoscale slots, and that this is advantageous for optical sensing; in this section we address the delivery of the sample solution to these sensors. One of the advantages of the slotted photonic crystal sensing architecture is the small footprint, on the order of 10–100 μm, thus potentially hundreds of independent sensors could be fabricated on a single chip. It therefore makes sense to control liquids at a similar scale such that samples can be directed to different sensors, whilst also minimizing their volume; something that can be important with precious samples and reagents. This is where microfluidics [84,85] comes into play; it being the control and manipulation of fluids in flow channels less than 1 mm wide. Microfluidics is a rapidly growing field in its own right, but is also being combined with other disciplines such as biophotonics [86]. One of the key features of fluid behaviour within microfluidic channels is that fluid flow is almost entirely laminar. The small size of microfluidic channels results in viscous forces dominating inertial ones, thus the Reynolds number is always much lower than 2,000, ensuring laminar flow [84].

One of the most common materials used in microfluidic chip fabrication is an elastomeric polymer called poly(dimethylsiloxane), or for short, PDMS. Using soft lithography, microfluidic devices can be rapidly fabricated in PDMS with feature sizes down to 30 nm [87]. PDMS is optically transparent and flexible. This flexibility can be exploited for making components such as pneumatic microvalves and pumps [88] in the microchannels. Here, microfluidic channels are integrated with the slotted photonic crystals as a way to supply the sensors with liquid samples of very low volume, whilst also preventing fast evaporation.

### Experiment

To form a master mould with the desired channel design (typically 100–200 μm wide channels), a piece of silicon is cleaned in acetone and IPA whilst inside an ultrasonic bath, and dried in dry nitrogen. SU-8 negative photoresist (2,050:2,000.5 in a ratio of 10:1) is then spun on top of the substrate at 1,000 rpm, giving a thickness of 40 microns. The sample is then baked on a hotplate at 65 °C for 5 min, followed by 95 °C for 15 min. To create the desired pattern, the sample is loaded into a mask aligner and exposed to UV light through a photomask of the channel pattern for 2 min. Following this, the sample is post exposure baked at 65 °C for 1 min and 95 °C for 10 min. The pattern is then developed in EC solvent for 3 min, and finally rinsed in IPA. To ensure good adhesion of the SU-8 to the silicon substrate, the sample is hard baked overnight in a 180 °C oven.

The PDMS base and curing agent are mixed in a ratio of 10:1.3 by volume, and degassed in a vacuum chamber to remove bubbles. The mixture is then carefully poured over the master mould inside a plastic Petri dish, and left it at 65 °C for over 4 hours to solidify. The solid PDMS is carefully pealed from the mould and cut to size with a blade. Access holes for tubing are punched through the microchannels where required. To seal the bottom of the channel, a coronal discharge gun is applied to the PDMS stamp with oxygen plasma [89] for 30 s, creating free OH groups on its surface [90]. When placed in contact with another surface with similar free groups, such as oxygen treated glass, PDMS, PMMA or silicon, the two surfaces can form covalent siloxane (Si–O–Si) bonds [90]. Each surface is treated with the oxygen plasma for 30 s, before rinsing in methanol; this allows the two surfaces to be placed in contact and aligned (the methanol acting as a lubricant). To ensure good bonding, this process must be carried out within a few minutes of oxygen treatment, thus methanol is used as it evaporates fast. After contact, the samples are placed in a 65 °C oven for over 4 hours. Better bonds result when using PDMS that is cured on the same day (ideally poured ~4 hours before bonding). The final sample has tubing of diameter slightly greater than the access holes, compressed into these holes to take liquids in and out of the microchannel. These moulds can be re-used numerous times to make PDMS microchannels. The full process is sketched in Figure 10.

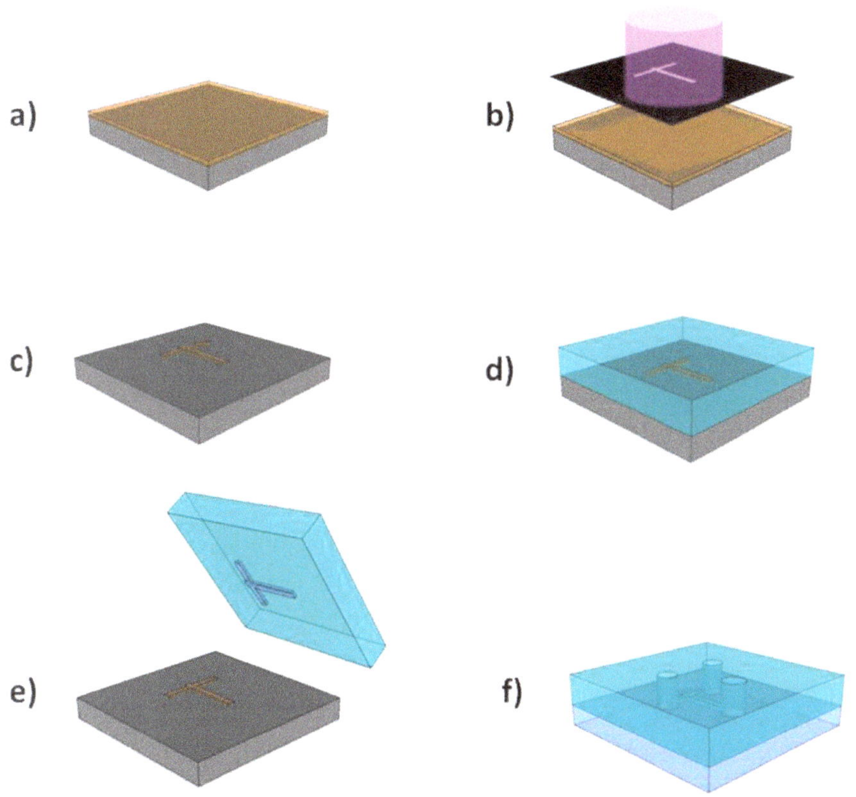


**Figure 10:** Soft lithography method of fabricating PDMS microfluidics. (**a**) SU-8 photoresist of thickness equal to desired channel height is spun and baked on clean silicon substrate; (**b**) Sample exposed to UV light through photomask of channel design; (**c**) Followed by, post-exposure bake, development and hard baking to complete mould; (**d**) Liquid PDMS base and curing agent poured on top of mould and allowed to solidify; (**e**) Solid PDMS containing imprint carefully peeled from mould. This mould can be used repeatedly for making PDMS imprints; (**f**) Access holes punched in to PDMS in order to connect pipes. PDMS and bottom cover (e.g., glass) then treated with oxygen plasma and bonded together.

In order to integrate the PDMS microchannels onto the silicon chip, a slightly different approach is adopted for bonding. Instead of plasma treating both surfaces, a piranha solution is used for the SOI containing the slotted photonic crystals. This method is used due to fears that arcs between a coronal discharge gun and a silicon surface could damage the membraned slotted photonic crystals. After cutting the PDMS chip to size, it is rinsed in IPA within an ultrasonic bath and dried with nitrogen gas in order to ensure the surface is clean. The piece of SOI containing the photonic crystals is then immersed in piranha solution for 15 min, before rinsing in DI water, acetone and IPA to create OH groups on the surface. The PDMS surface containing the microchannel is exposed to oxygen plasma on top of a plastic Petri dish for 30 s using the coronal discharge gun, before rinsing both the SOI and PDMS in methanol. The two surfaces are placed in contact under the microscope, and aligned using a pair of tweezers. The alignment of the microchannel to the photonic crystals has to be completed before the lubricating methanol evaporates, and shortly after oxygen treatment, in order to achieve good bonding at the desired location. Once aligned, the sample is left for a few minutes to allow the methanol to evaporate, before it is carefully transported to a 65 °C oven, and left to bond overnight. After removing from the oven, the sample is bonded to a glass sample holder using silver conductive paint, and tubing inserted into the access holes. Images of a finished chip are shown inFigure 11. No leaks were detected.

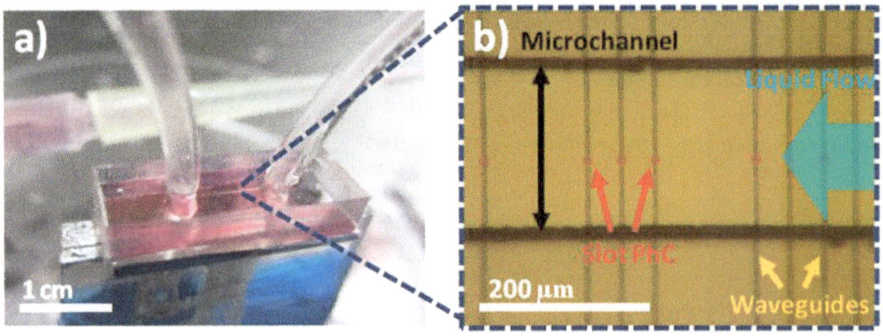


**Figure 11:** Slotted photonic crystals with integrated microfluidic channels. (**a**) PDMS microchannel bonded on top of SOI chip. Sample glued to glass holder. Tubing inserted into access holes, to which syringes can be connected. (**b**) Optical microscope image of channel with slotted photonic crystals inside. Access waveguides take light to and from photonic crystals. (Copyright © Elsevier. Adapted with permission from [42]).

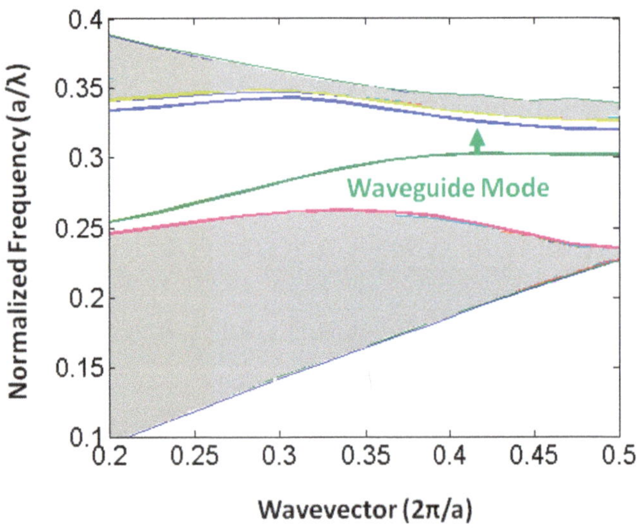


**Figure 12:** Bandstructure of slotted photonic crystal waveguide showing waveguide mode. A small local compression of the lattice period reduces the amount of dielectric, which in turn raises the operational frequency further into the forbidden bandgap region.

## BIOSENSING USING SLOTTED PHOTONIC CRYSTAL CAVITIES

### Cavities

Several cavity designs have been explored for slotted photonic crystals, including heterostructure [41,47], local line width modulated [91,92], L3 type [93], and a varying slot width [43] approach amongst others [50,51,94]. Amongst these, the heterostructure approach presents better fabrication tolerances, though all of these can, and have, been adopted for sensing. In the heterostructure cavity of Noda [95] for standard W1 photonic crystals, the lattice is locally expanded in one direction along the W1 (single row of missing holes) waveguide for a few lattice periods. The lattice remains unaltered on either side of this defect. Only light of a certain frequency can be supported in the defect region, and is confined by the surrounding lattice which acts as a mirror. As seen earlier, the different nature of a slotted defect versus a standard W1 defect photonic crystal results in different optical properties. To form a cavity in the slotted photonic crystal architecture requires the opposite approach to that of the W1. As shown in [41], heterostructure cavities within slotted photonic crystal require acompression, rather than expansion, of the lattice. As the waveguide mode of the slotted photonic crystal cuts-off at high frequency, a new forbidden frequency can be introduced by increasing the frequency slightly into the bandgap, one way being lattice compression. With this design our group has demonstrated theoretical quality factors of 80,000 in air, 6,000 in water, and experimental values of 50,000 in air and 4,000 in water [41].

Slotted photonic crystal cavities are advantageous for sensing applications as most of the field interacts with contents of the air slot. This is clear from the previous discussion and is highlighted by the FDTD simulation of Figure 13 for a heterostructure cavity. In conventional photonic crystal cavities and other devices such as ring resonators, most of the light is confined to the dielectric of the waveguide structure, and only the evanescent tail of the optical mode sees the analyte of interest. In the slotted photonic crystal case a change in the refractive index of the slot contents is felt by the majority of the modal field, altering the resonant wavelength of the system more strongly. This field overlap is reflected by the sensitivity 500 nm/RIU vs. 50–100 nm/RIU. We use an in-line coupling mechanism as demonstrated in Figure 14(c). Light of resonant wavelength couples from the mirror interface, into the cavity, and back out to the waveguide beyond the second mirror. This produces a peak in the transmission spectrum. The important difference with side coupling methods shown in Figure 14(a,b) is that other wavelengths of light which are guided, rather than reflected, by the wavelength dependent mirrors can pass also through. This produces a broad region of high signal within the output transmission spectrum, which can be used for alignment purposes. The full spectrum therefore shows a broad region of high transmission, with a stopband at the mirror wavelengths, within which lies the transmission peak of the cavity resonance. The disadvantage of this approach is that there are more interfaces to consider.

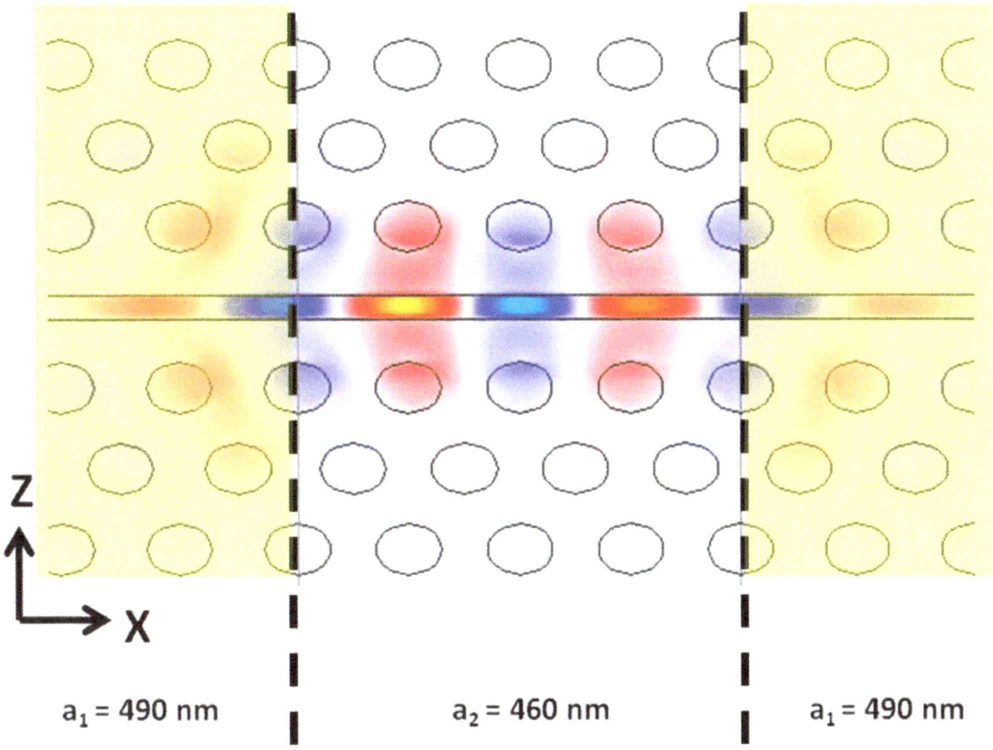


**Figure 13:** 2D FDTD simulation of an air slotted photonic crystal cavity mode showing Ez component. Mirrors are highlighted in yellow. The majority of the field intensity is confined to the slot, a property which lends itself to sensing because of the resulting strong light-matter interaction. (Copyright © Elsevier. Adapted with permission from [42]).

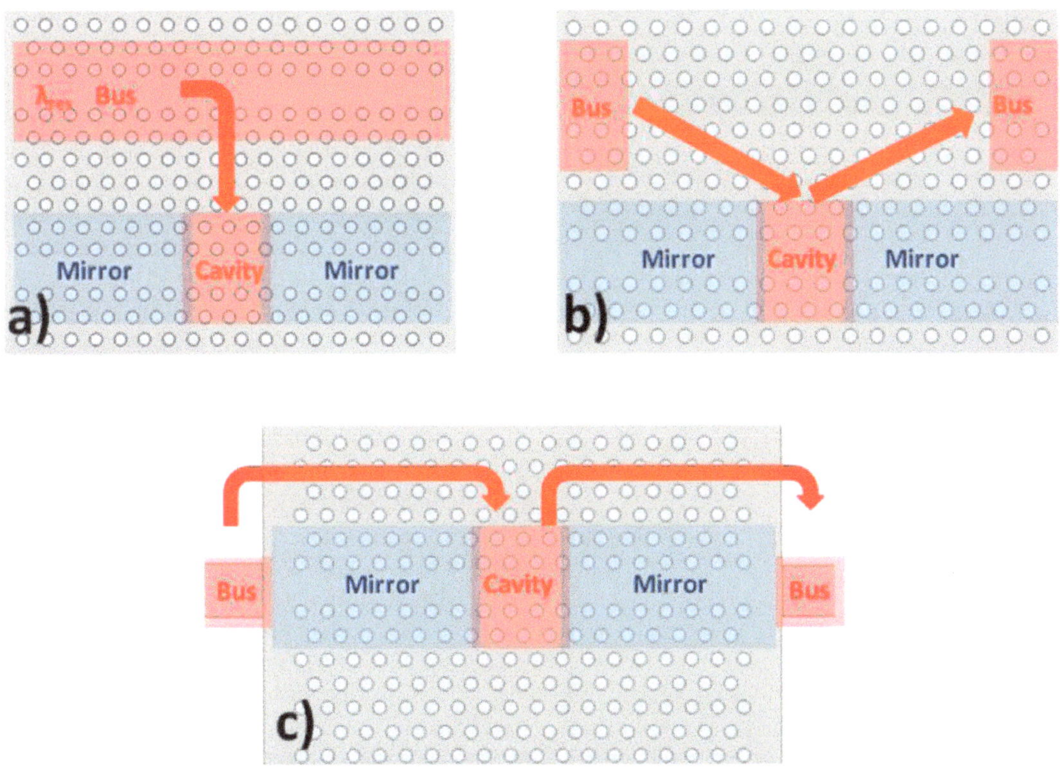


**Figure 14:** Common photonic crystal cavity coupling mechanisms. (**a**) Side coupling from bus waveguide produces dip in transmission spectrum; (**b**) Side coupling using input and output buses produces peak in transmission; (**c**) In-line coupling produces cavity peak plus guided wavelengths of mirror waveguide region.

Cavities were fabricated in SOI as outlined earlier. For operation at 1,550 nm for water filled devices, a hole radius of 150 nm, and slot width of 180–200 nm were chosen. PDMS microfluidic channels of 40 μm height and 200 μm width were integrated with the chip. We found experimentally that most repeatable results occurred when using the interface design shown in Section 4. An SEM image of the structure is shown in Figure 15. As a first test different alcohols were introduced into the microchannel, as shown in Figure 16 below. The cavity peak (around 1,532 nm for ethanol) is clearly distinguished from the bulk transmission of the mirror waveguide (above 1,550 nm for ethanol), and was found to shift significantly with the different alcohols. At 1,550 nm, Ethanol has a refractive index of approximately 1.354 and IPA has 1.378. Given that the shift in resonant wavelength is approximately 10 nm, this corresponds to a refractive index sensitivity of approximately 415 nm/RIU.

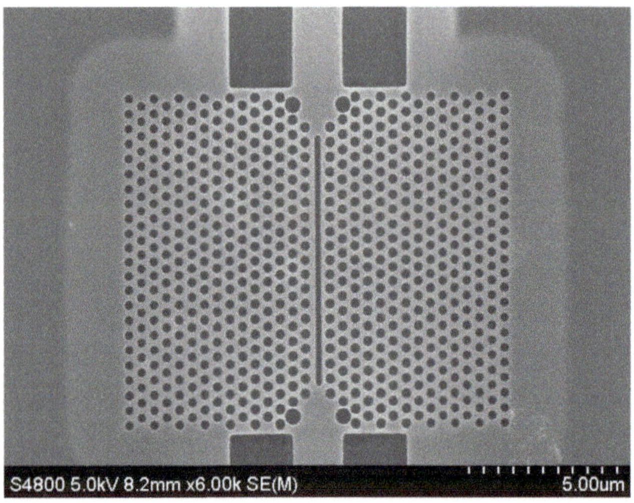


**Figure 15:** SEM image of slotted photonic crystal cavity fabricated in SOI.

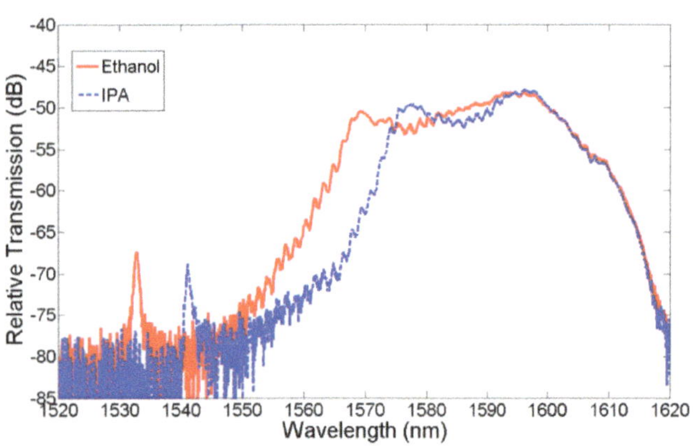


**Figure 16:** Transmission spectra of a slotted photonic crystal cavity with different alcohols injected into the integrated microchannel.

## Functionalization

In order to test the cavity's potential as a biosensor, suitable capture agents and targets had to be chosen. We initially decided to investigate the avidin-biotin system as it is often used as a model for studying protein interactions [64]. It can also be used to help immobilize a variety of antibodies on surfaces. Its use here is therefore of great interest as it could be adapted to suit most types of antibody, and thus different types of sensor. Biotin can be immobilized on silicon dioxide surfaces using a chemical known as aminopropyltriethoxysilane (APTES). This chemical binds to free OH groups on the surface of the glass, whilst presenting free $NH_2$ groups for further functionalization. NHS-linked Biotin can attach to these $NH_2$ groups [96]. In order to functionalize the photonic crystals with biotin receptors, a protocol similar to that of De Vos et al. [1] was adopted, as they demonstrated biotin functionalization of silicon microphotonic structures.

Before bonding the microfluidics, the SOI chips containing the photonic crystals were immersed in Piranha solution for greater than 20 min to ensure that the surface was clean, hydrophilic, and to expose OH groups on the surface (from the resulting oxidation) to aid functionalization. The microfluidic circuitry was attached as in Section 5. After bonding the chip to a sample holder with silver conductive paint, the chip was placed in the optical characterization setup. APTES was diluted in toluene to a concentration of 1% and was injected into the chip to sit for several hours. We found, however, that the toluene caused the PDMS microfluidics to swell up, and in some cases even melt, destroying both the microfluidic circuitry and the silicon surfaces. The later step of biotin involved DMF, which has similarly detrimental effects on PDMS. As noted by Whitesides' group [97], PDMS swells up in the presence of nonpolar organic solvents, one good example of which is toluene, with the amount of swelling varying between solvents.

In order to combat this problem, we tried a number of approaches. Liquid silica solutions such as Flowable Oxide (FOX) were coated on the PDMS chip through spinning, and baking on a hotplate. The problem with this method is that FOX 14 (Dow Corning) requires baking to temperatures in excess of 400 °C, which are not suitable for PDMS (PDMS is thermally stable only up to approximately 180 °C), thus the full glass transition could not be allowed to take place. Another approach was to sputter thin layers of gold into the channel itself, masking the surrounding PDMS surfaces with tape during the procedure so as to protect the parts to be placed in contact with the silicon. This appeared to reduce the swelling in toluene, but resulted in reduced light transmission, suggesting the gold was scattering some of the light at the edges of the channel in contact with the silicon or that the gold was detaching when solutions were flowed through the channel. As we did not succeed in adapting the PDMS to the functionalization chemicals, we decided to change the chemistry. Literature (for example [2,27,96]) shows that silicon dioxide surfaces can also be functionalized with APTES dissolved in many different solvents including ethanol and water. Some of these solvents only have very little effect on the PDMS. The immersion times also varied widely from a matter of minutes to 24 hours. After several test runs with different concentrations and timing, the protocol described below was adopted (see Figure 17).

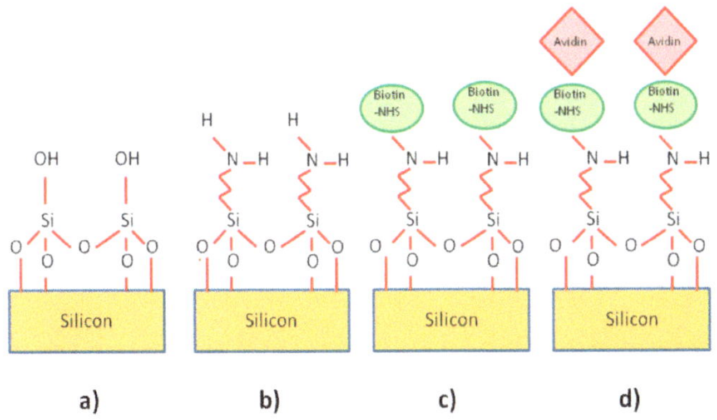


**Figure 17:** Biotin functionalization of silicon surface using APTES. (**a**) Silicon immersed in Piranha solution to expose OH groups on surface; (**b**) APTES links to OH groups on surface and presents $NH_2$ groups; (**c**) NHS linked biotin attaches to $NH_2$ groups; (**d**) Avidin captured by immobilized biotin.

Following the piranha and microfluidic bonding steps mentioned above, ethanol was diluted to a concentration of 95% in DI water (1 part DI water to 19 parts ethanol). Using this solution, APTES was diluted to a concentration of 2%. This was then injected into the microchannel and left overnight (typically 19 hours in total). No noticeable swelling or deterioration of the PDMS was detected. After this, the channel was flushed several times with ethanol and then DI water, and the whole chip was placed inside a 65 °C oven for over 5 hours in order to allow the APTES to cure. Biotinamidohexanoic acid N-hydoxysuccinimide (NHS linked biotin) ester powder (Sigma–Aldrich, Dorset, UK) was dissolved in dimethylformamide (DMF) to a concentration of 1 mg/mL. This was then further diluted in phosphate buffered saline (PBS) to a concentration of 100 µg/mL. This was flowed into the cured chip, and left to sit in the channel for over 12 hours in order to saturate the surface with biotin receptors. The chip was then flushed several times with PBS. As each step produced a shift in the cavity peak wavelength that did not recover after flushings, then the functionalization protocol was taken to be successful.

## Protein Sensing

To test the performance of the device, avidin from egg-white lyophilized powder (Sigma–Aldrich) was diluted down to 10 µg/mL in PBS. After functionalization and optical alignment, a blank PBS solution was first used, before injecting the avidin, via a syringe and inlet tubing into the chip as shown in Figure 18 (see Section 3 for full optical setup). During this time the transmission spectra of the slotted photonic crystal was recorded from the optical spectrum analyser (OSA). The high sensitivity setting on the OSA was used to minimize noise. The results are shown in Figure 19, which depicts the variation in the peak wavelength of the cavity versus time.

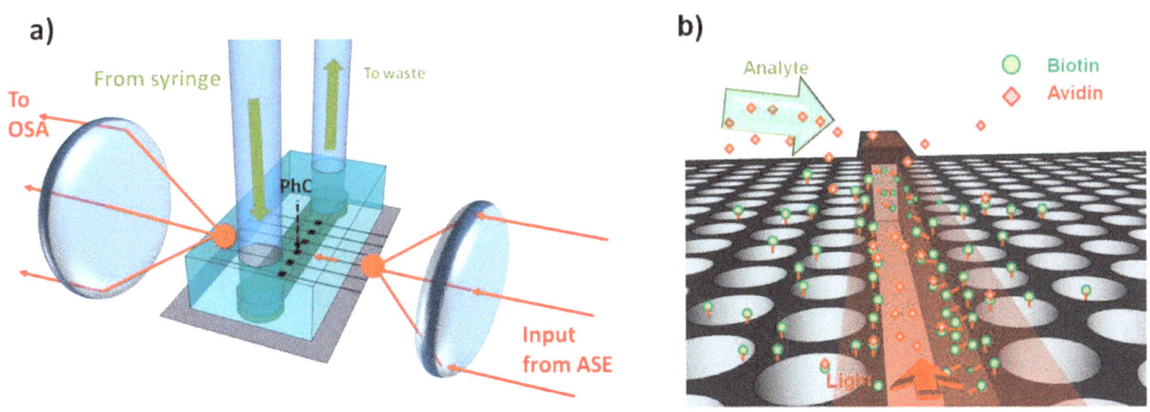


**Figure 18:** Biosensing experimental setup (not to scale). (**a**) Light focused into optical waveguide at front edge. This light is then taken through the photonic crystal (inside the microfluidic channel) and then to the back edge where it is collected. Liquids are passed through the microchannel via connected tubing and syringe; (**b**) Avidin binds to biotin coated sensor inducing a shift in the cavity peak wavelength.

The results of Figure 19 showed that avidin binding produced a large response, and that this response is shaped by diffusion. 30 minutes was also found to be sufficient to reach saturation. In order to test the lowest limit of detection another fresh chip was fabricated and functionalized as before. This time, avidin was serially diluted in PBS from 100 µg/mL down to 1 ng/mL. Each solution was introduced in order of increasing concentration into the chip and left to sit for typically 30 min each. Spectra were taken every minute, and the cavity peak wavelength extracted from a Lorentzian fit. Whilst some errors will result from the fit, given that the shifts in wavelength during binding were greater than the linewidth of the cavity, these fitting errors were assumed to be negligible. The results are shown in Figure 20. Note that all of these measurements were performed on a single photonic crystal.

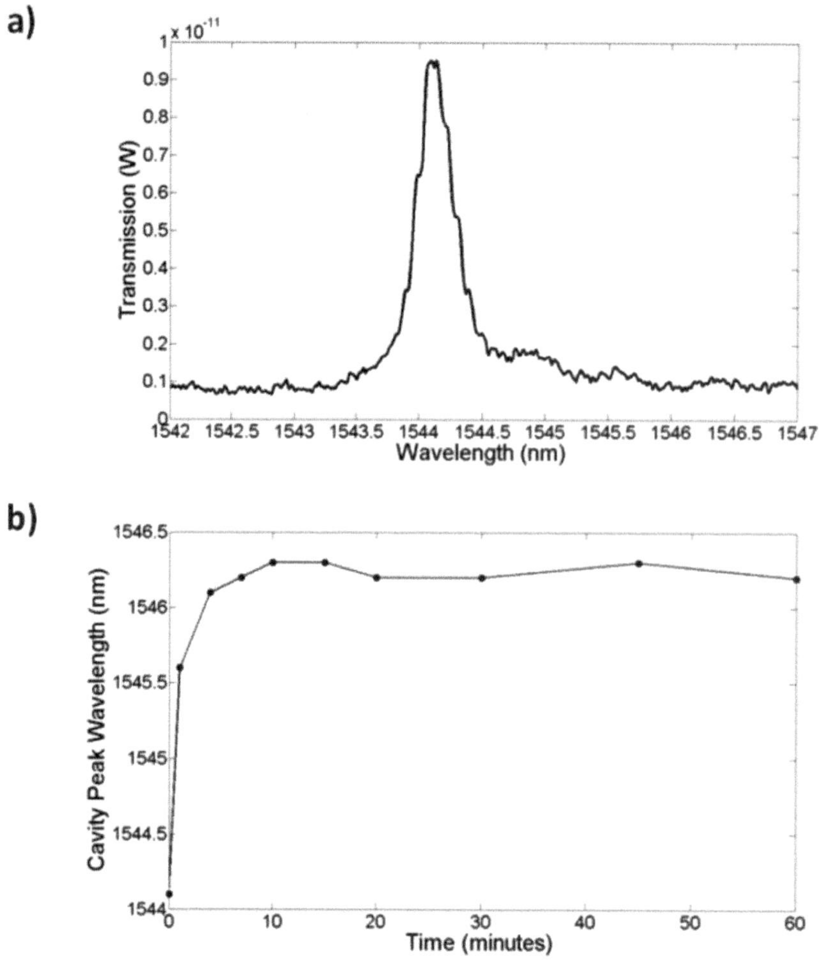


**Figure 19:** Initial avidin biosensing experiments. (**a**) Cavity spectrum; (**b**) Cavity peak wavelength as a function of time for 10 μg/mL concentration.

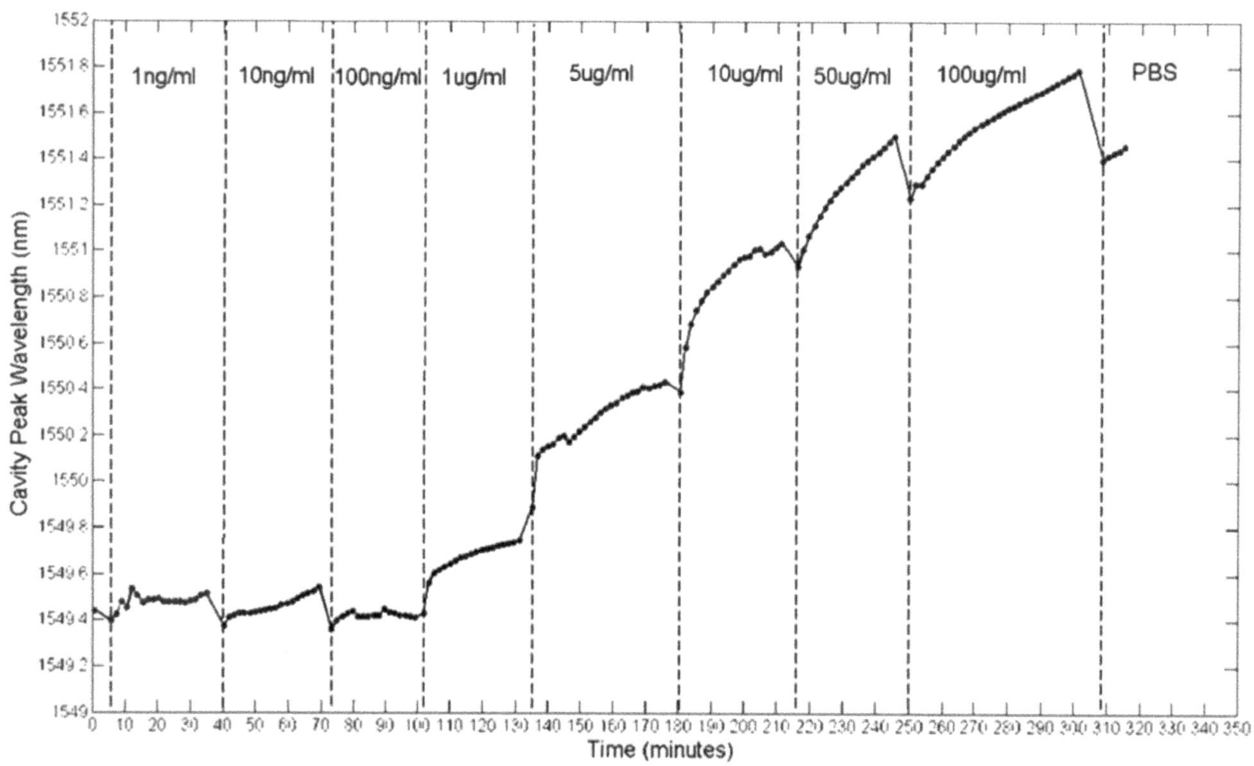


**Figure 20:** Cavity peak wavelength for a single slotted photonic crystal cavity as a function of time for different concentrations of avidin in PBS (Copyright © Elsevier. Adapted with permission from [42]).

Figure 20 highlights again that the time-response of the sensor is dominated by diffusion. Whilst the experiment was only run for 30 min per concentration, leaving it for longer would allow these diffusion curves to plateau at their respective saturation values. Each concentration has a different curve and as expected, lower concentrations have shallower gradients. The little dips in sensor response at each injection point can be attributed to the flushing of material that has not properly bonded to the biotin [18]. A better measure of the actual bound material are therefore these dip points, as here the unbound material has been discarded. Taking these points, Figure 21 shows how the sensor response for a thirty minutes reaction varies for different avidin concentrations. The lowest concentration that can be clearly observed is 1 μg/mL. The average quality factor was estimated to be 3,000 from the measurements.

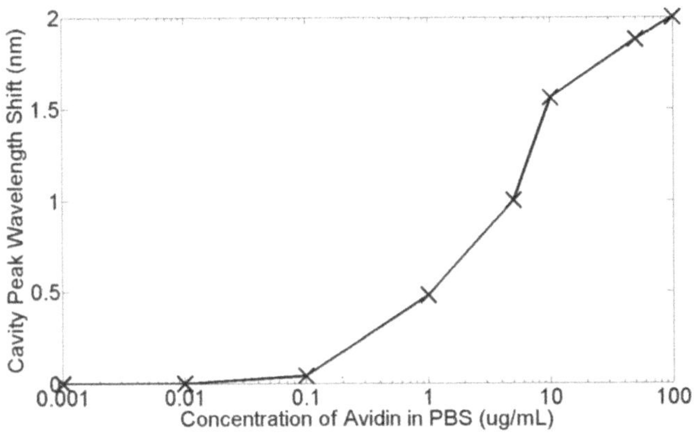


**Figure 21:** Cavity peak wavelength shift after binding for each avidin concentration. (Copyright © Elsevier. Adapted with permission from [42]).

## Analysis

From Figures 20 and 21, a dissolved avidin concentration of 1 μg/mL was clearly detected. As avidin has a molecular weight of 68 kDa [98], this corresponds to a molar concentration of 15 nM. Given that the sample stage was not temperature controlled, thermal noise must be taken into consideration. For de-ionised water at a temperature of 20 °C, light of wavelength 1,550 nm experiences a thermo-optic coefficient of approximately $10^{-4}$ RIU/°C [99]. Assuming a maximum temperature fluctuation of 0.5 °C within the lab over the time period for each concentration, and the refractive index sensitivity of the slotted photonic crystal cavity being 500 nm/RIU, this would correspond to a shift in resonant wavelength of 0.025 nm. Even assuming an order of magnitude higher thermo-optic coefficient of $10^{-3}$ RIU/°C would produce a shift of 0.25 nm. In both cases, the wavelength shift is smaller than the shifts observed for avidin concentrations above and including 1 μg/mL. To experimentally test the thermal stability of the PBS solutions, the cavity peak wavelength for a PBS filled channel was recorded over a 20 min period. The results shown in Figure 22 show fluctuations in wavelength of less than ~0.05 nm, which is consistent with a thermo-optic coefficient of $5 \times 10^{-4}$ RIU/°C. With these numbers, and given the linewidth of the cavity, it is fair to say that binding can be attributed to wavelength shifts of greater than 0.1 nm. Better environmental control should allow this number to be reduced in future. Comparing with Table 1 shows that a detection limit of 15 nM is impressive given the small size of the slotted photonic crystal cavity. To aid comparison further, an estimate of the surface density and bound mass is also made. For the smallest observed wavelength shift $\delta\lambda$ (in this case taking 0.4 nm for 1 μg/mL) the surface density of bound material $\sigma_p$ can be estimated from the equation [100]:

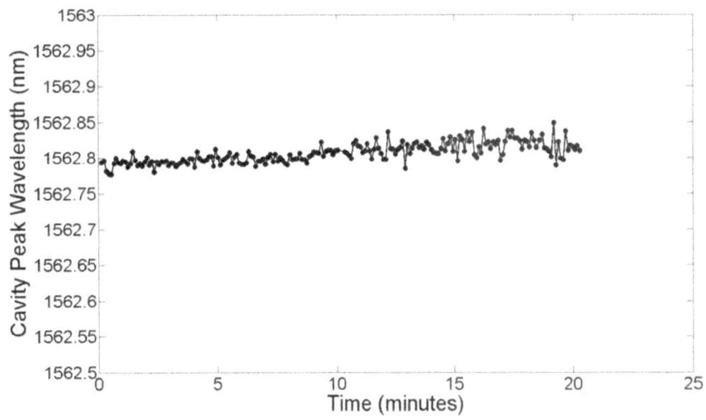


**Figure 22:** Cavity peak wavelength as a function of time for PBS filled microchannel.

$$\frac{\partial\lambda}{\lambda} = \sigma_p \alpha_{ex} \frac{2\pi\sqrt{n_m^2 - n_s^2}}{\varepsilon_0 \lambda^2} \frac{n_m}{n_s^2} S \tag{3}$$

where $\alpha_{ex}$ is the excess polarizability of the molecule, $n_m$ is the refractive index of the sensor substrate, $n_s$ is the refractive index of the analyte and S is the bulk refractive index sensitivity. Taking the values for this experiment: $\alpha_{ex} = 4\pi\varepsilon_0 (3.2 \times 10^{-21})$ $cm^3$[101], $n_m = 3.4$ (silicon), $n_s = 1.35$ (PBS), S = 500 nm/RIU [41], $\delta\lambda = 0.4$ nm and$\lambda = 1{,}550$ nm, the bound surface density is found to be $5 \times 10^8$ molecules/mm$^2$. For avidin, this corresponds to a surface mass density of 60 pg/mm$^2$. From FDTD simulations of the cavity, the sensing surface area is estimated to be as low as 2.2 μm$^2$ (based on the surface area of the slot walls within, and the six holes surrounding, the cavity), thus the bound mass detected is of the order of 100 ag. Such numbers should be used with caution, as they are based on assumptions. The concentration limit of detection is the most reliable quantity as this was actively controlled; it should be noted, however, that this value will differ for different antibody-antigen pairs. Although a biotin solution is left in contact with the substrate for several hours in order to ensure saturation, and the specificity of this protocol has already been established [1], non-specific binding could be eliminated by adding suitable blocking agents to the protocol.

This result highlights the increased light-matter interaction of slotted photonic crystal cavities. Wavelength shifts comparable to that of much larger structures can be observed in smaller devices. Compactness allows potential for hundreds of these devices to be integrated onto a centimetre scale chip. When combined antibody patterning techniques, this would allow hundreds of independent biological tests to be conducted simultaneously using a minimum amount of precious analyte. The disadvantage when compared with larger structures is that the target sensing area is much smaller. When dealing with very low concentrations, the target molecule may never see the 2.2 μm$^2$ sensing area on a reasonable timescale. This is a general limitation of all surface affinity based biosensors. Microfluidics has potential to help here. Water absorption at 1,550 nm also impacts on the performance of the device, by reducing the quality factor of the cavity. High quality factor cavities are advantageous for avidin sensing as higher Q-factor cavities allow smaller shifts to be detected. The slot engenders higher field overlap with the sample, which gives bigger shifts but also results in greater absorption from the solution in which the target molecules are contained when compared to standard ring resonators. Shifting to shorter wavelengths, where water absorption is lower, should reduce this problem. 1,550 nm was chosen here due to the wide availability of sources and detectors and the group›s familiarity with working at this wavelength. The lowest water absorption occurs at visible wavelengths, but this is inaccessible with slotted photonic crystals fabricated in silicon, due to strong absorption of silicon below 1,000 nm. Other materials (such as polymers) that are transparent at such wavelengths do not have as high a refractive index as silicon, and are thus not of sufficient index contrast necessary for the slot waveguide effect. Reducing the wavelength also requires downscaling the photonic crystal parameters (radius, slot width, period), which is harder to fabricate and increases potential for clogging. Despite these problems it is conceivable that performance could be improved by using wavelengths around 1,300 nm. The quality factor Q of a photonic crystal cavity is given by [102]:

$$\frac{1}{Q_{tot}} = \frac{1}{Q_{rad}} + \frac{1}{Q_{abs}} \quad (4)$$

where the subscripts refer to the total Q-factor, and the Q-factors describing radiation and absorption losses. At the 1/e point, the propagation limit L due to absorption is given by:

$$L = \frac{1}{\alpha} \quad (5)$$

where α is the absorption coefficient. The relationship between the quality factor and the lifetime $\tau_c$ of a cavity is given by:

$$Q_{abs} = \frac{2\pi f \tau_c}{\eta} \quad (6)$$

where η is the overlap of the mode with the absorbing medium and f is frequency. Combining Equations (5) and (6) yields the relationship between $Q_{abs}$ and α:

$$Q_{abs} = \frac{2\pi n}{\alpha \lambda \eta} \quad (7)$$

where n is the refractive index of the cavity medium. As $\alpha \approx 9.68\ cm^{-1}$ and $n \approx 1.319$ at 1,550 nm [103], and the overlap of the mode with water within the slot is approximately 0.85, then Equation (7) gives a value of $Q_{abs} = 6{,}500$. Taking $Q_{rad} \approx 50{,}000$ (highest Q in experiment without water [41]), then from Equation (4) $Q_{tot} = 5{,}700$, which is somewhat higher but of the same order, as the maximum experimental Q-factor we observe ($Q_{exp} \approx 4{,}000$ [41]). The water absorption at 1,300 nm is almost one order of magnitude lower as is evident from its absorption coefficient, $\alpha \approx 1.34\ cm^{-1}$ [103]. Using the same analysis as before, for 1,300 nm, given that $n \approx 1.324$, the quality factors are predicted to be $Q_{abs} = 56{,}000$ and $Q_{tot} = 26{,}000$. As this Q-factor is one order of magnitude greater than at 1,550 nm, working at 1,300 nm should allow smaller shifts to be detected.

## BARRIERS TO ENTRY

From the wide variety of applications, sensitivity and commercial success, surface plasmon resonance can be viewed as the technology to beat in terms of label-free biosensing. For any technology to replace the existing standard it must meet at least one of the following requirements:

Perform better: in the case of biosensing, having better sensitivity or limit of detection, and/or be easier to operate.

Cost less: For biosensors, not only the cost of the device, but also the cost per test in terms of materials and labour.

Do something new: can the biosensor investigate some problem that other devices cannot?

One of the main barriers for taking full advantage of the integrated optics (ring resonator, photonic crystal, slot waveguide etc.) approach is the sources and detectors available. Although the sensor chips can be very small, they often require a bulky spectrometer to function fully, and are thus more ‘chip-in-a-lab’ than ‘lab-on-a-chip’. This point is often brushed over and is one of the key limiting factors of many devices in the literature. Multiplexing at the same time can also be tricky. Either the source has to be split and the detector arrayed (e.g., [18]), which can be lossy, or each sensor has to be rapidly addressed individually as in the case of Genalyte [36]. Sensors based on silicon also suffer from a lack of integrated light sources due to the indirect bandgap of silicon, though solutions such as flip-chip bonding can be used to attach sources constructed in other material platforms. A further barrier comes from the fact that the machines sold by Biacore benefit from a research and commercial history stretching back more than 20 years, and possess the majority of the current market.

More generally the fragile and expensive nature of antibodies limits the vision of protein microarrays below their DNA counterparts. Whilst DNA can be copied rapidly in PCR, proteins are more vulnerable to harsh conditions. The use of surface based sensors also means that detection is based on diffusion and sedimentation, which limits both performance and speed. Antigens in solution may never see the capture antibody on the surface of the device within a reasonable timescale. Microfluidics has potential to help here. Many photonic sensors also rely on having prepared solutions involving steps such as centrifugation and dilution rather than work with unaltered samples (e.g., a drop of blood), and are thus not user-friendly to the non-specialist. Despite all of these problems, it is conceivable that integrated optical sensors will feature in future commercial biosensors due to the advantages of size, mass producibility and sensitivity they offer.

## SUMMARY AND OUTLOOK

Slotted photonic crystals have been developed towards a realistic and promising platform for lab-on-a-chip style biosensors. The great advantage over many other photonic devices is the strong confinement of light, not only spatially, but also temporally due to combination of slot waveguide and photonic crystal effects. To better understand the limitations of the system and to maximise the amount of light within the slot, we performed studies on the losses. The propagation losses were found to be comparable to standard photonic crystals in the fast regime as a result of the slot waveguide effect.

Coupling losses were reduced to 1.5 dB per interface over a large bandwidth of 78 nm using a resonant coupler design. This allowed full advantage of the slotted photonic crystal light confinement in air to be taken. The sensitivity of this system was shown in proof-of-principle experiments where the protein avidin was detected, at a lowest concentration of 15 nM, and where bigger shifts in resonant wavelength were observed for the same concentration as many larger photonic biosensors. When combined with the very small size of the device, this makes them attractive for the high throughputs, multiplexing and multifunctionality required by protein microarrays. Larger wavelength shifts also reduce the performance requirements for any on-chip spectrometer, which is necessary for a real lab-on-a-chip device. This work also showed that slotted photonic crystals can be integrated with microfluidic circuits.

Slotted photonic crystals therefore provide an entirely new platform for optical biosensing, allowing higher sensitivities in smaller structures, and great potential for on-chip spectrometry. Whilst it will take at timeframe of 5–10 years and more resources and engineering to make this a commercial reality, the basis of design, performance, detection, proof-of-principle and application have all been demonstrated within this body of work. Slotted photonic crystals therefore still have much more to give, and have strong potential for true 'lab-on-a-chip' devices. Our hope is that with this foundation they will shine new light on many of the mysteries of biology in years to come.

## ACKNOWLEDGMENTS

Andrea Di Falco is supported by an EPSRC Career Acceleration Fellowship (EP/I004602/1). We also like to thank Carlos A. Barrios of Universidad Politecnica de Madrid for fruitful discussions.

## REFERENCES

1. De Vos, K.; Bartolozzi, I.; Schacht, E.; Bienstman, P.; Baets, R. Silicon-on-insulator resonator for sensitive and label-free biosensing. Opt. Express 2007,15, 7610–7615.
2. Ksendzov, A.; Lin, Y. Integrated optics ring-resonator sensors for protein detection. Opt. Lett. 2005, 30, 3344–3346.
3. Iqbal, M.; Gleeson, M.A.; Spaugh, B.; Tybor, F.; Gunn, W.G.; Hochberg, M.; Baehr-Jones, T.; Bailey, R.C.; Gunn, L.C. Label-free biosensor arrays based on silicon ring resonators and high-speed optical scanning instrumentation. IEEE J. Sel. Top. Quantum Electrodyn. 2010, 16, 654–661.
4. Homola, J. Surface plasmon resonance sensors for detection of chemical and biological species. Chem. Rev. 2008, 108, 462–493.
5. Barlen, B.; Mazumdar, S.D.; Lezrich, O.; Kampfer, P.; Keusgen, M. Detection of salmonella by surface plasmon resonance. Sensors 2007, 7, 1427–1446.
6. Stevens, R.C.; Soelberg, S.D.; Eberhart, B.-T.L.; Spencer, S.; Wekell, J.C.; Chinowsky, T.M.; Trainer, V.L; Furlong, C.E. Detection of the toxin domoic acid from clam extracts using a portable surface plasmon resonance biosensor.Harmful Algae 2007, 6, 166–174.
7. Kausaite-Minkstimiene, A.; Ramanaviciene, A.; Ramanavicius, A. Surface plasmon resonance biosensor for direct detection of antibodies against human growth hormone. Analyst 2009, 134, 2051–2057.
8. Lee, J.W.; Sim, S.J.; Cho, S.M.; Lee, J. Characterization of a self-assembled monolayer of thiol on a gold surface and the fabrication of a biosensor chip based on surface plasmon resonance for detecting anti-GAD antibody. Biosens. Bioelectron. 2005, 20, 1422–1427.
9. Feltis, B.N.; Sexton, B.A.; Glenn, F.L.; Best, M.J.; Wilkins, M.; Davis, T.J. A hand-held surface plasmon resonance biosensor for the detection of ricin and other biological agents. Biosens. Bioelectron. 2008, 23, 1131–1136.
10. Homola, J.; Dostalek, J.; Chen, S.; Rasooly, A.; Jiang, S.; Yee, S.S. Spectral surface plasmon resonance biosensor for detection of staphylococcal enterotoxin B in milk. Int. J. Food Microbiol. 2002, 75, 61–69.
11. Mayer, K.M.; Lee, S.; Liao, H.; Rostro, B.C.; Fuentes, A.; Scully, P.T.; Nehl, C.L.; Hafner, J.H. A label-free immunoassay based upon localized surface plasmon resonance of gold nanorods. ACS Nano 2008, 2, 687–692.
12. Luff, B.J.; Wilkinson, J.S.; Piehler, J.; Hollenbach, U.; Ingenhoff, J.; Fabricius, N. Integrated optical Mach-Zehnder biosensor. J. Light. Tech. 1998, 16, 583–592.
13. Prieto, F.; Sepulveda, B.; Calle, A.; Llobera, A.; Dominquez, C.; Lechuga, L.M. Integrated Mach-Zehnder interferometer based on ARROW structures for biosensing applications. Sens. Actuators B 2003, 92, 151–158.
14. Wu, S.Y.; Ho, H.P.; Law, W.C.; Lin, C.; Kong, S.K. Highly sensitive differential phase-sensitive surface plasmon resonance biosensor based on the Mach-Zehnder configuration. Opt. Lett. 2004, 29, 2378–2380.
15. Ymeti, A.; Greve, J.; Lambeck, P.V.; Wink, T.; van Hovell, S.W.F.M.; Beumer, T.A.M.; Wijn, R.R.; Heideman, R.G.; Subramaniam, V.; Kanger, J.S. Fast, ultrasensitive virus detection using a Young interferometer sensor. Nano Lett.2007, 7, 394–397.
16. Barrios, C.A.; Gylfason, K.B.; Sanchez, B.; Griol, A.; Sohlstrom, H.; Holgado, M.; Casquel, R. Slot-waveguide biochemical sensor. Opt. Lett. 2007, 32, 3080–3082.
17. Banuls, M.-J.; Gonzalez-Pedro, V.; Barrios, C.A.; Puchades, R.; Maquieira, A. Selective chemical modification of silicon nitride/silicon oxide nanostructures to develop label-free biosensors. Biosens. Bioelectron. 2010, 25, 1460–1466.
18. Carlborg, C.F.; Gylfason, K.B.; Kazmierczak, A.; Dortu, F.; Banuls Polo, M.J.; Maquieira Catala, A.; Kresbach, G.M.; Sohlstrom, H.; Moh, T.; Vivien, L.; et al. A packaged optical slot-waveguide ring resonator sensor array for multiplex label-free assays in lab-on-chips. Lab Chip 2010, 10, 257–396.
19. Claes, T.; Molera, J.G.; de Vos, K.; Schacht, E.; Baets, R.; Bienstman, P. Label-free biosensing with a slot-waveguide-based ring resonator in silicon on insulator. IEEE Photonics J. 2009, 1, 197–204.
20. Toccafondo, V.; Garcia-Ruperez, J.; Banuls, M.J.; Griol, A.; Castello, J.G.; Peransi-Llopis, S.; Maquieira, A.

Single-strand DNA detection using a planar photonic-crystal-waveguide-based sensor. Opt. Lett. 2010, 35, 3673–3675.

21. Skivesen, N.; Tetu, A.; Kristensen, M.; Kjems, J.; Frandsen, L.H.; Borel, P.I. Photonic-crystal waveguide biosensor. Opt. Express 2007, 15, 3169–3176.

22. Lee, M.; Fauchet, P.M. Two-dimensional silicon photonic crystal based biosensing platform for protein detection. Opt. Express 2007, 15, 4530–4535.

23. Lee, M.; Fauchet, P.M. Nanoscale microcavity sensor for single particle detection. Opt. Lett. 2007, 32, 3284–3286.

24. Zlatanovic, S.; Mirkarimi, L.W.; Sigalas, M.M.; Bynum, M.A.; Chow, E.; Robotti, K.M.; Burr, G.W.; Esener, S.; Grot, A. Photonic crystal microcavity sensor for ultracompact monitoring of reaction kinetics and protein concentration. Sens. Actuators B 2009, 141, 13–19.

25. Dorfner, D.; Zabel, T.; Hurlimann, T.; Hauke, N.; Frandsen, L.; Rant, U.; Abstreiter, G.; Finley, J. Photonic crystal nanostructures for optical biosensing applications. Biosens. Bioelectron. 2009, 24, 3688–3692.

26. Kang, C.; Phare, C.T.; Vlasov, Y.A.; Assefa, S.; Weiss, S.M. Photonic crystal slab sensor with enhanced surface area. Opt. Express 2010, 18, 27930–27937.

27. Mandal, S.; Goddard, J.M.; Erickson, D. A multiplexed optofluidic biomolecular sensor for low mass detection. Lab Chip 2009, 9, 2924–2932.

28. Konopsky, V.N.; Alieva, E.V. Photonic crystal surface waves for optical biosensors. Anal. Chem. 2007, 79, 4729–4735.

29. Guo, Y.; Ye, J.Y.; Divin, C.; Huang, B.; Thomas, T.P.; Baker, J.R.; Norris, T.B. Real-time biomolecular binding detection using a sensitive photonic crystal biosensor. Anal. Chem. 2010, 82, 5211–5218.

30. Orosco, M.M.; Pacholski, C.; Miskelly, G.M.; Sailor, M.J. Protein-coated porous-silicon photonic crystals for amplified optical detection of protease activity. Adv. Mat. 2006, 18, 1393–1396.

31. Lin, B.; Li, P.; Cunningham, B.T. A label-free biosensor-based cell attachment assay for characterization of cell surface molecules. Sens. Actuators B 2006,114, 559–564.

32. Chan, L.L.; Gosangari, S.L.; Watkin, K.L.; Cunningham, B.T. A label-free photonic crystal biosensor imaging method for detection of cancer cell cytotoxicity. Apoptosis 2007, 12, 1061–1068.

33. Lidstone, E.A.; Chaudhery, V.; Kohl, A.; Chan, V.; Wolf-Jensen, T.; Schook, L.B.; Bashir, R.; Cunningham, B.T. Label-free imaging of cell attachment with photonic crystal enhanced microscopy. Analyst 2011, 136, 3608–3615.

34. Grepstad, J.O.; Kaspar, P.; Solgaard, O.; Johansen, I.-R.; Sudbo, A.S. Photonic-crystal membranes for optical detection of single nano-particles designed for biosensor application. Opt. Express 2012, 20, 7954–7965.

35. Biacore Life Sciences. Available online: http://www.biacore.com (accessed on 12 March 2013).

36. Genalyte. Available online: http://www.genalyte.com (accessed on 12 March 2013).

37. Manz, A.; Graber, N.; Widmer, H.M. Miniaturized total chemical analysis systems: A novel concept for chemical sensing. Sens. Actuators B 1990, 1, 244–248.

38. Melin, J.; Quake, S.R. Microfluidic large-scale integration: The evolution of design rules for biological automation. Annu. Rev. Biophys. Biomol. Struct.2007, 36, 213–231.

39. Weibel, D.B.; Di Luzio, W.R.; Whitesides, G.M. Microfabrication meets microbiology. Nat. Rev. Microbiol. 2007, 5, 209–218.

40. Di Falco, A.; O'Faolain, L.; Krauss, T.F. Dispersion control and slow light in slotted photonic crystal waveguides. Appl. Phys. Lett. 2008, 92.

41. Di Falco, A.; O'Faolain, L.; Krauss, T.F. Chemical sensing in slotted photonic crystal heterostructure cavities. Appl. Phys. Lett. 2009, 94, 063503.

42. Scullion, M.G.; Di Falco, A.; Krauss, T.F. Slotted photonic crystal cavities with integrated microfluidics for biosensing applications. Biosens. Bioelectron. 2011,27, 101–105.

43. Jagerska, J.; Zhang, H.; Diao, Z.; Le Thomas, N.; Houdre, R. Refractive index sensing with an air-slot photonic crystal nanocavity. Opt. Lett. 2010, 35, 2523–2525.

44. Lai, W.-C; Chakravarty, S.; Wang, X.; Lin, C.; Chen, R.T. On-chip methane sensing by near-IR absorption signatures in a photonic crystal slot waveguide.Opt. Lett. 2011, 36, 984–986.

45. Lai, W.-C; Chakravarty, S.; Wang, X.; Lin, C.; Chen, R.T. Photonic crystal slot waveguide absorption spectrometer for on-chip near-infrared spectroscopy of xylene in water. Appl. Phys. Lett. 2011, 98, 023304.

46. Wang, B.; Dundar, M.A.; Notzel, R.; Karouta, F.; He, S.; van der Heijden, R.W. Photonic crystal slot nanobeam slow light waveguides for refractive index sensing. Appl. Phys. Lett. 2010, 97, 151105.

47. Kwon, S.-H.; Sunner, T.; Kamp, M.; Forchel, A. Optimization of photonic crystal cavity for chemical sensing. Opt. Express 2008, 16, 11709–11717.

48. Lin, S.; Hu, J.; Kimerling, L.; Crozier, K. Design of nanoslotted photonic crystal waveguide cavities for single nanoparticle trapping and detection. Opt. Lett.2009, 34, 3451–3453.

49. Kurt, H.; Erim, M.N.; Erim, N. Various photonic crystal bio-sensor configurations based on optical surface modes. Sens. Actuators B 2012, 165, 68–75.

50. Kita, S.; Hachuda, S.; Otsuka, S.; Endo, T.; Imai, Y.; Nishijima, Y.; Misawa, H.; Baba, T. Super-sensitivity in label-free protein sensing using a nanoslot nanolaser. Opt. Express 2011, 19, 17683–17690.

51. Lee, M.R.; Miller, B.L.; Fauchet, P.M. Two-dimensional photonic crystal slot microcavity sensor for virus-sized particle detection. Proceedings of Integrated Photonics and Nanophotonics Research and Applications Conference, Boston, MA, USA, 13 July 2008.

52. Almeida, V.R.; Xu, Q.; Barrios, C.A.; Lipson, M. Guiding and confining light in void nanostructure. Opt. Lett. 2004, 29, 1209–1211.

53. Jackson, J.D. Plane Electromagnetic Waves and Wave

Propagation. In Classical Electrodynamics, 3rd ed.; John Wiley & Sons: New York, NY, USA, 1999; Chapter 7; pp. 302–306.

54. Yang, A.H.J.; Moore, S.D.; Schmidt, B.S.; Klug, M.; Lipson, M.; Erickson, D. Optical manipulation of nanoparticles and biomolecules in sub-wavelength slot waveguides. Nature 2009, 457, 71–75.

55. Robinson, J.T.; Chen, L.; Lipson, M. On-chip gas detection in silicon optical microcavities. Opt. Express 2008, 16, 4296–4301.

56. Yablonovitch, E. Inhibited spontaneous emission in solid-state physics and electronics. Phys. Rev. Lett. 1987, 58, 2059–2062.

57. John, S. Strong localization of photons in certain disordered dielectric superlattices. Phys. Rev. Lett. 1987, 58, 2486–2489.

58. Krauss, T.F.; De La Rue, R.M.; Brand, S. Two-dimensional photonic-bandgap structure operating at near-infrared wavelengths. Nature 1996, 383, 699–702.

59. Krauss, T.F. Slow light in photonic crystal waveguides. J. Phys. D 2007, 40, 2666–2670.

60. Krauss, T.F. Why do we need slow light? Nat. Phot. 2008, 2, 448–450.

61. Jane, A.; Dronov, R.; Hodges, A.; Voelcker, N.H. Porous silicon biosensors on the advance. Trends Biotech. 2009, 27, 230–239.

62. Fan, S.; Joannopoulos, J.D. Analysis of guided resonances in photonic crystal slabs. Phys. Rev. B 2002, 65, 235112.

63. SRU Biosystems. Available online: http://www.srubiosystems.com (accessed on 12 March 2013).

64. Wilchek, M.; Bayer, E.A.; Livnah, O. Essentials of biorecognition: The (strept)avidin-biotin system as a model for protein-protein and protein-ligand interaction. Immun. Lett. 2006, 103, 27–32.

65. MacBeath, G. Protein microarrays and proteomics. Nat. Gen. 2002, 32, 526–532.

66. Reardon, C.; Rey, I.H.; Welna, K; Krauss, T.F. Fabrication and characterization of photonic crystal slow light waveguides and cavities. J. Vis. Expr. 2012, 69, E50216.

67. Wang, Z.; Zhu, N.; Tang, Y.; Wosinski, L.; Dai, D.; He, S. Ultracompact low-loss coupler between strip and slot waveguides. Opt. Lett. 2009, 34, 1498–1500.

68. Brosi, J.-M; Koos, C.; Andreani, L.C.; Waldow, M.; Leuthold, J.; Freude, W. High-speed low-voltage electro-optic modulator with a polymer-infiltrated silicon photonic crystal waveguide. Opt. Express 2008, 16, 4177–4191.

69. Gersen, H.; Karle, T.J.; Engelen, R.J.P.; Bogaerts, W.; Korterik, J.P.; van Hulst, N.F.; Krauss, T.F.; Kuipers, L. Real-space observation of ultraslow light in photonic crystal waveguides. Phys. Rev. Lett. 2005, 94, 073903.

70. Scullion, M.G.; Krauss, T.F.; Di Falco, A. High efficiency interface for coupling into slotted photonic crystal waveguides. IEEE Photonics J. 2011, 3, 203–208.

71. Chen, X.; Jiang, W.; Chen, J.; Gu, L.; Chen, R.T. 20 dB-enhanced coupling to slot photonic crystal waveguide using multimode interference coupler. Appl. Phys. Lett. 2007, 91, 091111.

72. Moreno, E.; Garcia-Vidal, F.J.; Martin-Moreno, L. Enhanced transmission and beaming of light via photonic crystal surface modes. Phys. Rev. B 2004, 69, 121402.

73. Morrison, S.; Kivshar, Y.S. Engineering of directional emission from photonic-crystal waveguides. Appl. Phys. Lett. 2005, 86, 081110.

74. Frei, W.R.; Tortorelli, D.A.; Johnson, H.T. Topology optimization of a photonic crystal waveguide termination to maximize directional emission. Appl. Phys. Lett.2005, 86, 111114.

75. Chen, C.-C.; Pertsch, T.; Iliew, R.; Lederer, F.; Tunnermann, A. Directional emission from photonic crystal waveguides. Opt. Express 2006, 14, 2423–2428.

76. Kurt, H. Theoretical study of directional emission enhancement from photonic crystal waveguides with tapered exits. IEEE Photon. Tech. Lett. 2008, 20, 1682–1684.

77. Bauer, J.; John, S. Broadband optical coupling between microstructured fibers and photonic band gap circuits: Two-dimensional paradigms. Phys. Rev. A 2008,77, 013819.

78. Di Falco, A.; Massari, M.; Scullion, M.G.; Schulz, S.A.; Romanato, F.; Krauss, T.F. Propagation Losses of Slotted Photonic Crystal Waveguides. IEEE Photonics J. 2012, 4, 1536–1541.

79. Notomi, M.; Tanabe, T.; Shinya, A.; Kuramochi, E.; Taniyama, H.; Mitsugi, S.; Morita, M. Nonlinear and adiabatic control of high-Q photonic crystal nanocavities. Opt. Express 2007, 15, 17458–17481.

80. Baehr-Jones, T.; Hochberg, M.; Walker, C.; Scherer, A. High-Q optical resonators in silicon-on-insulator-based slot waveguides. Appl. Phys. Lett. 2005,86, 081101.

81. Spott, A.; Baehr-Jones, T.; Ding, R.; Liu, Y.; Bojko, R.; O'Malley, T.; Pomerene, A.; Hill, C.; Reinhardt, W.; Hochberg, M. Photolithographically fabricated low-loss asymmetric silicon slot waveguides. Opt. Express 2011, 19, 10950–10958.

82. Zhang, H.; Zhang, J.; Chen, S.; Song, J.; Kee, J.S.; Yu, M.; Lo, G.-Q. CMOS-compatible fabrication of silicon-based sub-100-nm slot waveguide with efficient channel slot coupler. IEEE Photon. Tech. Lett. 2012, 24, 10–12.

83. Alasaarela, T.; Korn, D.; Alloatti, L.; Saynatjoki, A.; Tervonen, A.; Palmer, R.; Leuthold, J.; Honkanen, S. Reduced propagation loss in silicon strip and slot waveguides coated by atomic layer deposition. Opt. Express 2011, 19, 11529–11538.

84. Squires, T.M.; Quake, S.R. Microfluidics: Fluid physics at the nanoliter scale.Rev. Mod. Phys. 2005, 77, 977–1026.

85. Whitesides, G.M. The origins and the future of microfluidics. Nature 2008, 442, 368–373.

86. Schmidt, H.; Hawkins, A.R. The photonic integration of non-solid media using optofluidics. Nat. Phot. 2011, 5, 598–604.

87. Xia, Y.; Whitesides, G.M. Soft lithography. Annu. Rev. Mater. Sci. 1998, 28, 153–184.

88. Unger, M.A.; Chou, H.-P.; Thorsen, T.; Scherer, A.; Quake, S.R. Monolithic microfabricated valves and pumps by multilayer soft lithography. Science 2000,288, 113–116.

89. Haubert, K.; Drier, T.; Beebe, D. PDMS bonding by means of a portable, low-cost corona system. Lab Chip 2006, 6, 1548–1549.

90. Eddings, M.A.; Johnson, M.A.; Gale, B.K. Determining the optimal PDMS-PDMS bonding technique for microfluidic devices. J. Micromech. Microeng. 2008, 18, 067001.

91. Yamamoto, T.; Notomi, M.; Taniyama, H.; Kuramochi, E.; Yoshikawa, Y.; Torii, Y.; Kuga, T. Design of a high-Q air-slot cavity based on a width-modulated line-defect in a photonic crystal slab. Opt. Express 2008, 16, 13809–13817.

92. Gao, J.; McMillan, J.F.; Wu, M.-C.; Zheng, J.; Assefa, S.; Wong, C.W. Demonstration of an air-slot mode-gap confined photonic crystal slab nanocavity with ultrasmall mode volumes. Appl. Phys. Lett. 2010, 96, 051123.

93. Gao, J.; Green, W.M.; Vlasov, Y.; Assefa, S.; Yang, X.; Wong, C.W. Demonstrations of an air-slot photonic crystal nanocavity with ultrasmall mode volumes for enhanced light-matter interactions. Proceedings of Conference on Lasers and Electro-Optics, Baltimore, MD, USA, 2–4 June 2009.

94. Foubert, K.; Lalouat, L.; Cluzel, B.; Picard, E.; Peyrade, D.; de Fornel, F.; Hadji, E. An air-slotted nanoresonator relying on coupled high Q small V Fabry-Perot nanocavities. Appl. Phys. Lett. 2009, 94, 251111.

95. Song, B.-S.; Noda, S.; Asano, T.; Akahane, Y. Ultra-high-Q photonic double-heterostructure nanocavity. Nat. Mat. 2005, 4, 207–210.

96. Lapin, N.A.; Chabal, Y.J. Infrared characterization of biotinylated silicon oxide surfaces, surface stability, and specific attachment of streptavidin. J. Phys. Chem. B. 2009, 113, 8776–8783.

97. Lee, J.N.; Park, C.; Whitesides, G.M. Slovent compatibility of poly(dimethylsiloxane)-based microfluidic devices. Anal. Chem. 2003, 75, 6544–6554.

98. Green, N.M. The molecular weight of avidin. Biochem. J. 1964, 92, 16–17.

99. Kim, Y.H.; Park, S.J.; Jeon, S.-W.; Ju, S.; Park, C.-S.; Han, W.-T.; Lee, B.H. Thermo-optic coefficient measurement of liquids based on simultaneous temperature and refractive index sensing capability of a two-mode fiber interferometric probe. Opt. Express 2012, 20, 23744–23754.

100. White, I.M.; Fan, X. On the performance quantification of resonant refractive index sensors. Opt. Express 2008, 16, 1020–1028.

101. Arnold, S.; Khoshsima, M.; Teraoka, I.; Holler, S.; Vollmer, F. Shift of whispering-gallery modes in microspheres by protein adsorption. Opt. Lett.2003, 28, 272–274.

102. Xu, T.; Wheeler, M.S.; Ruda, H.E.; Mojahedi, M.; Aitchison, J.S. The influence of material absorption on the quality factor of photonic crystal cavities. Opt. Express 2009, 17, 8343–8348.

103. Palmer, K.F.; Williams, D. Optical properties of water in the near infrared. J. Opt. Soc. Am. 1974, 64, 1107–1110.

CHAPTER 3

# A PROGRAMMABLE NANOREPLICA MOLDING FOR THE FABRICATION OF NANOPHOTONIC DEVICES

Longju Liu[1], JingxiangZhang[1], MohsinAli Badshah[2], Liang Dong[1], Jingjing Li[3], Seok-min Kim[2] & Meng Lu[1,4]

[1] Department of Electrical and Computer Engineering, Iowa State University, Ames, Iowa 50011, USA

[2]School of Mechanical Engineering, Chung-Ang University, Seoul 156-756, Republic of Korea

[3]Department of Mechanical Engineering, University of Hawaii, Honolulu, Hawaii 96822, USA

[4]Department of Mechanical Engineering, Iowa State University, Ames, Iowa 50011, USA

## ABSTRACT

**The ability to fabricate periodic structures with sub-wavelength features has a great potential for impact on integrated optics, optical sensors, and photovoltaic devices. Here, we report a programmable nanoreplica molding process to fabricate a variety of sub-micrometer periodic patterns using a single mold. The process utilizes a stretchable mold to produce the desired periodic structure in a photopolymer on glass or plastic substrates. During the replica molding process, a uniaxial force is applied to the mold and results in changes of the periodic structure, which resides on the surface of the mold. Direction and magnitude of the force determine the array geometry, including the lattice constant and arrangement. By stretching the mold, 2D arrays with square, rectangular, and triangular lattice structures can be fabricated. As one example, we present a plasmonic crystal device with surface plasmon resonances determined by the force applied during molding. In addition, photonic crystal slabs with different array patterns are fabricated and characterized. This unique process offers the capability of generating various periodic nanostructures rapidly and inexpensively.**

## INTRODUCTION

Periodic nanostructures, such as 1D and 2D sub-wavelength gratings, are critical to a broad range of optical applications as they control light propagation and can enhance light-matter interactions1. They are exploited in various photonic devices, including diffraction gratings, wire grid polarizers, grating couplers, distributed feedback lasers, and photonic crystals2·3·4·5. Fabrication of periodic nanostructures has been limited by the need to work on a sub-micrometer scale; conventional lithography methods, such as using e-beam and deep ultraviolet, are either too expensive or have insufficient throughput for wafer-scale fabrications. To address this issue, interference lithography and soft lithography have been successfully applied6·7·8·9·10·11. Soft lithography provides the capability of inexpensive, roll-to-roll fabrication of periodic nanostructures for example, the pattern of a mold can be transferred to a photo-curable polymer material at room temperature by the nanoreplica molding process, without the need for large mechanical forces12·13·14.

Although soft lithography has been successful, a drawback is the high cost of molds. To introduce or modify a feature, a new mold must be fabricated. To facilitate the production of different patterns, and periodic structures in particular, programmable approaches have been developed. For example, thermal tuning of a thermoplastic substrate bearing a nanopattern can generate a variety of patterns from a single mold15. Alternatively, Pokroy et al. took advantage of

**Citation:** Longju Liu, Jingxiang Zhang, Mohsin Ali Badshah, Liang Dong, Jingjing Li, Seok-min Kim & Meng Lu, A programmable nanoreplica molding for the fabrication of nanophotonic devices, doi:10.1038/srep22445.

elastomeric and flexible molds to generate arrays of nanoposts with a variety of micrometer-scale periods16.

In this paper, we present various plasmonic crystals and photonic crystal slabs fabricated by a programmable nanoreplica molding process that uses a single mold. This technique uses mechanical stretching of an elastic polydimethylsiloxane (PDMS) mold to create periodic structures with various periods and lattice arrangements. At the same time, it maintains the high-throughput and low-cost features of the conventional nanoreplica molding approach. When a force is applied, the surface of the PDMS mold adjusts itself to a negative volume profile of the desired periodic structure. Replicating the stretched mold shape in a UV-curable polymer (UVCP) yields programmable nanostructures in a process that is inexpensive and amenable to scale-up. Following replica molding, the periodic structures produced may be coated with a dielectric or thin metal film; some examples are titanium dioxide ($TiO_2$), gold, and silver. With a 100 nm-thick silver coating, the fabricated 2D plasmonic crystals exhibit surface plasmon resonances (SPRs) in the spectral range 410 nm to 570 nm. Using the same PDMS mold, we fabricated photonics crystal slabs with three different lattice arrangements, namely, square, rectangular, and triangular structures. The photonic crystal slabs use a 160 nm-thick $TiO_2$ film as the light-confinement layer. Finally, band diagrams of the fabricated photonic crystal slabs are experimentally determined and compared to simulations from electromagnetic theory.

## RESULTS AND DISCUSSION

Figure 1 summarizes the major steps of the programmable nanoreplica molding process. The major fabrication procedures include mold preparation, mold stretching, pattern transfer, and mold release. As the initial step, a PDMS mold is replicated from a rigid glass stamp carrying a 2D array of nanoposts with a square lattice arrangement and periods $\Lambda_x = \Lambda_y = 300$ nm. The glass stamp was produced using a glass thermal imprinting process, reported previously17,18. Next, the PDMS mold is precisely stretched to obtain a lattice pattern for a particular device. A uniaxial force is applied to the PDMS mold in the plane parallel to its surface. Along the direction of the force (y-axis), the array is stretched and its period increases. Consequently, the array is also compressed in the perpendicular x-direction and the array period is reduced. Likewise, stretching the PDMS mold in other directions also allows us to program the lattice arrangement. As shown in Fig. 1, the PDMS mold can be pulled along a diagonal direction to convert the original square lattice into a triangular lattice. From a stretched PDMS mold, the modified periodic pattern is replicated onto a glass or plastic substrate by the nanoreplica molding process. The details of the process are described below in the experimental section. Briefly, a layer of liquid UVCP material is squeezed between the stretched PDMS mold and a glass or plastic substrate. On exposure to UV illumination, the UVCP solidifies and subsequently releases from the PDMS mold.

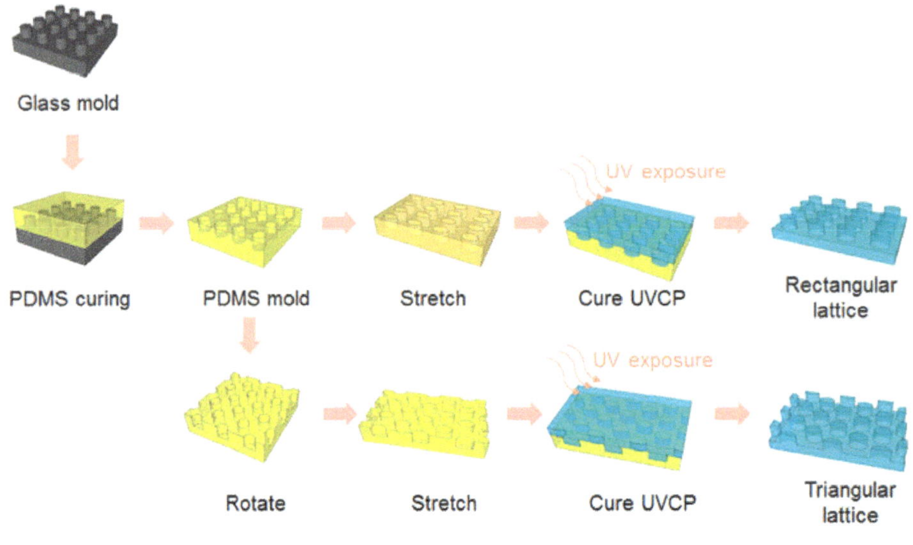


**Figure 1:** Schematic of the programmable nanoreplica molding process.

Figure 2(a) schematically illustrates how the geometry of the periodic nanostructure is tuned to produce different types of arrays. The blue dots represent the nanoposts of the un-stretched square lattice with a period of 300 nm. When a uniaxial force is applied along the x-axis, the square lattice is changed to a rectangular one, as shown by the array of cherry-colored dots. Figure 2(b–d) are the scanning electron microscopy (SEM) images of replicated triangular, square, and rectangular arrays, respectively. The rectangular array shown in Fig. 2(d) was fabricated when the PDMS mold was under a strain of 40%. The period of the array increases to $\Lambda_x = 420$ nm along the direction of the applied force. The lattice shows a slight shrinkage in the direction perpendicular to the force, with a period of $\Lambda_y = 258$ nm. Meanwhile, the cross section of the nanoposts changes from circular to elliptical; however the duty cycles of the periodic structure (post size/period) remain unchanged. When the PDMS mold is stretched along the diagonal direction, it is possible to produce periodic arrays with a triangular lattice, which is illustrated by the array of orange dots in Fig. 2(b). Figure 2(b) shows the replication fabricated when the PDMS mold was under a 40% stretch along the 45° direction to the x-axis. With precise control of the degree and direction of the stretch, the triangular lattice could be tuned into a hexagonal one.

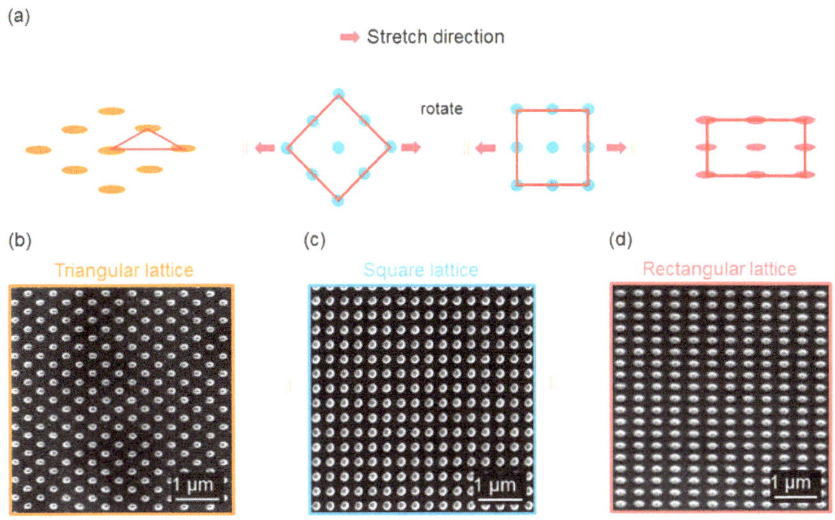


**Figure 2:** Nanopost arrays replicated using a single PDMS mold.

Scheme of the mold programming process, beginning at the third figure from the left (**a**). SEM images of nanopost arrays with triangular lattice (**b**), square lattice (**c**), and rectangular lattice (**d**).

The capability of programming the array lattice is particularly useful for some nanophotonic devices. As an example, we studied plasmonic crystals with different periodic arrays fabricated using the process described above. As a result of the different geometries, these arrays exhibit distinct plasmonic resonances. The plasmonic crystal structure, shown in Fig. 3(a), consists of a replicated 2D array of nanoposts with a 100 nm-thick silver coating, and supports grating- coupled surface plasmon resonance (SPR) modes. Excitation light that meets the resonance condition can be coupled into an SPR mode and may be strongly absorbed. As a result, the reflection spectrum exhibits a dip with the minimum reflectance at the SPR wavelength ($\lambda_{SP}$). The resonance wavelength can be estimated using the equation:

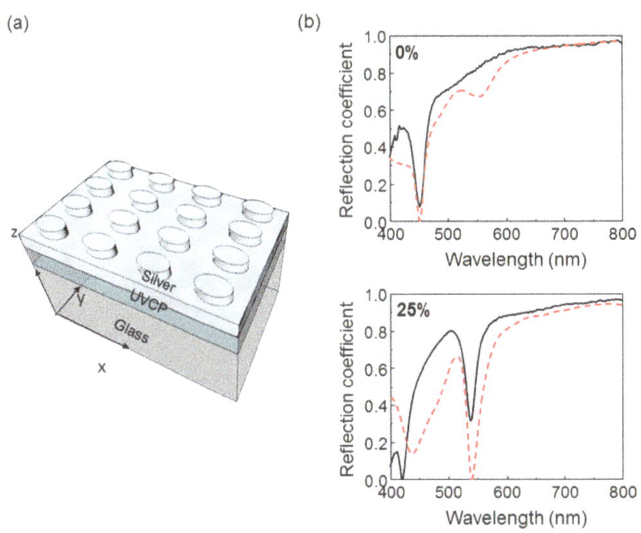


**Figure 3:** Optical characterization of 2D plasmonic crystals.

(**a**) Schematic of a silver-coated plasmonic slab. (**b**) Simulations for 0% and 25% strain, compared to experimental data. The black solid lines represent the measurement results, and the red dashed lines represent the simulations.

$$\lambda_{sp} = \sqrt{\frac{\varepsilon_d \varepsilon_m}{\varepsilon_d + \varepsilon_m}} \cdot \frac{\Lambda}{\sqrt{i^2 + j^2}}, \qquad (1)$$

where Λ is the period of the grating, (i, j) represent the Bragg resonance orders, and $\varepsilon_m$ and $\varepsilon_d$ are the dielectric constants of metal and surrounding medium, respectively[19,20]. Because the SPRs of plasmonic crystals are determined by the grating period shown in Eq. 1, the programmable replica molding process can be used to finely tune the periodic array and obtain the desired SPR mode wavelength.

In fact, the resonance wavelength varies proportionally to the strain (∈) that is generated in the PDMS mold by the uniaxial stretch. In order to experimentally investigate the correlation between $\lambda_{SP}$ and ϵ, we replicated 10 different arrays of nanoposts by placing the PDMS mold under a range of strains (∈=0%, 2.5%, 5%, 7.5%, 10%, 15%, 20%, 25%, 30%, and 35%). The replicated 2D gratings were subsequently coated with a 100 nm-thick silver film to form the plasmonic crystals. The same PDMS mold was reused 10 times to produce the plasmonic devices. Reflection spectra of the plasmonic crystals were measured to identify their resonance wavelengths. Rigorous coupled wave analysis (RCWA) was used to simulate these reflection spectra. Details of the simulation and measurement are discussed in the experimental section. Figure 3(b)compares the measured and simulated reflection spectra of devices fabricated with 0% and 25% strain. With 0% strain, the device exhibits a reflection dip at $\lambda_{SP}$=456 nm; the x- and y-polarized SPR modes coincide in wavelength because the array structure is symmetric. As shown in Fig. 4(a), the stretch applied during molding splits the differently polarized resonance modes. The SPR mode with the electric field polarized along the x-axis shifts to longer wavelengths with increased stretch; the shift is proportional to the degree of stretching. The corresponding compression along the y-axis causes the y-polarized SPR mode to move to shorter wavelengths. For ϵ=35%, the plasmonic resonances are shifted by 111 nm and −41 nm for the x- and y- polarizations, respectively. Figure 4(b) summarizes $\lambda_{SP}$ as a function of strain for both polarizations. The measured resonance wavelengths are fitted by straight lines with slopes of $\lambda_{SP}/\epsilon$=3.13 and −1.14 nm/(% strain) for the x- and y-polarizations, respectively.

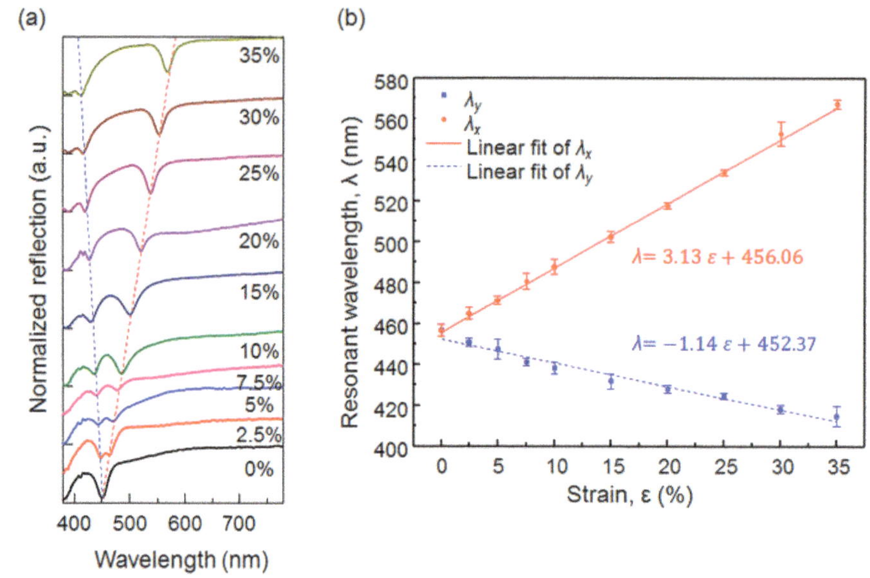


**Figure 4:** (**a**) The SPR reflection dips for plasmonic slabs fabricated with 0%, 2.5%, 5%, 7.5%, 10%, 15%, 20%, 25%, 30%, and 35% strain. The split between the dips for x- and y-polarizations follows the lattice period deviations in the x-direction (elongation) and y-direction (compression). (**b**) Plasmonic resonance wavelength with respect to applied strain. The SPR dips in the stretched and compressed directions from the experimental data in (**a**) are shown by the red dots and blue squares, respectively; linear fits to the resonance wavelengths, by the solid red and dashed blue lines. The error bars represent the standard deviation of 10 measurements taken at different locations on the sample.

The array control in the nanoreplica molding method can also be used to examine and modify the photonic band diagram of photonic crystal slabs. In this study, the photonic crystal slab is based on a grating coupled waveguide, which is also known as a leaky-mode waveguide. The guided-mode resonance (GMR) phenomenon supported by this structure provides narrowband optical resonances3·21·22. As shown inFig. 5(a), the photonic crystal slab consists of a replicated 2D grating on a glass substrate, which can couple light into and out of a thin-film dielectric waveguide coated on the surface of the grating. At a specific combination of wavelength and angle of incidence, the GMR mode can be excited with nearly 100% reflectance. Based on the diffraction grating equation, the GMR wavelength and the resonant angle can be calculated from the equations: Finally, $n_1$, $n_2$, and $n_3$ represent the refractive indices of the superstrate, the grating layer material, and the substrate, respectively23·24. Like plasmonic crystals, the resonance characteristics of a photonic crystal slab also depend on the array geometry.

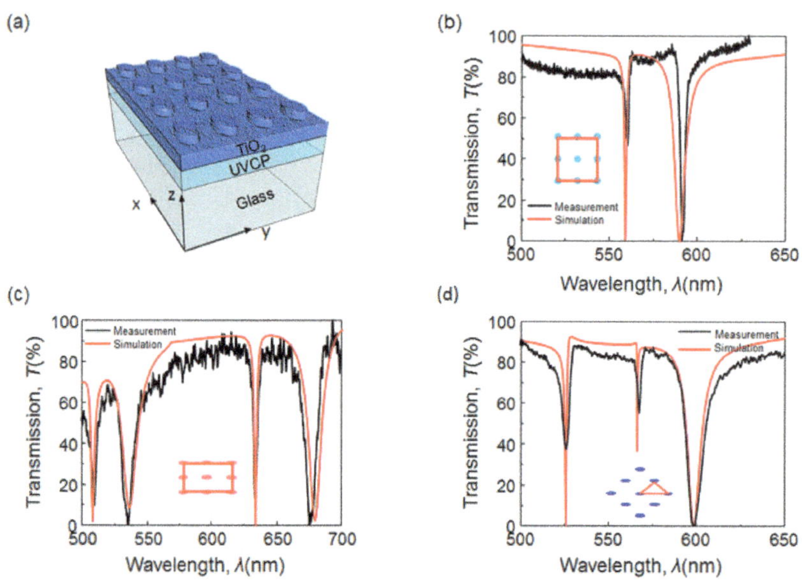


**Figure 5:** Reflection spectra of plasmonic crystals.

(**a**) Schematic diagram of a photonic crystal slab with a rectangular lattice. Experimental and simulated transmission spectra of photonic crystal slabs when the resonance modes are excited from the normal direction, for a square (**b**), rectangular (**c**), and triangular lattice (**d**).

Here, we demonstrate photonic crystal slabs with three different lattice arrangements, namely, the square, the rectangular, and the triangular lattice. The photonic crystal slabs were fabricated using the programmable replica molding approach, followed by coating with a 160 nm-thick $TiO_2$ layer (refractive index n=2.0). The transmission spectra were measured using broadband light from the direction normal to the sample surface. As shown in Fig. 5(b), the photonic crystal slab with a square lattice ($\Lambda_x = \Lambda_y = 300$ nm) shows two GMR modes at 560.2 nm and 591.6 nm, representing the TM-polarized and TE-polarized modes, respectively. For the TM-polarized modes, the electric field components are perpendicular to the periodic modulation, while for the TE-polarized modes, they are parallel to the modulation. Because of the symmetry of the square lattice, the reflection measurement is independent of the excitation light polarization. Figure 5(b) also compares the measurement to the RCWA simulation. The discrepancy between measured and simulated resonance characteristics likely arises because the simulation does not take into account the slight divergence of the incident light.

To generate a rectangular lattice, the PDMS mold was stretched with a uniaxial force along the x-axis to generate a strain of 25%. The replicated array structure shows two distinct periods ($\Lambda_x = 375$ nm and $\Lambda_y = 280$ nm) along the x-axis and y-axis, respectively. As shown in Fig. 5(c), the rectangular lattice slab exhibits two TM-polarized modes at 634.5 nm and 510.1 nm corresponding to the modified array periods. At the same time, there are also two TE-polarized modes locating at 677.2 nm and 535.5 nm. Next, to fabricate a photonic crystal slab with a triangular lattice (Fig. 2(d)), the PDMS mold was stretched along its diagonal direction to achieve a strain of 35%. The measured transmission of the triangular lattice photonic crystal is shown in Fig. 5(d). For comparison, the red curves in Fig. 5(c,d) represent the transmission spectra calculated using an RCWA simulation.

To further investigate the effect of lattice arrangement, the dispersion of GMR modes as a function of incident angle was studied. Transmission spectra were recorded when the incident angle was scanned along the high-symmetry directions, as described in the experimental section. For the square lattice PC slab, the measured and simulated photonic band structures are shown in Fig. 6(a). When the angle of incidence ($\theta_x$) increases from the Γ point toward the X point, both TE-polarized and TM-polarized modes form three bands. The center "flat" band represents the GMR coupled with the excitation light through the grating modulation along the y-direction. In contrast, the GMR modes coupled through the grating modulation along the x-direction split into two bands, an upper band and a lower band with respect to their resonant wavelength. The lower TE-polarized band and upper TM-polarized band intersect at $\theta_x = 15°$. When the angle of incidence changes along the Γ-M, the TE-polarized GMR modes form an upper band and a lower band. The TM-polarized modes branch into two upper bands and two lower bands.

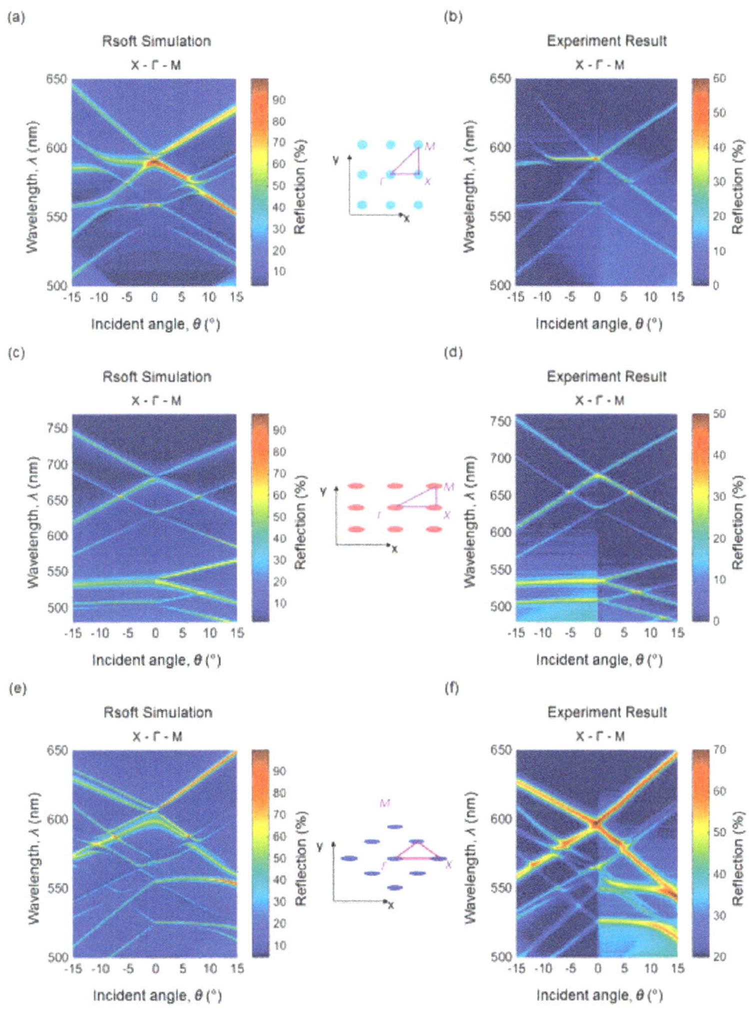


**Figure 6:** Photonic band diagrams derived from the transmission spectra.

**(a,b)** Simulated and measured photonic band diagrams for the square lattice structure. **(c,d)** Simulated and measured photonic band diagrams for the rectangular lattice structure. **(e,f)** Simulated and measured photonic band diagrams for the triangular lattice structure.

Figure 6(b) shows the measured and simulated band structures for the rectangular photonic crystal slab with the angle of incidence scanned from the Γ point to the X and M points, respectively. When the angle of incidence increases from the Γ point towards the X point, two flat bands (one for a TE band and one for a TM) are observed. The flat bands are the GMR modes coupled via the grating modulation along the y-direction. The GMR modes coupled via the x-direction modulation are located in a longer wavelength region because the rectangular grating is not symmetric and $\Lambda_x$ is larger than $\Lambda_y$. For both TE and TM polarizations, the GMR modes split into an upper band and a lower band. From the Γ point to the M point, all four GMR modes form two bands, resulting in eight different bands. Likewise, Fig. 6(c) shows the experimental and calculated photonic band diagrams for the device with a triangular array, where the angle of incidence is scanned from the Γ point to the X and M points.

## CONCLUSION

In summary, this paper reports a programmable nanoreplica molding method that facilitates fabrication of grating-based nanophotonic devices. Using a single PDMS mold, this fabrication method can produce sub-wavelength structures with various lattices. We adopted the stretchable PDMS mold, which was replicated from a glass stamp, to fabricate periodic nanostructures using UVCP material on a glass or a plastic substrate. During nanoreplica molding, the PDMS mold was stretched precisely to produce the desired lattice geometry. Nanophotonic devices, including plasmonic and photonic crystals, were formed by coating metal or dielectric thin films on the replicated UVCP. The optical resonances of these devices were characterized experimentally and compared to simulation results. For the plasmonic crystals, an SPR wavelength of 456 nm increased to 566 nm with application of 35% uniaxial strain during molding. The range of the resonance wavelengths could be even greater with increased stretch, since the PDMS can endure as high as 100% strain25·26. For the photonic crystal slabs, reflection filters with three different lattice geometries were fabricated. Rectangular and triangular lattices were successfully obtained from the original square lattice. The photonic band diagrams of all three devices were measured and showed good agreement with simulations.

## MATERIALS AND METHODS

### Nanoreplica Molding Process

The master stamp with a 2D square array (Λ = 300 nm) of 150 nm diameter posts and an overall dimension of 50 mm × 50 mm was fabricated using the glass thermal imprinting method with vitreous mold prepared by carbonization of replicated furan precursor17·18. The glass master stamp was cleaned and treated using an anti-adhesion silane (Repel Silane, GE Healthercare) in order to facilitate the replications. The PDMS mold was fabricated from the master stamp by the thermally curing of a mixture of PDMS elastomer and curing agent (in a volume ratio of 1:10) on the master stamp. The thickness of PDMS molds was controlled to be 2 mm. After curing at 100 °C for 4 hours, the solidified PDMS was peeled away from the master stamp. Then the PDMS mold was cut into a rectangle that has a length of 50 mm and a width of 15 mm. The orientation of the lattice on the PDMS mold was identified and marked. During the UV-based replica molding process, the PDMS mold was placed between two grips that were separated by L ~ 45 mm on a customized stage. One of the grips was fixed and the other one was pulled horizontally (along x-direction) by a linear translation stage. The translation stage induced strain values, $\varepsilon_x = \Delta x/L$, where Δx is the amount of stretch. The PDMS mold was held at the desired length during the replica molding process. A layer of liquid UVCP (NOA 88, Norland Product Inc.) was squeezed between the PDMS mold and a glass coverslip. The UV-curing process took place by exposing the coverslip/liquid UVCP/PDMS stack to UV illumination for 300 s. After curing, the replica of a 2D array of nanoposts and the PDMS mold were separated by peeling the coverslip away from the PDMS mold. The fully cured polymer preferentially adheres to the glass substrate without leaving any residue on the PDMS mold. Subsequent to the replica molding, a layer of dielectric ($TiO_2$) or silver thin film was deposited over the surface relief 2D grating by electron-beam evaporation to complete the device fabrication.

### Reflection Measurement for Plasmonic Crystals

Reflection spectra of plasmonic crystals were measured using a white light reflection setup. A halogen lamp was used as a broadband excitation source, and was coupled into a bifurcated fiber (BFY50HS02, Thorlabs), with a fiber tip collimator at the exit. An iris and a linear polarizer were placed in front of the collimator to control the spot size and polarization of the incident beam. The illumination assembly was attached to a kinematic mount for precise adjustment of the angle of incidence. The reflected light was coupled into a spectrometer (USB2000, OceanOptics) through the same bifurcated fiber. For measurement of its reflection spectrum, a plasmonic crystal sample was mounted on a motorized x-y translation stage and immersed in deionized water. A silver-coated mirror was used as the reference for reflectance. Software developed using C# was used to control movement of the translation stage and collect spectra from the spectrometer for chosen sampling locations on the plasmonic crystals. The measured spectra were fitted using a second-order polynomial function to find the resonance wavelength of the plasmonic modes.

### Photonic Crystal Transmission and Band Diagram Measurements

The dispersion band diagrams of the photonic crystal slabs were measured by recording the transmission spectra for multiple incident angles. The experimental setup for transmission measurement has three main parts: a halogen lamp for white broadband illumination, a sample mount, and a spectrometer. The white light was collimated before exiting the coupling fiber. The light beam passed through the photonic crystal slab and the transmitted light was collected using a multimode fiber collector, which was connected to the spectrometer. The instruments were aligned horizontally and the plane of incidence was chosen to be the horizontal plane. The sample mount was carefully designed, with two perpendicular rotation stages and a kinematic mount, providing sufficient degrees of freedom for complicated and precise angle adjustments. At the beginning of each measurement, a sample was mounted with its top face perpendicular to the incoming light (at the Γ-Point).

Then, based on the direction to be measured, i.e., Γ-X or Γ-M, the sample was rotated vertically so that the direction being measured was parallel with the horizontal plane. Next, the sample was rotated horizontally so that the incident angle θ was scanned accordingly from 0° to 15° by increments of 0.5°. For each θ the transmission spectrum was recorded, for both Γ-X and Γ-M directions, to form a data cube. The photonic band diagram was plotted using θand λ as x-axis and y-axis, respectively.

## Electromagnetic Modeling

The RCWA simulations were performed to model the reflection and transmission spectra of the plasmonic crystals and photonic crystal slabs. The simulation region was setup to the unit volume of the periodic structures. Periodic boundary conditions were applied to truncate the calculation domain in x-y plane. Ten spatial harmonics were used at both x- and y-directions. For plasmonic crystals, the devices were illuminated from the direction normal to the surface by a plane wave. The incident wave was linearly polarized with the electric field oriented at 45° to x-axis in x-y plane. The material properties of silver were taken from the Palik's handbook, and then fitted by the multi-coefficient model in the wavelength range from 300 nm to 1000 nm. Reflection spectra of plasmonic crystals were calculated in the wavelength range from 375 nm to 775 nm. The simulation of photonic crystal slabs generated their dispersion diagrams. The transmission spectra of a photonic crystal slab were recorded when the angle of incidence (θ) was scanned from 0° to 15° with an incremental of 0.5°. The dispersion diagrams for the Γ-X directions were shown by plotting the transmission spectra as a function of θ when $\varphi = 0°$. The dispersion diagrams for the Γ-M directions were plotted when $\varphi = 45°$, 37.12°, and 90°, for the square, the rectangular, and the triangular lattices, respectively.

## REFERENCES


1. Busch, K. et al. Periodic nanostructures for photonics. Phys. Rep. 444, 101–202 (2007).
2. Collin, S. Nanostructure arrays in free-space: optical properties and applications. Rep. Prog. Phys. 77, 126402 (2014).
3. Johnson, S. G., Fan, S. H., Villeneuve, P. R., Joannopoulos, J. D. & Kolodziejski, L. A. Guided modes in photonic crystal slabs. Phys. Rev. B 60, 5751–5758 (1999).
4. Carroll, J. E., Whiteaway, J. & Plumb, D. Distributed feedback semiconductor lasers. (The Institution of Electrical Engineers: SPIE Optical Engineering Press, 1998).
5. Taillaert, D. et al. Grating couplers for coupling between optical fibers and nanophotonic waveguides. Jpn. J. Appl. Phys. 45, 6071–6077 (2006).
6. Chou, S. Y., Krauss, P. R. & Renstrom, P. J. Nanoimprint lithography. J. Vac. Sci. Technol. B 14, 4129–4133 (1996).
7. Zhao, X. M., Xia, Y. N. & Whitesides, G. M. Soft lithographic methods for nano-fabrication. J. Mater. Chem. 7, 1069–1074 (1997).
8. George, M. C., Liu, J. N., Farhang, A., Williamson, B., Black, M., Wangensteen, T., Fraser, J., Petrova, R. & Cunningham, B. T. Waferscale plasmonic and photonic crystal sensors, In SPIE Nanoscience + Engineering, 95471F, doi: 10.1117/12.2188631 (2015).
9. Lu, C. & Lipson, R. H. Interference lithography: a powerful tool for fabricating periodic structures. Laser Photonics Rev. 4, 568–580 (2010).
10. Pokhriyal, A. et al. Photonic crystal enhanced fluorescence using a quartz substrate to reduce limits of detection. Opt. Express 18, 24793–24808 (2010).
11. Lu, M., Choi, S. S., Irfan, U. & Cunningham, B. T. Plastic distributed feedback laser biosensor. Appl. Phys. Lett. 93, 111113 (2008).
12. Lee, H. S., Kim, D. S. & Kwon, T. H. UV nano embossing for polymer nano structures with non-transparent mold insert. Microsyst. Technol.
13. 13, 593–599 (2007). 13. Kooy, N., Mohamed, K., Pin, L. T. & Guan, O. S. A review of roll-to-roll nanoimprint lithography. Nanoscale Res. Lett. 9, 320 (2014).
14. Ahn, S. H. & Guo, L. J. Large-Area Roll-to-Roll and Roll-to-Plate Nanoimprint Lithography: A Step toward High-Throughput Application of Continuous Nanoimprinting. Acs Nano 3, 2304–2310 (2009).
15. Lee, M. H., Huntington, M. D., Zhou, W., Yang, J. C. & Odom, T. W. Programmable Soft Lithography: Solvent-Assisted Nanoscale Embossing. Nano. Lett. 11, 311–315 (2011).
16. Pokroy, B., Epstein, A. K., Persson-Gulda, M. C. M. & Aizenberg, J. Fabrication of Bioinspired Actuated Nanostructures with Arbitrary Geometry and Stiffness. Adv. Mater. 21, 463–469 (2009).
17. Ju, J., Han, Y., Seok, J. & Kim, S. M. Development of Low-cost and Large-area Nanopatterned Vitreous Carbon Stamp for Glass Nanoreplication. 2012 12th IEEE Conference on Nanotechnology, Birmingham, UK. (doi: 10.1109/NANO.2012.6322076) (2012, August).
18. Ju, J., Lim, S., Seok, J. & Kim, S. M. A Method to Fabricate Low-Cost and Large Area Vitreous Carbon Mold for Glass Molded Microstructures. Int. J. Precis. Eng. Man. 16, 287–291 (2015).
19. Ghaemi, H. F., Thio, T., Grupp, D. E., Ebbesen, T. W. & Lezec, H. J. Surface plasmons enhance optical transmission through subwavelength holes. Phys. Rev. B 58, 6779–6782 (1998).
20. Chang, S. H., Gray, S. K. & Schatz, G. C. Surface plasmon generation and light transmission by isolated nanoholes and arrays of nanoholes in thin metal films. Opt. Express 13, 3150–3165 (2005).
21. Wang, S. S., Magnusson, R., Bagby, J. S. & Moharam, M. G. Guided-Mode Resonances in Planar Dielectric-Layer Diffraction Gratings. JOSA A 7, 1470–1474 (1990).
22. Magnusson, R. & Shokooh-Saremi, M. Widely tunable guided-mode resonance nanoelectromechanical RGB pixels. Opt. Express 15, 10903–10910 (2007).
23. Liu, J. N., Schulmerich, M. V., Bhargava, R. & Cunningham, B. T. Sculpting narrowband Fano resonances inherent in the large-area mid-infrared

photonic crystal microresonators for spectroscopic imaging. Opt. Express 22, 18142–18158 (2014).

24. Ondic, L., Babchenkol, O., Varga, M., Kromka, A., Ctyroky, J. & Pelant, I. Diamond photonic crystal slab: Leaky modes and modified photoluminescence emission of surface-deposited quantum dots. Sci. Rep. 2, 00914 (2012).

25. Zhao, S., Xia, H., Wu, D., Lv, C., Chen, Q. D., Ariga, K., Liu, L. Q. & Sun, H. B. Mechanical stretch for tunable wetting from topological PDMS film. Soft Matter 9, 4236–4240 (2013).

26. Lin, P. C., Vajpayee, S., Jagota, A., Hui, C. Y. & Yang, S. Mechanically tunable dry adhesive from wrinkled elastomers. Soft Matter 4, 1830–1835 (2008).

CHAPTER 4

# PHOTONIC CRYSTALS CAUSE ACTIVE COLOUR CHANGE IN CHAMELEONS

Je´re´mie Teyssier[1], Suzanne V. Saenko[2], Dirk van der Marel[1] & Michel C. Milinkovitch[2]

[1]Department of Quantum Matter Physics, University of Geneva, Geneva 1211, Switzerland

[2] Laboratory of Artificial and Natural Evolution (LANE), Department of Genetics and Evolution, University of Geneva, Sciences III, 30, Quai Ernest-Ansermet, Geneva 1211, Switzerland

## ABSTRACT

**Many chameleons, and panther chameleons in particular, have the remarkable ability to exhibit complex and rapid colour changes during social interactions such as male contests or courtship. It is generally interpreted that these changes are due to dispersion/aggregation of pigment-containing organelles within dermal chromatophores. Here, combining microscopy, photometric videography and photonic band-gap modelling, we show that chameleons shift colour through active tuning of a lattice of guanine nanocrystals within a superficial thick layer of dermal iridophores. In addition, we show that a deeper population of iridophores with larger crystals reflects a substantial proportion of sunlight especially in the near-infrared range. The organization of iridophores into two superposed layers constitutes an evolutionary novelty for chameleons, which allows some species to combine efficient camouflage with spectacular display, while potentially providing passive thermal protection.**

## INTRODUCTION

Ever since their description by Aristotle, chameleons have populated myths and legends because of a number of features such as a long projectile tongue, independently movable eyes, zygodactylous feet, a very slow pace and the striking capacity of some species to rapidly shift from one vivid colour to another[1, 2, 3]. Many vertebrates can rapidly change colour for camouflage, communication and thermoregulation[2, 4, 5, 6, 7], but these so-called physiological colour changes are generally mediated by modifications of skin brightness (that is, diffuse and/or specular reflectivity) through dispersion/aggregation of pigment-containing organelles, especially melanosomes, within dermal chromatophores[6, 7]. On the other hand, rapid active tuning of skin hue has been described in only a handful of species and generally involves structural, rather than pigmentary, components, that is, multilayer nano-reflectors with alternating high and low refractive indices that generate interference of light waves. For example, some species of squid can rapidly tune skin iridescence through periodical invaginations of plasma membrane deep into specialized cells called iridophores, generating arrays of alternating cytoplasmic protein-rich lamellae and extracellular channels[8, 9]. In fish, amphibians and reptiles, iridophores containing transparent guanine nanocrystals generate a large variety of structural colours, and modifications of the multilayer reflector geometry has been suggested to generate colour change in a few species[10,11, 12, 13, 14]. Finally, it must be emphasized that the colour of a reptile skin patch is often the result of interactions among pigmentary and structural elements[14, 15].

**Citation:** Je´re´mie Teyssier, Suzanne V. Saenko, Dirk van der Marel & Miche C. Milinkovitch, Photonic crystals cause active colour change in chameleons, DOI: 10.1038/ncomms7368.

Combining histology, electron microscopy and photometric videography techniques with numerical band-gap modelling, here we show that chameleons have evolved two superimposed populations of iridophores with different morphologies and functions: the upper multilayer is responsible for rapid structural colour change through active tuning of guanine nanocrystal spacing in a triangular lattice, whereas the deeper population of cells broadly reflects light, especially in the near-infrared range. This combination of two functionally different layers of iridophores constitutes an evolutionary novelty that allows some species of chameleons to combine efficient camouflage and dramatic display, while potentially moderating the thermal consequences of intense solar radiations.

## RESULTS

### Colour-Change Abilities of Panther Chameleons

We study the panther chameleon (Furcifer pardalis), a lizard from Madagascar, capable of dynamic colour change. Our Raman spectroscopy analyses indicate the presence of two types of dark chromatophores containing respectively melanin[16] and an unidentified dark-blue pigment (Supplementary Fig. 1). Although panther chameleons of both sexes and all ages can strongly modulate the brightness of the skin through dispersion of these pigments (for example, in response to stress), adult males are additionally characterized both by exceptionally large intraspecific colour variation (with various combinations of white, red, green and blue skin) among geographic locales within Madagascar and their ability to rapidly change colour (hue). Indeed, when encountering a male competitor or a potentially receptive female, a mature male panther chameleon can shift the background colour of its skin (Fig. 1a) from green to yellow or orange, whereas blue patches turn whitish and red becomes brighter with less conspicuous hue modifications. This process occurs within a couple of minutes and is fully reversible (Supplementary Movies 1,2,3).

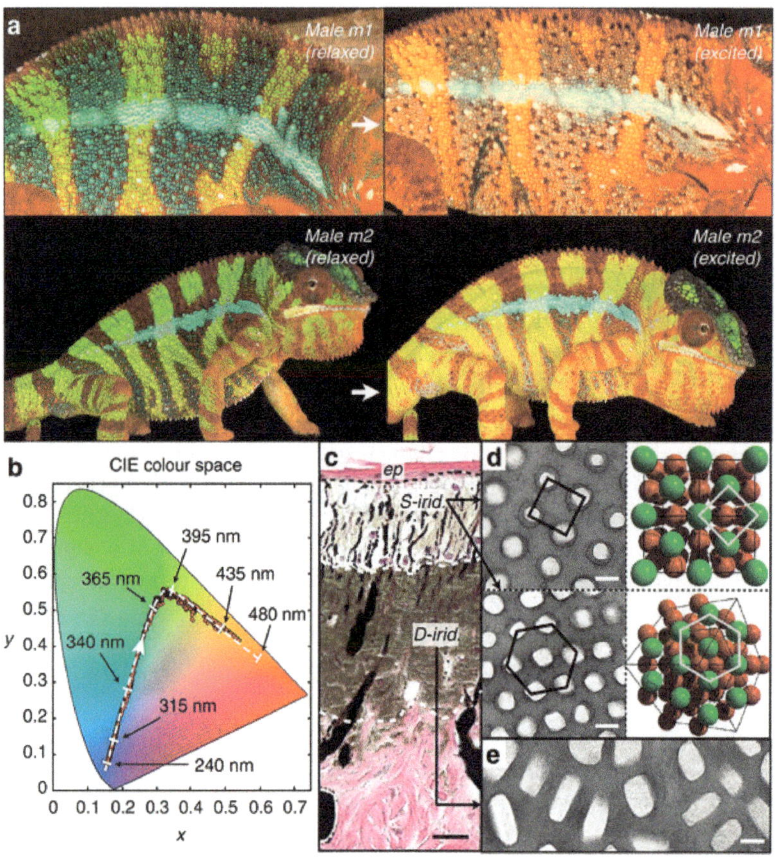


**Figure 1:** Colour change and iridophore types in panther chameleons.

(**a**) Reversible colour change is shown for two males (m1 and m2): during excitation (white arrows), background skin shifts from the baseline state (green) to yellow/orange and both vertical bars and horizontal mid-body stripe shift from blue to whitish (m1). Some animals (m2 and Supplementary Movie 2) have their blue vertical bars covered by red pigment cells. (**b**) Red dots: time evolution in the CIE chromaticity chart of a third male with green skin in a high-resolution video (Supplementary Movie 3); dashed white line: optical response in numerical simulations using a face-centred cubic (FCC) lattice of guanine crystals with lattice parameter indicated with black arrows. (**c**) Haematoxylin and eosin staining of a cross-section of white skin showing the epidermis (ep) and the two thick layers of iridophores (see alsoSupplementary Fig. 1). (**d**) TEM images of guanine nanocrystals in S-iridophores in the excited state and three-dimensional model of an FCC lattice (shown in two orientations). (**e**) TEM image of guanine nanocrystals in D-iridophores. Scale bars, 20 μm (**c**); 200 nm (**d**,**e**).

## *In-vivo* Photometry and Skin Structure

We used red–green–blue (RGB) photometry on high-resolution videos (Supplementary Fig. 2) to analyse in vivo the optical response of the skin of male F. pardalis individuals during male–male contests. The time evolution of the skin colour in the CIE (International Commission on Illumination) chromaticity chart indicates a gradual spectral weight transfer from the blue to the green to the red portions of the visible electromagnetic spectrum (Fig. 1b). This phenomenon is difficult to explain by dispersion/aggregation of pigments within chromatophores alone and is likely to additionally involve tuning of a structural colour mechanism such as, for example, multilayer interference[17]. Our histological and transmission electron microscopy (TEM) analyses in five adult males, four adult females and four juveniles revealed that, unlike other lizards[7, 14, 18]the skin of panther chameleons consists of two superposed thick layers of iridophore cells (Fig. 1c) containing guanine crystals of different sizes, shapes and organizations (Fig. 1d,e). The upper layer is fully developed only in the skin of adult males (and reduced in the skin of female and juveniles; Supplementary Fig. 1b) and contains iridophores (hereafter called superficial (S-) iridophores) with small close-packed guanine crystals[19] (diameter 127.4±17.8 nm;Supplementary Table 1) organized in a triangular lattice (Fig. 1d). This arrangement of high and low refractive index materials ($n_{guanine}$=1.83, $n_{cytoplasm}$=1.33; ref. 13) has the potential of behaving as a so-called photonic crystal[20, 21], such as those that generate bright colours in some birds and insects[22, 23, 24, 25]. Comparing multiple TEM images from F. pardalis samples of blue or green skin (resting state) with those obtained on yellow or white skin (excited state) of the same individuals, we find that crystal size in S-iridophores does not vary, but the distance among guanine crystals is on average 30% smaller in the resting than in the excited skin (Fig. 2a andSupplementary Table 1). Given that even slight alterations of geometry in a photonic crystal can generate dramatic changes in colour[17], we hypothesized that panther chameleons shift from one vibrant colour to another by modifying guanine crystal spacing in their S-iridophores. As in other lizards[14], the green background skin of panther chameleons contains chromatophores with yellow pigments (xanthophores, Supplementary Fig. 1a). Hence, it is likely to be that the increase in mean distance among nanocrystals in excited male panther chameleons causes S-iridophores to shift their selective reflectivity from short (blue) to long (red or infrared) wavelengths, causing the corresponding shift from green to yellow/orange skin. Note that in red skin, the upper layer ofS-iridophores remains well developed but a large proportion of iridophores is replaced by red-pigment chromatophores (that is, erythrophores; Supplementary Fig. 1a), explaining that red skin hue does not change dramatically during excitation, but its brightness increases.

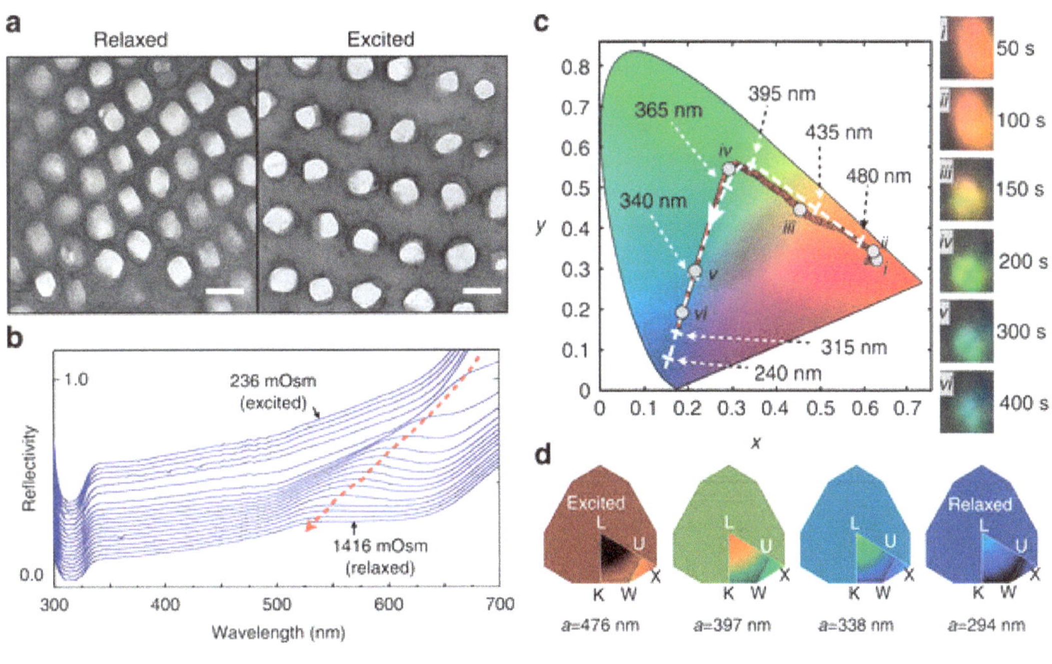


**Figure 2:** In-vivo **skin** colour change in chameleons is reproduced ex vivo.

(**a**) TEM images of the lattice of guanine nanocrystals in S-iridophores from the same individual in a relaxed and excited state (two biopsies separated by a distance <1 cm, scale bar, 200 nm). This transformation and corresponding optical response is recapitulated ex vivo by manipulation of white skin osmolarity (from 236 to 1,416 mOsm): (**b**) reflectivity of a skin sample (for clarity, the 19 reflectivity curves are shifted by 0.02 units along the y axis) and (**c**) time evolution (in the CIE chromaticity chart) of the colour of a single cell (insets i–vi; Supplementary Movie 4); both exhibit a strong blue shift (red dotted arrow in **b**) as observed in vivo during behavioural colour change. Dashed white line: optical response in numerical simulations (cf. Fig. 1b) with lattice parameter indicated with dashed arrows. Note that increased osmotic pressure corresponds to behavioural relaxation; hence, the reverse order (white arrowhead in CIE colour chart) of red to green to blue time evolution in comparison with Fig. 1b. (**d**) Variation of simulated colour photonic response for each vertex of the irreducible first Brillouin zone (colour outside of the Brillouin zone indicates the average among all directions) shown for four lattice parameter values (from Supplementary Movie 5) of the modelled photonic crystal. L-U-K-W-X are standard symmetry points.

## Osmotic Pressure Experiments and Optical Modelling

To test whether indeed period modulation of the lattice of guanine nanocrystals explains colour change in chameleons, we subjected samples of excited skin (white/yellow) to hypertonic solutions to generate osmotic pressure likely to cause the guanine crystal lattice to shrink to its resting state. This treatment indeed results in a blue shift in the reflectivity of S-iridophores (Fig. 2b). Furthermore, cell tracking during increase of extracellular osmolarity indicates that individual cells experience a gradual shift in colour across the whole visible spectrum (Fig. 2c andSupplementary Movie 4). Hence, expansion/contraction of the photonic crystal lattice in S-iridophores is sufficient to explain the reversible shifts of colours observed in vivo.

Using band-gap modelling of the photonic crystal optical response[25, 26], we simulated the colours generated by a face-centred cubic lattice of close-packed guanine crystals for a fixed crystal size and a range of lattice parameter (distance) values measured on TEM images of various excited and unexcited male panther chameleon skin samples of different colours (Supplementary Table 1). The irreducible Brillouin zone was meshed (Fig. 2d) and the photonic band structure was computed for each vertex using block-iterative frequency-domain methods[26](Supplementary Fig. 3). As no preferential orientation of photonic crystals relative to skin surface was observed in S-iridophores, we also computed the average among all directions. Reflectivity was set to unity in the gapped region and convolution with standard X, Y, Z spectral functions returned simulated colours (Supplementary Movie 5) that closely match those observed in vivo(Fig. 1b) and during osmotic pressure experiments (Fig. 2c).

## Function of D-Iridophores

In addition, we investigated the second thick layer of iridophores (Fig. 1e), hereafter called deep (D-) iridophores, which contain larger brick-shaped and somewhat disorganized guanine crystals (length 200–600 nm, height 90–150 nm). This population of D-iridophores is present in all panther chameleons (regardless of sex or age) and in the three distantly related chameleon species we investigated (Figs 1e and 3a), and is particularly thick in comparison with the layer of iridophores observed in other (non-chameleonid) lizards. In chameleons, we never found this layer to change colour (in the visible range) during osmotic pressure experiments, suggesting that the main function of D-iridophores is not associated to shifts in hue. Our measurements indicate that the reflectivity (R) in the near-infrared region (700–1,400 nm) is particularly high (Fig. 3b), causing a substantial decrease in the absorption of sunlight. Multiplying the sun radiance[27] (blue curve inFig. 3b) by $1-R$, to yield the amount of light transmitted by the dermis (hence absorbed by the peritoneum or deeper tissues; red curve in Fig. 3b), indicates that ~45% of the radiation energy in that spectral range is screened in panther chameleons by reflection on the dermis. To test whether this infrared reflectivity is probably due to coherent scattering on guanine crystals in D-iridophores, we generated two-dimensional Fourier spectra[28] on extensive TEM image assemblies of panther chameleon D-iridophores (see online Methods). Note that the disorder of guanine crystals insideD-iridophores prevents the use of more rigorous modelling. We then used the computed Fourier power spectrum as an estimate of the spectral shape (red curve in Fig. 3c) of the light back-scattered by deep iridophores. This shows that the D-iridophore layer is a broad-band reflector in the near infrared region, as the power spectrum is essentially a step function going from 0 below 400 nm to a plateau above 900 nm. Multiplying the power spectrum by the transmittance of a 150-μm-thick layer of skin[29] (identical, in this spectral range, to that of water[30]), we produce a reflectance spectrum (green curve in Fig. 3c) that matches the shape of the measured reflectivity spectrum (black dashed curve in Fig. 3c) in the range 900–2,500 nm. The match below wavelengths of 900 nm is substantially less good, as we exclusively consider the D-iridophore crystals in our Fourier power spectrum analysis, that is, we ignore pigments and S-iridophores, which both strongly influence the measured reflectivity in the visible range. Hence, the thick layer of D-iridophores has the potential to play in some species, such as the panther chameleon, a substantial role in thermal protection. Comparative analyses with similar measurements in chameleonid and non-chameleonid species (for example, see Supplementary Fig. 4 and refs 31,32) is warranted to identify whether reflectivity in the near-infrared range is substantially and systematically higher in chameleons than in other lizards. It is noteworthy that the iridophores found in non-chameleonid lizards can exhibit guanine crystals with diverse sizes, shapes and organizations (some of which generate structural colours[14]) but are not organized into two superposed layers of functionally different iridophores (Fig. 3a).

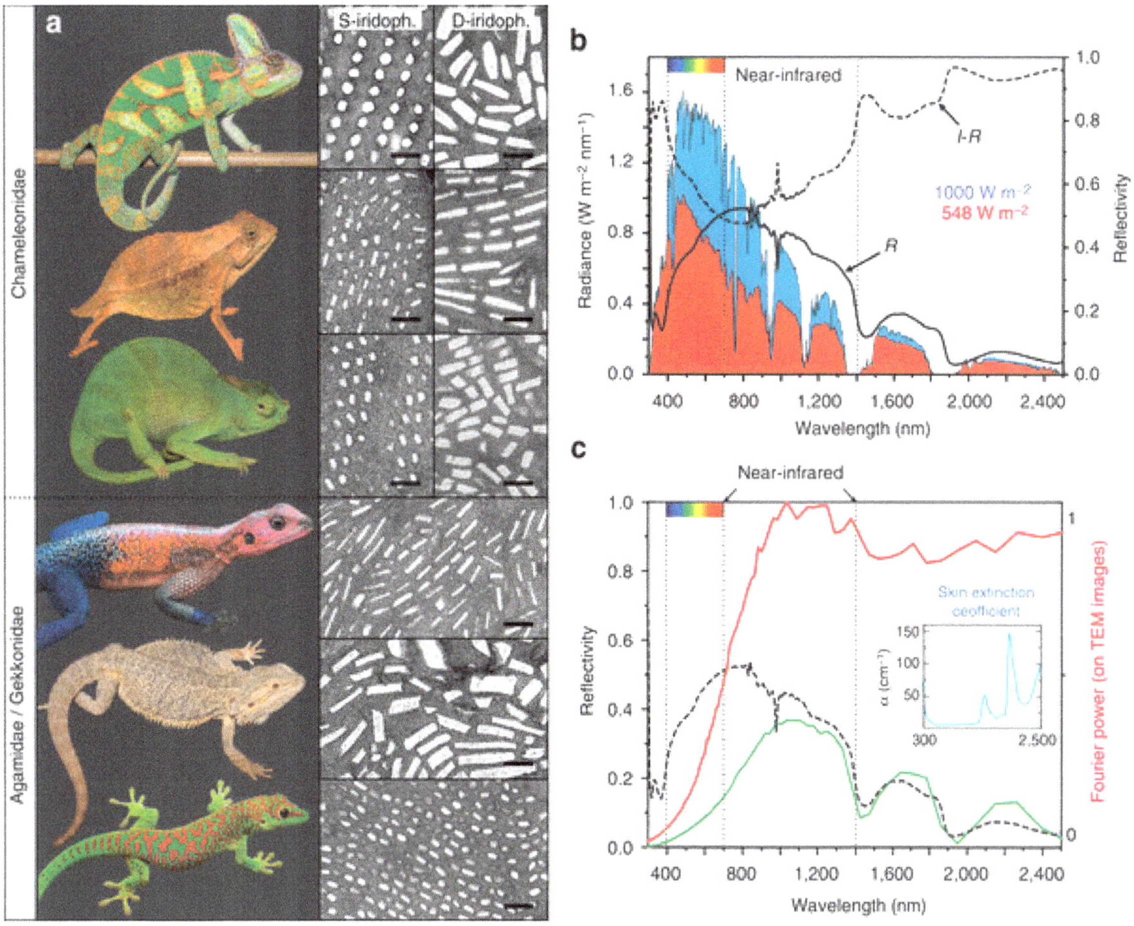


**Figure 3:** Iridophore types in lizards and function of D-iridophores in chameleons.

(**a**) In addition to F. pardalis (Fig. 1), other chameleonidae (top to bottom: Chamaeleo calyptratus,Rhampholeon spectrum and Kinyongia matschiei) exhibit two superposed layers of (S- and D-) iridophores, whereas agamids (the sister group to chameleons) and gekkonids have a single-type iridophore layer (top to bottom: Agama mwanzae, Pogona vitticeps and Phelsuma grandis). Scale bar, 500 nm. (**b**) Reflectivity (R) of a panther chameleon white skin sample and solar radiation spectrum (blue curve) at sea level (1,000 W $m^{-2}$); the product of the solar radiation spectrum and (1−R) yields the amount of sun radiation absorbed by deep tissues (red curve, 548 W $m^{-2}$). (**c**) The product of the Fourier power spectrum (red curve, computed from TEM images of D-iridophores) and the extinction coefficient of skin[29](blue inset) yields a predicted reflectivity distribution (green curve) similar to the measured reflectivity spectrum (black dashed curve) of a panther chameleon red skin sample.

## DISCUSSION

Combining experimental methods from biology and physics, as well as optical modelling, we have shown that panther chameleons rapidly change colour (hue) by actively tuning the photonic response of a lattice of small guanine nanocrystals in S-iridophores. The molecular mechanisms involved in this process remain to be determined; however, given that iridophores share the same neural-crest origin as pigmented chromatophores, the active tuning of guanine crystal spacing we describe here could be considered analogous to movements of pigment-containing organelles in other types of chromatophores, possibly through similar neural or hormonal mechanisms[33]. In chameleons, these S-iridophores are positioned on the top of a second thick layer of D-iridophores, with larger, flatter and somewhat disorganized guanine crystals, which reflects a substantial proportion of direct and indirect sun radiations, especially in the near-infrared range.

Chameleons form a highly derived monophyletic group of iguanian lizards that originated in post-Gondwanan Africa around 90 million years ago[34, 35]. Undoubtedly, some species of chameleons occupy quite open environments where they are exposed to high levels of sunlight. In particular, panther chameleons and veiled chameleons (studied here) occur in dry, hot environments (Northern Madagascar and Yemen, respectively) and are highly exposed to sunlight such that the 45% decrease in sunlight absorption caused by D-iridophores (Fig. 3b,c) is likely to be advantageous for survival. However, the ancestral function of D-iridophores might not be associated with passive thermal protection, because extant species of the basal lineages in the phylogeny of chameleons[34] are dense-forest dwellers (that is, not exposed to a dry and sunny environment), suggesting that the common ancestor of chameleons might have exhibited a similar ecology (but see alternative evolutionary scenarios in Supplementary Discussion).

The organization of iridophores into two superposed layers constitutes an evolutionary novelty for chameleons that allows some species to combine efficient camouflage with spectacular display. Additional analyses are warranted to identify whether the deep layer of iridophores in chameleons further provide them with improved resistance to variable sunlight exposure.

## METHODS

### Animals

Maintenance of and experiments on animals were approved by the Geneva Canton ethical regulation authority (authorization 1008/3421/1R) and performed according to the Swiss law.

### Skin Structure and Raman Spectroscopy

We examined the skin (ultra)structure by histology and TEM, using standard procedures, for example, as described in ref. 14. Samples were taken with biopsy pinches (diameter 2 mm) from male skin patches: when comparing skin before and after excitation, biopsies were separated by a maximum distance of 1 cm. For the relaxed state, the biopsy was taken within a few seconds after taking the animal from its cage and immediately placed in fixative. For the excited state, the animal was engaged in a male–male combat and a biopsy was taken again. The colour of skin biopsies was checked after fixation, to ensure that only those samples with well-preserved colours were used for TEM.

Semi-thin (2 μm) and ultra-thin (80–90 nm) cross-sections were cut with a diamond knife on a Leica UCT microtome. Ultra-thin sections were viewed with a Tecnai G2 Sphera (FEI) TEM at 120 kV before and after staining with uranyl acetate and lead citrate. Raman spectroscopy[16] was performed on melanophores directly on 2 μm cross-sections of skin samples with a home-made micro-Raman system composed of a 50-cm focal length spectrometer coupled to a nitrogen-cooled Princeton charge-coupled device detector and an argon laser (wavelength 514.5 nm) as the excitation source.

### Nanocrystal Measurements

For S-iridophores, crystal height, length and spacing between nearest crystals were measured on unstained and stained sections (original magnification × 19,000) of the same skin samples. Distances between nearest crystals were similar for unstained (179.9±30.2, N=85) and stained (180.3±26.9, N=94) sections. A first set of experiments indicated that the average diameter of the more or less spherical 'holes' remaining after staining (124.2 nm, N=103) reasonably approximates the average size of intact crystals (length=149.7±15.3 nm, height=94.7±11.5 nm,N=145). Hence, all Supplementary Data were collected on multiple TEM images of stained sections (performed on samples obtained from skin of various colours; Supplementary Table 1).

For D-iridophores, white 'rectangular holes' (Fig. 1e; corresponding to guanine crystals dissolved during post staining) on × 800 magnification TEM images were fitted (in JMicroVision[36]) with ellipses and geometric parameters (length, height and orientation) were computed subsequently. We performed Fourier transform analyses (as described in ref. 28), for each skin sample, on large assemblies of 20–30 high-resolution TEM images (1 pixel=15 nm, typical size of a guanine crystal=200 nm) spanning over 100 × 100 μm, that is, about 50 times the length of the longest wavelength investigated inside the material (corresponding to 2.5 μm in vacuum). Each assembly included more than 100,000 crystals.

### Photometry

High-quality photographs and movies were obtained with a high-resolution digital single-lens reflex camera (Nikon D800) and a Panasonic HDC-HS700 video camera, respectively. To analyse each video frame (Supplementary Fig. 2a), RGB band-pass filters were applied (Supplementary Fig. 2b) to select a colour window through which the variations of RGB channels were monitored (Supplementary Fig. 2c). The corresponding RGB numbers of each frame were averaged across the picture (Supplementary Fig. 2d) and normalized over the sum (R+G+B) (Supplementary Fig. 2e), to remove fluctuations of illumination as well as potential global variation in skin reflectivity caused by movements of melanosomes. Next, each channel was normalized from 0 to 1 (Supplementary Fig. 2f), to exclusively measure the variation occurring (in S-iridophores) over a relatively constant background colour (generated by D-iridophores and/or pigments). After transformation of RGB numbers from video into CIE XYZ tristimulus values, we derived the final x and y values that define colour irrespective of its luminance. Experimental traces (dots) are plotted and compared with the model (dashed line) on the CIE chromaticity chart (Figs 1b and 2c).

Skin reflectivity in the ultraviolet range, highly relevant for colour perception in reptiles[37, 38], is not recorded by RGB photometry. This does not have an impact on our conclusions, as the excellent matching between the modelled photonic response and photometry analyses validates our conclusions. In addition, photometric measurements have been validated with accurate spectroscopic measurements ex vivo (Fig. 2b) and show that XYZ/RGB photometric videography is sufficient to detect the wavelength shift in the reflectivity spectrum. In-vivo measurements of skin reflectivity (including in the ultraviolet range) with spectrometers directly on the animal skin are difficult, mainly because the animals move and because chameleon skin darkens very rapidly when it comes in contact with the optical probe.

### Optical Modelling

The symmetry of the close-packed photonic crystals present in the top layer (S-iridophores) of the skin was deduced from direct observations, under TEM, of crystals sectioned in different planes (Fig. 1d). The structural element common to all close-packed structures is a triangular two-dimensional arrangement. Studies of the effect of symmetry variations (pp. 307–308 in ref. 39) have indicated that, for $a/\lambda<1$ the optical response is mostly sensitive to the first Fourier component of the dielectric modulation. All close-packed structures have the same first-order Fourier component. To model the photonic structure effect that corresponds to our samples, it is therefore sufficient to choose face-centred cubic crystal symmetry. Crystal diameter (d) and distance between centres of nearest crystals (s) were measured (on about 1,200 individual crystals, Supplementary Table 1) on TEM images as described above and the lattice parameter awas computed as $s\sqrt{2}$. The first irreducible Brillouin zone (red lined contour in Supplementary Fig. 3b) was meshed and the band energies were computed for each vertex centre using block-iterative frequency-domain methods as implemented in the Massachusetts Institute of Technology photonic band package[26]. For each direction of light propagation in the structure (that is, each mesh triangle centre at the surface of the Brillouin zone), the reflectivity was set to unity in gapped frequency regions. As no preferential orientation of photonic crystals relative to the skin surface was observed in S-iridophores, the weighted average among all directions was computed using the dimensionless $a/\lambda$ parameter, where $\lambda$ refers to the corresponding light wavelength in air and arepresents the lattice parameter. The refractive indices of guanine and cytoplasm were set to 1.83 and 1.33, respectively[13]. Convolution of local and average reflectivity with standard spectral functions returns X, Y and Z colour numbers. Colour of each vertex and the average colour are plotted inside and outside of the irreducible Brillouin zone, respectively. Supplementary Movie 5shows the evolution of local and average colours as the lattice parameter varies.

## Reflectivity Measurements

Reflectivity of skin samples was measured using a monochromatic source and the locked-in detection of a spectroscopic Woollam ellipsometer from 300 to 2,500 nm and with a resolution of 5 nm. Monochromatic light of the source was injected in one channel of a reflection fibre probe (QR400-7-VIS-NIR). The light reflected on the sample (near-normal incidence) was collected by the second optical fibre channel and guided to a silicon photodiode (for visible range) and constrained InGaAs detectors (for the near-infrared range). A white diffuse standard (Ocean Optics WS-1-SL) was used as reference. This set up allowed measurements in air or in a solution with adjustable osmolarity (from Ringer's 1 × to 6 ×, that is, from 236 to 1,416 mOsm).

## Single-Cell Videography

Photometric videography on individual S-iridophores, were measured with adjustable osmolarity. Osmotic pressure was applied to fresh skin samples taken with biopsy pinches (diameter 2 mm) from white stripes of male panther chameleons. Samples were placed in a Ludin chamber where Ringer's 1 × solution (236 mOsm) was slowly replaced with Ringer's 4 × solution (944 mOsm). Full high-definition videos (1,920 × 1,080 pixels) were recorded with a Nikon D800 camera attached to a Leica MZ16 stereoscope. XYZ/RGB evolution of individual S-iridophores were obtained following the normalization procedure described in the 'Photometry' section.

## REFERENCES


1. Ligon, R. A. & McGraw, K. J. Chameleons communicate with complex colour changes during contests: different body regions convey different information. Biol. Lett. 9, 20130892 (2013).
2. Stuart-Fox, D. & Moussalli, A. Selection for social signalling drives the evolution of chameleon colour change. PLoS Biol. 6, e25 (2008).
3. Ferguson, G. W., Murphy, J. B., Ramanamanjato, J.-B. & Raselimanana, A. P. The Panther Chameleon: Color Variation, Natural History, Conservation, and Captive Management. 118 (Krieger Publishing Company, 2004).
4. Stuart-Fox, D. & Moussalli, A. Camouflage, communication and thermoregulation: lessons from colour changing organisms. Philos. Trans. R. Soc. Lond. B. Biol. Sci. 364, 463–470 (2009).
5. Messenger, J. B. Cephalopod chromatophores: neurobiology and natural history. Biol. Rev. Camb. Philos. Soc. 76, 473–528 (2001).
6. Nilsson Skold, H., Aspengren, S. & Wallin, M. Rapid color change in fish and amphibians-function, regulation, and emerging applications. Pigment Cell Melanoma Res. 26, 29–38 (2013).
7. Taylor, J. D. & Hadley, M. E. Chromatophores and color change in the lizard, Anolis carolinensis. Z. Zellforsch. Mikrosk. Anat. 104, 282–294 (1970).
8. DeMartini, D. G., Krogstad, D. V. & Morse, D. E. Membrane invaginations facilitate reversible water flux driving tunable iridescence in a dynamic biophotonic system. Proc. Natl Acad. Sci. USA 110, 2552–2556 (2013).
9. Mathger, L. M., Denton, E. J., Marshall, N. J. & Hanlon, R. T. Mechanisms and behavioural functions of structural coloration in cephalopods. J. R. Soc. Interface 6(Suppl 2): S149–S163 (2009).
10. Yoshioka, S. et al. Mechanism of variable structural colour in the neon tetra: quantitative evaluation of the Venetian blind model. J. R. Soc. Interface 8, 56–66 (2011).
11. Amiri, M. H. & Shaheen, H. M. Chromatophores and color revelation in the blue variant of the Siamese fighting fish (Betta splendens). Micron 43, 159–169 (2012).
12. Mathger, L. M., Land, M. F., Siebeck, U. E. & Marshall, N. J. Rapid colour changes in multilayer reflecting stripes in the paradise whiptail, Pentapodus paradiseus. J. Exp. Biol. 206, 3607–3613 (2003).
13. Morrison, R. L., Sherbrooke, W. C. & Frost-Mason, S. K. Temperaturesensitive, physiologically active iridophores in the lizard Urosaurus ornatus: an ultrastructural analysis of color change. Copeia 1996, 804–812 (1996).
14. Saenko, S. V., Teyssier, J., van der Marel, D. & Milinkovitch, M. C. Precise colocalization of interacting structural and pigmentary elements generates extensive color pattern variation in Phelsuma lizards. BMC Biol. 11, 105 (2013).
15. Grether, G. F., Kolluru, G. R. & Nersissian, K. Individual colour patches as multicomponent signals. Biol. Rev. (Camb. Philos. Soc.) 79, 583–610 (2004).
16. Huang, Z. et al. Raman spectroscopy of in vivo cutaneous melanin. J. Biomed. Opt. 9, 1198–1205 (2004).
17. Arsenault, A. C., Puzzo, D. P., Manners, I. & Ozin, G. A. Photonic-crystal full-colour displays. Nat. Photon. 1, 468–472 (2007).
18. Kuriyama, T., Miyaji, K., Sugimoto, M. & Hasegawa, M. Ultrastructure of the dermal chromatophores in a lizard (Scincidae: Plestiodon latiscutatus) with conspicuous body and tail coloration. Zoolog. Sci. 23, 793–799 (2006).
19. Scott, D. G. Packing of spheres: packing of equal spheres. Nature 188, 908–909 (1960).

20. Vukusic, P. & Sambles, J. R. Photonic structures in biology. Nature 424, 852–855 (2003).

21. Welch, V. L. & Vigneron, J. P. Beyond butterflies - the diversity of biological photonic crystals. Opt. Quant. Electron 39, 295–303 (2007).

22. Galusha, J. W., Richey, L. R., Gardner, J. S., Cha, J. N. & Bartl, M. H. Discovery of a diamond-based photonic crystal structure in beetle scales. Phys. Rev. E. Stat. Nonlin. Soft Matter Phys. 77, 050904 (2008).

23. Eliason, C. M., Bitton, P. P. & Shawkey, M. D. How hollow melanosomes affect iridescent colour production in birds. Proc. Biol. Sci. 280, 20131505 (2013).

24. Yin, H. et al. Amorphous diamond-structured photonic crystal in the feather barbs of the scarlet macaw. Proc. Natl Acad. Sci. USA 109, 10798–10801 (2012).

25. Stavenga, D. G., Leertouwer, H. L., Pirih, P. & Wehling, M. F. Imaging scatterometry of butterfly wing scales. Opt. Exp. 17, 193–202 (2009).

26. Johnson, S. & Joannopoulos, J. Block-iterative frequency-domain methods for Maxwell's equations in a planewave basis. Opt. Exp. 8, 173–190 (2001).

27. Bird, R. E. & Riordan, C. Simple solar spectral model for direct and diffuse irradiance on horizontal and tilted planes at the Earth's surface for cloudless atmospheres. J. Climate Appl. Meteorol. 25, 87–97 (1986).

28. Prum, R. O., Torres, R., Williamson, S. & Dyck, J. Coherent light scattering by blue feather barbs. Nature 396, 28–29 (1998).

29. Anderson, R. R. & Parrish, J. A. The optics of human skin. J. Invest. Dermatol. 77, 13–19 (1981).

30. Palmer, K. F. & Williams, D. Optical properties of water in the near infrared. J. Opt. Soc. Am. 64, 1107–1110 (1974).

31. Walton, B. M. & Bennett, A. F. Temperature-dependent color-change in kenyan chameleons. Physiol. Zool. 66, 270–287 (1993).

32. Fan, M., Stuart-Fox, D. & Cadena, V. Cyclic colour change in the bearded dragon Pogona vitticeps under different photoperiods. PLoS One 9, e111504 (2014).

33. Bagnara, J. T. & Matsumoto, J. Comparative Anatomy and Physiology of Pigment Cells in Nonmammalian Tissues 2nd edn 11–59 (Blackwell Pub., 2006).

34. Tolley, K. A., Townsend, T. M. & Vences, M. Large-scale phylogeny of chameleons suggests African origins and Eocene diversification. Proceedings 280, 20130184 (2013).

35. Townsend, T. M. et al. Phylogeny of iguanian lizards inferred from 29 nuclear loci, and a comparison of concatenated and species-tree approaches for an ancient, rapid radiation. Mol. Phylogenet. Evol. 61, 363–380 (2011).

36. Roduit, N. JMicroVision: Image analysis toolbox for measuring and quantifying components of high-definition images. Version 1.2.5. www.jmicrovision.com.

37. Thorpe, R. S. & Richard, M. Evidence that ultraviolet markings are associated with patterns of molecular gene flow. Proc. Natl Acad. Sci. USA 98, 3929–3934 (2001).

38. Fleishman, L. J., Loew, E. R. & Whiting, M. J. High sensitivity to short wavelengths in a lizard and implications for understanding the evolution of visual systems in lizards. Proceedings 278, 2891–2899 (2011).

39. Limonov, M. F. & De La Rue, R. M. Optical Properties of Photonic Structures: Interplay of Order and Disorder (CRC Press, 2012).

CHAPTER 5

# ENHANCED BROADBAND ELECTROMAGNETIC ABSORPTION IN SILICON FILM WITH PHOTONIC CRYSTAL SURFACE AND RANDOM GOLD GROOVES REFLECTOR

Zhi-Hui Chen[1,2], Na Qiao[1], YibiaoYang[1] , HanYe[2], Shaoding Liu[1], Wenjie Wang[1] & Yuncai Wang[1]

[1] Key Lab of Advanced Transducers and Intelligent Control System, Ministry of Education and Shanxi Province, College of Physics and Optoelectronics, Taiyuan University of Technology, Taiyuan 030024, China

[2] State Key Laboratory of Information Photonics and Optical Communications, Beijing University of Posts and Telecommunications, Beijing 100876, China.

## ABSTRACT

**We show a hybrid structure consisting of Si film with photonic crystal surface and random triangular gold grooves reflector at the bottom, which is capable of realizing efficient, broad-band, wide-angle optical absorption. It is numerically demonstrated that the enhanced absorption in a broad wavelength range (0.3–9.9 μm) due to the scattering effect of both sides of the structure and the created resonance modes. Larger thickness and period are favored to enhance the absorption in broader wavelength range. Substantial electric field concentrates in the grooves of surface photonic crystal and in the Si film. Our structure is versatile for solar cells, broadband photodetection and stealth coating.**

## INTRODUCTION

Efficient broadband electromagnetic absorption is significant in solar energy harvesting1·2, infrared photodetection3, and stealth coating4.

In 1982, Yablonovitch limit (Y-limit) was proposed to show a light trapping limit in homogeneous semiconductor films5. However, it is only applicable to geometrical optics and could be surpassed by the light trapping in sub-wavelength nanostructures by combining the reflection, diffraction, and refraction effects. A lot of optical coupling structures have been designed to enhance the electromagnetic absorption2·3·6·7·8·9·10·11·12·13·14·15·16·17·18·19·20·21·22, and various optimized light trapping structures are explored by optimization algorithms23·24. Most of these structures acted on either visible wavelength range or infrared wavelength range. However, simultaneously large absorption in both visible and infrared wavelength ranges is also essential for solar cells, broadband photodetectors and stealth applications.

In this work, we propose a Si film with one dimensional (1D) photonic crystal (PC) on the surface and 1D random triangular gold grooves reflector (RTGGR) at the bottom to improve the electromagnetic absorption efficiency within the wavelength range from 0.3 μm to 9.9 μm. The surface PC is capable of reducing the reflection from the interface and increasing the optical path lengths in the Si film. Meanwhile, enhanced scattering and absorption of light is obtained by introducing RTGGR at the bottom. The absorption is close to the Y-limit at visible wavelengths and is also large in infrared wavelength range, especially for the transverse magnetic (TM) polarized light. Furthermore, the enhanced absorption efficiency is insensitive to the incidence angle.

**Citation:** Zhi-Hui Chen, Na Qiao, Yibiao Yang, Han Ye, Shaoding Liu, Wenjie Wang & Yuncai Wang, Enhanced Broadband Electromagnetic Absorption in Silicon Film with Photonic Crystal Surface and Random Gold Grooves Reflector, doi:10.1038/srep12794.

In the following, we take ultrathin silicon solar cell as an example, and the wavelength range of 0.3–2.0 μm is studied firstly. The cross-sectional view of the proposed structure with denoted geometrical parameters is shown in Fig. 1(a). The calculations are conducted within one period by using rigorous coupled wave analysis (RCWA)25,26 and finite difference time domain (FDTD) method3,27,28. The model parameters and calculation details can be found in Method section.

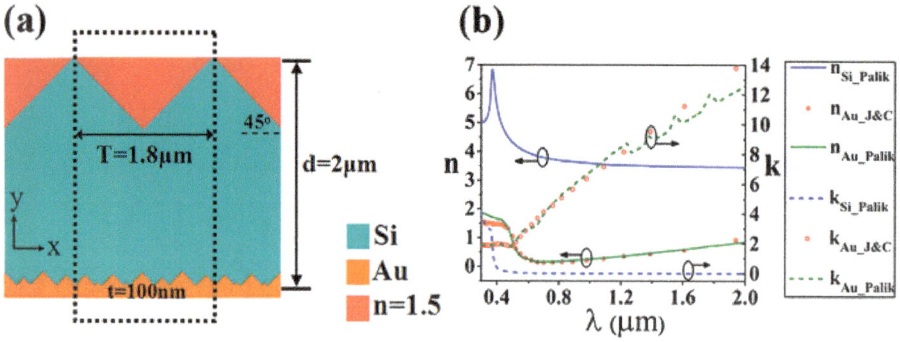


**Figure 1**: (**a**) The cross-sectional views of the proposed structure. One period is shown in dotted region. Polymer can be filled in the surface PC (n = 1.5). (**b**) The refractive indices of Si and gold in the wavelength range from 0.3 μm to 2 μm.

## RESULTS AND DISCUSSION

### Effects of Period and Angle on the Absorption Spectra

Figure 2 (a,b) displays the effect of period (T) of our structure (Fig. 1(a)) on the absorption efficiency at TM and TE polarization, respectively. It is observed that for TM polarization, the mean absorption efficiency of the structure exceeds 70% within the whole wavelength range from 0.3 μm to 2 μm when T is larger than 1.7 μm. The high absorption efficiencies of TE polarization are within the wavelength range from 300 nm to 700 nm. According to Fig. 2(a), the optimal periodicity should be close to the near-infrared wavelength for enhancing the electromagnetic absorption of TM polarization in both visible and near-infrared wavelength range. This is because for longer wavelength, a larger period is preferred to create more guided resonances which lead to higher absorption. In the following study, we choose T = 1.8 μm.

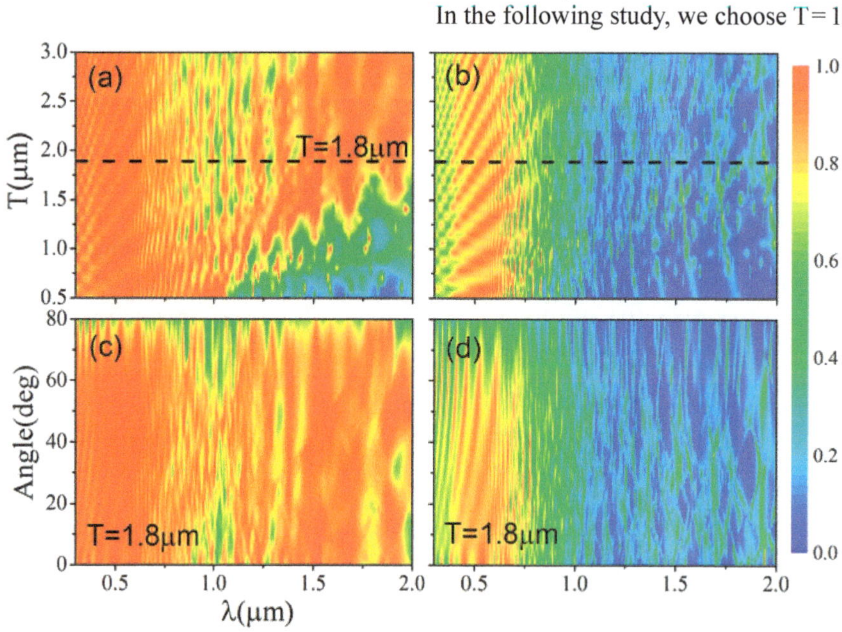


**Figure 2:** (**a,b**) are the absorption efficiencies of TM and TE polarization at normal incidence respectively when T changes from 0.5 μm to 3.0 μm. Figure 2 (**c,d**) are the absorption efficiencies of TM and TE polarization respectively at T = 1.8 μm when the incidence angle changes from 0° to 80°.

Figure 2(c,d) shows the angular absorption spectra of TM and TE polarization in our proposed structure. It is confirmed that the absorption of structure is insensitive to the angular changes when the incidence angle changes from 0° to 80°.

### Effects of Coupling Structures on the Absorption Spectra

In order to investigate the physical mechanism of light absorption enhancement, the absorption spectra of five structures compared with the Y-limit are shown in Fig. 3.

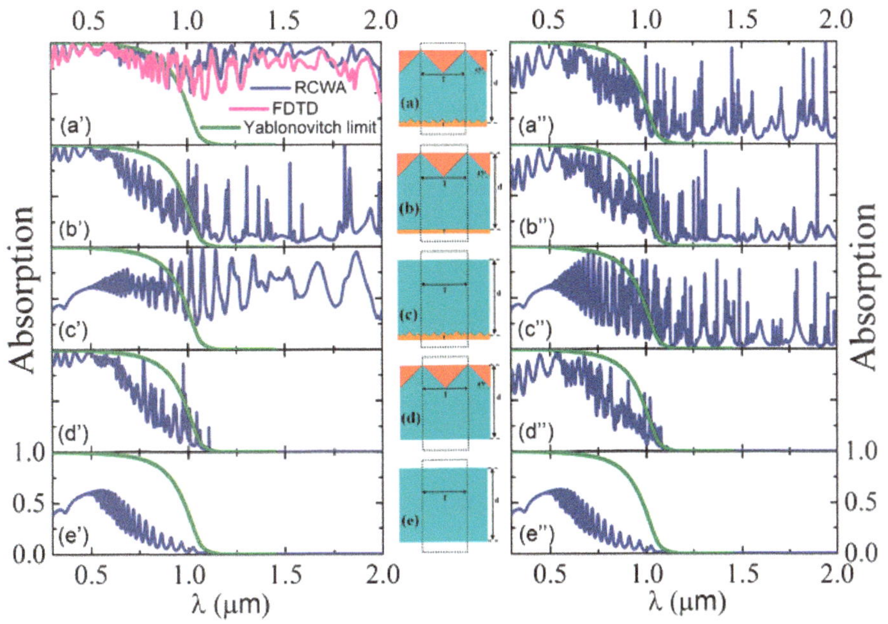


**Figure 3:** Absorption spectra of (a'–e') TM and (a"–e") TE polarizations at normal incidence in five structures (a–e) when T = 1.8 **μm**.

In all subfigures, blue curves are obtained by RCWA method, green curves are Y-limit, the pink curve in **(a')** is obtained by FDTD method.

From Fig. 3(a'), we can see that the result from RCWA method is in accordance with that from FDTD method. The Y-limit with perfect antireflection and perfect light trapping is calculated by:

$$A_{Yablonovitch} = 1 - 1/(1 + 4n^2\alpha d) \qquad (1)$$

which is shown in green curves2. ("n" is the real part of the refractive index of Si, "α" is the absorption coefficient, "d" is the thickness of Si film.)

When there is a bare Si film (structure "e"), the absorption is much smaller than the Y-limit. When only the surface PC is introduced (structure "d"), the absorption is enhanced due to the scattering and diffraction, but it is still below the Y-limit. When only the RTGGR is added (structure "c"), the absorption at infrared wavelengths is enhanced a lot, but the absorption at the visible wavelengths is not large. When both the surface PC and flat gold reflector are added (structure "b"), the absorption at infrared wavelengths is large at individual wavelengths due to some high Q-factor resonance modes. When both surface PC and RTGGR are introduced (structure "a"), obvious optical absorption enhancement is observed at both visible and infrared bands, and the absorption spectrum of our structure is very close to the Y-limit at visible wavelength and the absorption can even maintain large within infrared wavelengths range. It is noted that the absorption efficiencies of TM polarization in our proposed structure (structure "a") are higher than other four structures, especially in infrared wavelength range, the high absorption bands cover both visible and infrared bands. Similarly, gold reflector can also enhance the absorption of TE polarization at infrared wavelengths, while the enhancement of TE polarization in our structure is not as good as TM polarization, which is due to the 1D PC surface and 1D gold grooves. The surface PC and RTGGR work together on visible and infrared bands. Whether the top surface of Si is flat or structured, there will be resonance modes between the top surface and gold reflector. Stronger resonance modes can lead to more absorption by the Si and gold material. We show the electric field distributions corresponding to individual absorption peaks (TE polarization) of the structures (a–c) inFig. S1 of the Supplementary information file. There are no individual peaks in absorption spectra in the infrared wavelength range for structure (d) and (e) because they do not contain gold film and the Si film has little absorption in the infrared wavelength range.

## Spatial Distributions of the Electric Field

In order to study the spatial distributions of electric field in the proposed structure (Fig. 1(a)), we focused on TM polarization. Figure 4displays distribution of the electric field ($\mathbf{E}^2 = E_x^2 + E_y^2 + E_z^2$) calculated by RCWA at wavelengths of $\lambda_1 = 0.4\,\mu m$, $\lambda_2 = 1.2\,\mu m$, and $\lambda_3 = 1.8\,\mu m$, respectively. It can be seen from Fig. 4 that the electric field enhancement mainly concentrates in triangular grooves of surface PC at short wavelength. And at longer wavelength, the electric field distributes in both triangular grooves of surface PC and the Si film between PC and gold reflector. The distribution of the electric field has been confirmed by FDTD method. In solar cells or other photovoltaic applications, different type or different size of colloidal quantum dots (QDs) can be embedded in the PC's grooves to further enhance the broadband absorption29, because QDs have broadband absorption from the UV to the visible and even into the infrared wavelength regions mainly depending on the type and size of QDs30·31·32. Additionally, the optical resonance modes in the PC's grooves will enhance the interaction between light and QDs. When QDs are introduced into the surface PC's grooves, the increase of absorption and the wavelengths for the absorption enhancement towards the whole structure are related to the refractive index (n) and extinction coefficient (k) of the surface PC's grooves region, which are depended on the type, size and concentration (volume fraction ratio) of the QDs, and the QDs are usually in the status of film in a device33. In order to estimate the increase of absorption after embedding QDs in the grooves, we take $(PbSe)_{85}Cd_{15}$ QD as an example, we calculated the absorption efficiency (at TE polarization) of the structure with $(PbSe)_{85}Cd_{15}$ QDs film embedded in the surface PC's grooves (please see the model in Fig. S2 of the Supplementary information file). The refractive index (n=1.7) and extinction coefficient (k=0.42) of $(PbSe)_{85}Cd_{15}$ QDs thin film are obtained at the wavelength of $\lambda$=885 nm ($h$ =1.4 eV)34. According to the RCWA calculation results, the absorption efficiency of our proposed structure with polymer film (n=1.5, k=0) in the surface PC's grooves is about 44.3%, when $(PbSe)_{85}Cd_{15}$ QDs thin film (n=1.7, k=0.42) is introduced into the surface PC's grooves, the absorption efficiency of the structure can be enhanced to 92.6%.

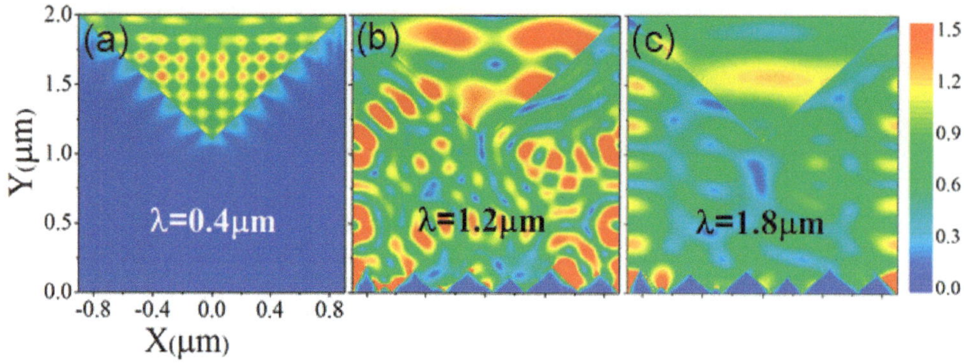


**Figure 4:** Electric field distribution ($E^2$) of TM polarization at **(a)** $\lambda_1$=0.4 μm, **(b)** $\lambda_2$=1.2 μm, and **(c)** $\lambda_3$=1.8 μm, respectively, in the structure of Fig. 1(a).

## Time Resolved Electric Field

We set three point monitors which are positioned in the surface PC's groove, Si film far from gold, Si film near the gold, respectively, in structure "a" and structure "e". The fold of time integral $\mathbf{E}^2$ between structure "a" and structure "e" is defined as:

$$N = \int E_a^2(t)dt / \int E_e^2(t)dt \qquad (2)$$

The N is "3.6", "2.6" and "4.7" in Fig. 5 (a–c), respectively. Time integral $\mathbf{E}^2$in structure "a" is larger than that of structure "e", which could confirm the results obtained in the frequency-domain (Fig. 3). The enhancement of electric field is because of light scattering and resonance modes in structure "a".

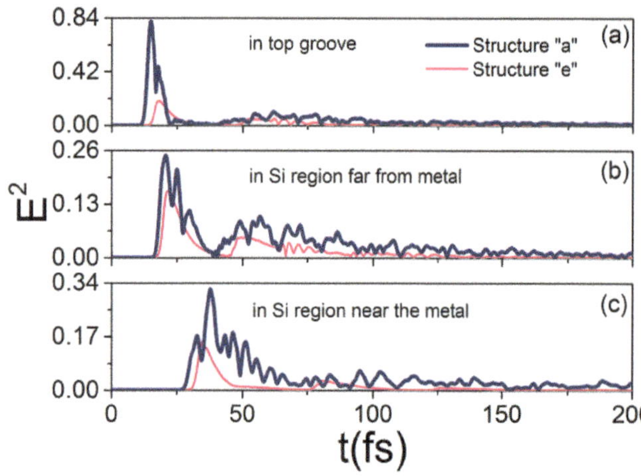


**Figure 5:** $E^2(t)$ in the surface PC's groove, Si film far from metal, Si film near the metal in structure **"a"**and structure **"e"**. The source is a broadband pulse covering 0.3–2.0 μm.

## Enhanced Broadband (0.3–9.9 μm) Absorption

As the optimal periodicity for light trapping should be close to the target wavelength range, absorption enhancement in broader wavelength range can be realized by using larger thickness and period. From Fig. 6, we can see that the large absorption (above 60%) wavelength range is more related to the T than the d. For different T, the absorption is similar at shorter wavelength, while is different at longer wavelength, and the drop point of absorption is at around the wavelength near the value of the T. Large d can only result in more resonance modes than the small d. When T=10 μm and d=10 μm, the absorption is larger at the wavelength from 0.3 μm to 9.9 μm which covers both the visible wavelength and infrared wavelength compared to the other smaller T and d. In our structure, T can be even larger and hold more guided resonances without obvious optical coupling and leaking to external channels.

The dark yellow line at 60% absorption is just for eye guide.

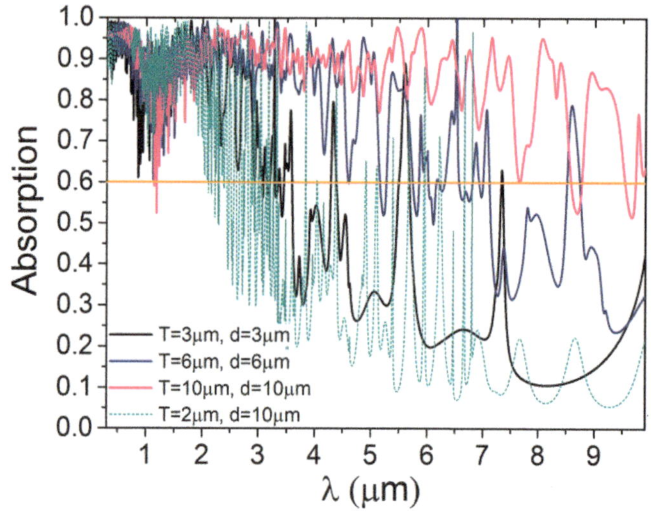


**Figure 6:** Absorption spectra of TM polarization at normal incidence in our structure with different periods (T) and thicknesses (d) of Si.

## CONCLUSIONS

In summary, we have numerically demonstrated a hybrid structure consisting of Si film with a photonic crystal surface and a random triangular gold groove reflector at the bottom, which is capable of realizing efficient (>60%), broad-band (0.3 μm–9.9 μm), wide-angle (0°–80°) electromagnetic absorption. It is observed that the absorption efficiencies of TM polarization in our proposed structure is close to Y-limit within the visible wavelength range and maintain large within the infrared wavelength range, such large absorption is insensitive to the incidence angle. Thus is mainly attributed to the enhancement of light scattering and light trapping in our structure. The surface PC and RTGGR work together on visible and infrared bands. The results are confirmed by the FDTD method and RCWA method. Colloidal quantum dots can be embedded in the surface PC's grooves to further enhance the absorption. The multifunctional structure provides a great potential for the application in solar cells, photodetectors and stealth coating.

## METHODS

### Proposed Structure

In the whole study, the calculations are conducted within one period (the dotted region) as shown in Fig. 1(a). The period (T) of the surface PC is 1.8 μm, and the cross-sectional views of one period of the PC is isosceles right triangle. We use eight isosceles right triangles with random sizes in sub-wavelength scale in one period to represent the quasi-random distribution of gold grooves. The thickness of Si (d) is 2 μm, and the thickness of gold substrate (t) is 100 nm which is thick enough to reflect the light in the wavelength range from 0.3 μm to 9.9 μm. The absorption efficiencies of TM and TE polarization are studied when T and incidence angle change. Here, the size of every gold triangle is in a certain proportion to T. The results are confirmed by the RCWA method and FDTD method.

### RCWA and FDTD Simulations

The Rigorous Coupled Wave Analysis (RCWA) method, which is an efficient method to calculate electromagnetic wave propagation in periodic structures26, is adopted in our work. In this method, Maxwell's equations are solved in the Fourier domain. The absorption can be calculated by the reflected, transmitted and diffracted power. In this work, the absorption efficiency has been computed in one unit of the periodic structure within the space domain (x, y) = (−0.9:0.9, −1.0:4.5) μm, the wavelength, incidence angle and period are scanned. The near-field electric field distributions are calculated by using a normal-direction incident plane wave source at specific wavelengths.

The finite-difference time-domain (FDTD) method can numerically solve two or three dimensional time-dependent Maxwell's equations in time domain without any approximations or theoretical restrictions35. In addition, FDTD can also obtain frequency-dependent solutions by using fast Fourier transform (FFT) or discrete Fourier transform (DFT). Thus the transmission and reflection of light can be calculated, the absorption can be obtained according to the law of energy conservation. In this work, we perform a two dimensional FDTD simulation in the region (x, y) = (−0.9:0.9, −1.0:4.5) μm with Bloch boundaries in the x direction and perfectly matched layers (PMLs) absorbing boundaries in the y direction. The simulation time is 20000 fs, and the spatial grid size is 4nm. The incident plane wave is simulated by a broadband pulse.

In both simulation methods, we take intrinsic losses of Si and gold into consideration. In the RCWA method and FDTD method, the wavelength dependent refractive indices of Si are obtained from Palik's book36, while the refractive indices of gold are from Palik's book36 and Johnson and Christy's data37 respectively. The complex refractive index of Si and gold are shown in Fig. 1(b). Plane wave is injected from the top of the structure, and the wavelength range we studied firstly in this work is from 0.3 μm to 2.0 μm. Then a broader band (0.3 μm–9.9 μm) is studied.

### Feasibility of Experimental Realization

Our structure is experimentally feasible. The grooves in the upper and bottom surface of Si film can be fabricated by various etching techniques38·39. Then the grooves at two sides of Si can be filled by deposited polymer and gold film, respectively. The structure is expected to enhance both TM and TE polarization obviously if it includes 2D patterned grooves in the upper and bottom surface of the Si film.

### Additional INFORMATION

**How to cite this article**: Chen, Z.-H. *et al.* Enhanced Broadband Electromagnetic Absorption in Silicon Film with Photonic Crystal Surface and Random Gold Grooves Reflector. *Sci. Rep.* **5**, 12794; doi: 10.1038/srep12794 (2015).

## REFERENCES

1. Brongersma, M. L., Cui, Y. & Fan, S. Light management for photovoltaics using high-index nanostructures. Nat. Mater. 13, 451–460 (2014).
2. Wang, K. X. Z., Yu, Z. F., Liu, V., Cui, Y. & Fan, S. H. Absorption Enhancement in Ultrathin Crystalline Silicon Solar Cells with Antireflection and Light-Trapping Nanocone Gratings. Nano Lett. 12, 1616–1619 (2012).
3. Chen, Z.-H., Hellström, S., Yu, Z.-Y., Qiu, M. & Fu, Y. Time-resolved photocurrents in quantum well/dot infrared photodetectors with different optical coupling structures. Appl. Phys. Lett. 100, 043502 (2012).
4. Shi, H., Ok, J.G., Won Baac, H. & Jay Guo, L. Low density carbon nanotube forest as an index-matched and near perfect absorption coating. Appl. Phys. Lett. 99, 211103 (2011).
5. Yablonovitch, E. & Cody, G. D. Intensity enhancement in textured optical sheets for solar cells. IEEE Trans. Elec. Dev. 29, 300–305 (1982).
6. Zhu, J., Hsu, C.-M., Yu, Z., Fan, S. & Cui, Y. Nanodome Solar Cells with Efficient Light Management and Self-Cleaning. Nano Lett. 10, 1979–1984 (2009).
7. Zhu, J. et al. Optical Absorption Enhancement in Amorphous Silicon Nanowire and Nanocone Arrays. Nano Lett. 9, 279–282 (2008).
8. Hsu, C.-M. et al. High-Efficiency Amorphous Silicon Solar Cell on a Periodic Nanocone Back Reflector. Adv. Energy Mater. 2, 628–633 (2012).

9. Jeong, S. et al. Hybrid Silicon Nanocone–Polymer Solar Cells. Nano Lett. 12, 2971–2976 (2012).

10. Jeong, S., McGehee, M. D. & Cui, Y. All-back-contact ultra-thin silicon nanocone solar cells with 13.7% power conversion efficiency. Nat. Commun. 4, 2950 (2013).

11. Yu, Z., Raman, A. & Fan, S. Fundamental limit of nanophotonic light trapping in solar cells. P. Natl. Acad. Sci. USA. 107, 17491–17496 (2010).

12. Chen, Z.-H., Hellström, S., Yu, Z.-Y. & Fu, Y. Comb-shaped photonic crystal structure for efficient broadband light diffraction and funnelling in solar cells. Sol. Energy Mater. Sol. Cells 99, 316–320 (2012).

13. Meng, L., Zhao, D., Li, Q. & Qiu, M. Polarization-sensitive perfect absorbers at near-infrared wavelengths. Opt. Express 21, A111–A122 (2013).

14. Xie, S. et al. Transparent gold nano-membranes for the enhanced light trapping of the indium tin oxide films. Opt. Mater. Express 4, 321–328 (2014).

15. Adachi, M. M. et al. Broadband solar absorption enhancement via periodic nanostructuring of electrodes. Sci. Rep. 3, 2928 (2013).

16. Ganesh, N. et al. Enhanced fluorescence emission from quantum dots on a photonic crystal surface. Nat. Nanotechnol. 2, 515–520 (2007).

17. Yu, R., Lin, Q., Leung, S.-F. & Fan, Z. Nanomaterials and nanostructures for efficient light absorption and photovoltaics. Nano Energy 1, 57–72 (2012).

18. Hellström, S. et al. Increased photocurrent in quantum dot infrared photodetector by subwavelength hole array in metal thin film. Appl. Phys. Lett. 96, 231110 (2010).

19. Zhang, Y., Stokes, N., Jia, B., Fan, S. & Gu, M. Towards ultra-thin plasmonic silicon wafer solar cells with minimized efficiency loss. Sci. Rep. 4, 4939 (2014).

20. Fang, J., Liu, B., Zhao, Y. & Zhang, X. Two-dimensional high efficiency thin-film silicon solar cells with a lateral light trapping architecture. Sci. Rep. 4, 6169 (2014).

21. Chen, X. et al. Graphenized Carbon Nanofiber: A Novel Light-Trapping and Conductive Material to Achieve an Efficiency Breakthrough in Silicon Solar Cells. Adv. Mater. 27, 849–855 (2015).

22. Varghese, L. T. et al. Enhanced Photon Management of Thin-Film Silicon Solar Cells Using Inverse Opal Photonic Crystals with 3D Photonic Bandgaps. Adv. Opt. Mater. 1, 692–698 (2013).

23. Wang, C., Yu, S., Chen, W. & Sun, C. Highly Efficient Light-Trapping Structure Design Inspired By Natural Evolution. Sci. Rep. 3, 1025 (2013).

24. Wang, D. & Su, G. New strategy to promote conversion efficiency using high-index nanostructures in thin-film solar cells. Sci. Rep. 4, 7165 (2014).

25. Chen, Z.-H. et al. Enhanced normal-direction excitation and emission of dual-emitting quantum dots on a cascaded photonic crystal surface. Nanoscale 6, 14708–14715 (2014).

26. Liu, V. & Fan, S. S4: A free electromagnetic solver for layered periodic structures. Comput. Phys. Commun. 183, 2233–2244 (2012).

27. Taflove, A. & Hagness, S. C. Computational electrodynamics, Vol. 160. (Artech house Boston, 2000).

28. Chen, Z.-H., Yu, Z.-Y., Liu, Y.-M., Lu, P.-F. & Fu, Y. Multiple beam splitting to free space from a V groove in a photonic crystal waveguide. Appl. Phys. B 102, 857–861 (2011).

29. Chen, Z.-H., Hellström, S., Ning, Z.-J., Yu, Z.-Y. & Fu, Y. Exciton polariton contribution to the stokes shift in colloidal quantum dots. J. Phys. Chem. C 115, 5286–5293 (2011).

30. Coe-Sullivan, S., Steckel, J.S., Woo, W.K., Bawendi, M.G. & Bulović, V. Large-Area Ordered Quantum-Dot Monolayers via Phase Separation During Spin-Casting. Adv. Funct. Mater. 15, 1117–1124 (2005).

31. Bera, D., Qian, L., Tseng, T.-K. & Holloway, P. H. Quantum dots and their multimodal applications: a review. Materials 3, 2260–2345 (2010).

32. Dai, Q. et al. Size-dependent composition and molar extinction coefficient of PbSe semiconductor nanocrystals. ACS nano 3, 1518–1524 (2009).

33. Li, J., Jia, B., Zhou, G. & Gu, M. Fabrication of three-dimensional woodpile photonic crystals in a PbSe quantum dot composite material. Opt. Express 14, 10740–10745 (2006).

34. Alvi, M. & Khan, Z. H. Synthesis and characterization of nanoparticle thin films of a-(PbSe)100-x Cdx lead chalcogenides. Nanoscale Res. Lett. 8, 1–10 (2013).

35. Chen, Z. Light manipulation in micro and nano photonic materials and structures. (KTH Royal Institute of Technology, Sweden, 2012).

36. Palik, E. D. Handbook of optical constants of solids, Vol. 3. (Academic press, 1998).

37. Johnson, P. B. & Christy, R. W. Optical Constants of the Noble Metals. Phys. Rev. B 6, 4370–4379 (1972).

38. Olzierski, A., Nassiopoulou, A., Raptis, I. & Stoica, T. Two-dimensional arrays of nanometre scale holes and nano-V-grooves in oxidized Si wafers for the selective growth of Ge dots or Ge/Si hetero-nanocrystals. Nanotechnology 15, 1695 (2004).

39. Xia, Y. & Whitesides, G. M. Shadowed sputtering of gold on V -shaped microtrenches etched in silicon and applications in microfabrication. Adv. Mater. 8, 765–768 (1996).

CHAPTER 6

# MACROSCOPIC OPTICAL RESPONSE AND PHOTONIC BANDS

J S Perez-Huerta[1], Guillermo P Ortiz[2], Bernardo S Mendoza[3] and W Luis Mochan[1]

[1] Instituto de Ciencias F´ısicas, Universidad Nacional Autonoma de M ´ exico, ´ Apartado Postal 48-3, 62251 Cuernavaca, Morelos, Mexico

[2] Departamento de F´ısica, Facultad de Ciencias Exactas, Naturales y Agrimensura, Universidad Nacional del Nordeste, Avenida Libertad 5400 Campus-UNNE, W3404AAS Corrientes, Argentina

[3] Division of Photonics, Centro de Investigaciones en Optica, Leon, Guanajuato, ´ Mexico

## ABSTRACT

**We develop a formalism for the calculation of the macroscopic dielectric response of composite systems made of particles of one material embedded periodically within a matrix of another material, each of which is characterized by a well-defined dielectric function. The nature of these dielectric functions is arbitrary, and could correspond to dielectric or conducting, transparent or opaque, absorptive and dispersive materials. The geometry of the particles and the Bravais lattice of the composite are also arbitrary. Our formalism goes beyond the long-wavelength approximation** as it fully incorporates retardation effects. **We test our formalism through the study of the propagation of electromagnetic waves in two-dimensional photonic crystals made of periodic arrays of cylindrical holes in a dispersionless dielectric host. Our macroscopic theory yields a spatially dispersive macroscopic response which allows the calculation of the full photonic band structure of the system, as well as the characterization of its normal modes, upon substitution into the macroscopic field equations. We can also account approximately for the spatial dispersion through a local magnetic permeability and analyze the resulting dispersion relation, obtaining a region of left handedness.**

## INTRODUCTION

The propagation of light within homogeneous materials is completely characterized by using their electromagnetic response. In the optical regime, magnetic effects are typically negligible so that knowledge of the dielectric response is sufficient for the study of wave propagation [1]. However, when the system has spatial inhomogeneities, the behavior of light is not trivial and this has captured the attention of many researchers [2–11]. As an example, plasmon–polariton spectra of quasiperiodic superlattices have been investigated [12, 13] and have been shown to have a fractal character. Novel effects have been obtained, for example, negative refraction, inverse Doppler effect, optical invisibility cloaking and optical magnetism, etc [14–17].4

Effective medium theories [18–26] have been proposed for describing inhomogeneous systems, such as diluted colloidal suspensions in the quasistatic or long-wavelength limit, in terms of a homogeneous macroscopic response. Further developments on homogenization of composites have been proposed [27–35] and their limits of validity have been discussed [2, 36, 37].

In this paper, we obtain the macroscopic optical response using a computationally efficient reformulation of a procedure [5, 6] originally developed for accounting for

**Citation:** J S Perez-Huerta, Guillermo P Ortiz, Bernardo S Mendoza and W Luis Mochan, Macroscopic optical response and photonic bands, doi:10.1088/1367-2630/15/4/043037.

the local field effect of systems with spatial fluctuations. Our main result is that for any system, the macroscopic response to an external excitation is simply the average projection of the corresponding microscopic response. We apply this general result to Maxwell equations in order to obtain an explicit expression for the macroscopic dielectric response of binary composites. We illustrate its use with numerical results for a two-dimensional (2D) periodic photonic crystal. As the wavelength of the electromagnetic field becomes comparable with the characteristic microscopic spatial lengthscale of the system, retardation effects become important, yielding a non-local [38, 39] macroscopic response, with a strong dependence on the wavevector besides its usual dependence on frequency. When the full spatial dispersion is taken into account, the photonic band structure of the system can be obtained and analyzed from its macroscopic response.

This paper is organized as follows. In section 2, we formulate a general theory for the macroscopic dielectric response of composite systems. In section 3, we adapt our formulation to periodic systems made of two alternating materials, each characterized by well-defined dielectric functions. In section 4, we develop an efficient computational scheme which allows the numerical calculation of the macroscopic response without recourse to explicit operations between large matrices. In section 5, we obtain the non-local macroscopic dielectric tensor and use it to calculate macroscopically the full band structure of a 2D photonic crystal made up of non-dispersive dielectric components; we compare it with exact results as well as with results within a local approximation that partially accounts for spatial dispersion through an effective magnetic permeability. Finally, in section 6, we present the conclusions.

## GENERAL THEORY

Consider a non-magnetic inhomogeneous system characterized by its dielectric response $\hat{\epsilon}$, defined through the constitutive equation $\mathbf{D} = \hat{\epsilon}\mathbf{E}$, where **D** and **E** are the microscopicdisplacement and electric fields, respectively. We designate these fields as microscopic as they have spatial fluctuations due to the inhomogeneities of the material. They are not microscopic in the atomic scale, but rather on the scale of the spatial inhomogeneities of the system. We use a caret ($\hat{}$) over a symbol to denote its operator nature and leave implicit the dependence of the dielectric response on position as well as on frequency. Our purpose is to obtain the macroscopic dielectric response of the system, relating the macroscopic displacement and electric fields, from which the fluctuations have been removed.

We start from Maxwell equations for monochromatic microscopic fields. We follow the usual procedure [40] to decouple the magnetic field from the electric field to obtain the second-order wave equation

$$\hat{\mathcal{W}}\mathbf{E} = \frac{4\pi c}{iq}\mathbf{J}, \tag{1}$$

where

$$\hat{\mathcal{W}} = \hat{\epsilon} - \frac{1}{q^2}\nabla\times\nabla\times = \hat{\epsilon} + \frac{1}{q^2}\nabla^2\hat{\mathcal{P}}^{\mathrm{T}} \tag{2}$$

is the wave operator, q = ω/c is the wavenumber of light in vacuum, ω is the frequency, c is the speed of light in vacuum and **J** is the external electric current density. Here we introduced the transverse projector $\hat{\mathcal{P}}^{\mathrm{T}}$ such that the transverse projection of a field **F** is $\mathbf{F}^{\mathrm{T}} = \hat{\mathcal{P}}^{\mathrm{T}}\mathbf{F}$. For completeness, we also introduce the longitudinal projector $\hat{\mathcal{P}}^{\mathrm{L}}$ such that $\hat{\mathcal{P}}^{\mathrm{L}} + \hat{\mathcal{P}}^{\mathrm{T}} = \hat{1}$, with $\hat{1}$ the identity operator. Note that equation (1) contains both a longitudinal and a transverse part, so it describes not only transverse waves, but also allows for the possible excitation of longitudinal waves such as plasmons [41].

We solve equation (1) formally for **E** to obtain

$$\mathbf{E} = \frac{4\pi c}{iq}\hat{\mathcal{W}}^{-1}\mathbf{J}. \tag{3}$$

As we are interested on the macroscopic response of the system, we introduce the average $\hat{\mathcal{P}}_{\mathrm{a}}$ and fluctuation $\hat{\mathcal{P}}_{\mathrm{f}}$ projectors, such that an arbitrary field **F** can be written as $\mathbf{F} = \mathbf{F}_{\mathrm{a}} + \mathbf{F}_{\mathrm{f}}$ with $\mathbf{F}_{\mathrm{a}} = \hat{\mathcal{P}}_{\mathrm{a}}\mathbf{F}$ its average and $\mathbf{F}_{\mathrm{f}} = \hat{\mathcal{P}}_{\mathrm{f}}\mathbf{F}$ its fluctuating part, and we identify the average field with the macroscopic field $\mathbf{F}_{\mathrm{a}} \equiv \mathbf{F}_{\mathrm{M}}$. Later, we will provide appropriate explicit definitions for $\hat{\mathcal{P}}_{\mathrm{a}}$ and $\hat{\mathcal{P}}_{\mathrm{f}}$; here we remark that they must be idempotent $\hat{\mathcal{P}}_{\mathrm{a}}^2 = \hat{\mathcal{P}}_{\mathrm{a}}$, $\hat{\mathcal{P}}_{\mathrm{f}}^2 = \hat{\mathcal{P}}_{\mathrm{f}}$, i.e. the average of the average is the average, and the fluctuations of the fluctuations are the fluctuations [5]. Furthermore, $\hat{\mathcal{P}}_{\mathrm{a}}\hat{\mathcal{P}}_{\mathrm{f}} = \hat{\mathcal{P}}_{\mathrm{f}}\hat{\mathcal{P}}_{\mathrm{a}} = 0$ and $\hat{\mathcal{P}}_{\mathrm{a}} + \hat{\mathcal{P}}_{\mathrm{f}} = \hat{1}$.

As the macroscopic response would be useless unless the external excitations are devoid of microscopic spatial fluctuations, we assume that the external current has no fluctuations, so$\mathbf{J}_{\mathrm{f}} = \mathbf{0}$ and $\mathbf{J} = \mathbf{J}^{\mathrm{M}} = \hat{\mathcal{P}}_{\mathrm{a}}\mathbf{J}^{\mathrm{M}}$, where we used the idempotency of $\hat{\mathcal{P}}_{\mathrm{a}}$. Thus, acting on both sides of equation (3) with $\hat{\mathcal{P}}_{\mathrm{a}}$ we obtain

$$\mathbf{E}^{\mathrm{M}} = \frac{4\pi c}{iq}\hat{\mathcal{W}}_{\mathrm{aa}}^{-1}\mathbf{J}^{\mathrm{M}}. \tag{4}$$

Here we define $\hat{O}_{\alpha\beta} \equiv \hat{\mathcal{P}}_{\alpha}\hat{O}\hat{\mathcal{P}}_{\beta}$, with α,β = a,f for any operator $\hat{O}$. As equation (4) relates the macroscopic external electric current to the macroscopic electric field, we may identify the macroscopic inverse wave operator $\hat{\mathcal{W}}_{\mathrm{M}}^{-1}$,

$$\mathbf{E}^{\mathrm{M}} = \frac{4\pi c}{iq}\hat{\mathcal{W}}_{\mathrm{M}}^{-1}\mathbf{J}^{\mathrm{M}}, \tag{5}$$

where

$$\hat{\mathcal{W}}_{\mathrm{M}}^{-1} = \hat{\mathcal{W}}_{\mathrm{aa}}^{-1}. \tag{6}$$

We summarize this result stating that the macroscopic inverse wave operator is simply the average of the microscopic inverse wave operator. This is a particular case of a more general result: the macroscopic response to an external excitation is simply the average of the corresponding microscopic response.

From the macroscopic Maxwell equations, we may further relate the macroscopic wave operator in equations (5) and (6) to the macroscopic dielectric response of the system $\hat{\epsilon}^{\mathrm{M}}$ through

$$\hat{\mathcal{W}}^{\mathrm{M}} = \hat{\epsilon}^{\mathrm{M}} + \frac{1}{q^2}\nabla^2\hat{\mathcal{P}}_{\mathrm{a}}\hat{\mathcal{P}}^{\mathrm{T}} \tag{7}$$

in analogy to equation (2). Thus, we have to invert the wave operator, average it and invert it again to finally identify the macroscopic dielectric response. Note that we have made no approximation whatsoever. We remark that as $\hat{\epsilon}$ relates to two fields, **E** and **D**, that have spatial fluctuations, we may not simply average $\hat{\epsilon}$ to obtain its macroscopic counterpart, i.e. $\hat{\epsilon}^{\mathrm{M}} \neq \hat{\epsilon}_{\mathrm{aa}}$. The difference constitutes the local field effect [5].

Our result may easily be generalized to other situations and other response functions. The procedure consists of first identifying the response (in our case $\hat{\mathcal{W}}^{-1}$) to the externalperturbation (i.e. **J**) and then averaging it to yield the corresponding macroscopic response (i.e. $\hat{\mathcal{W}}_{\mathrm{M}}^{-1} = \hat{\mathcal{W}}_{\mathrm{aa}}^{-1}$), which may further be related to the desired macroscopic response operator (i.e. $\hat{\epsilon}^{\mathrm{M}}$).

## PERIODIC BINARY SYSTEMS

In this section, we use equation (6) to obtain the optical properties of an artificial binary crystal made of two materials A and B with dielectric functions $\epsilon A$ and $\epsilon B$. We assume that both media are local and isotropic so that $\epsilon A$ and $\epsilon B$ are simply complex functions of the frequency. For convenience, we will further assume that $\epsilon A$ is real, although this assumption is easily relaxed [42].

We introduce the characteristic function B(**r**) of the inclusions, such that B(**r**) ≡ 1 whenever **r**is on the region B occupied by the inclusions, and B(**r**) ≡ 0 otherwise. Thus, we may write the microscopic dielectric response as

$$\epsilon(\mathbf{r}) = \frac{\epsilon_A}{u}\left(u - B(\mathbf{r})\right), \tag{8}$$

where we defined the spectral variable $u \equiv 1/(1 - \epsilon B/\epsilon A)$ [19]. The microscopic wave operator of equation (2) may be written as

$$\hat{\mathcal{W}} = \frac{\epsilon_A}{u}(u\hat{g}^{-1} - \hat{\mathcal{B}}), \tag{9}$$

where the characteristic operator $\hat{\mathcal{B}}$ corresponds to multiplication by B(**r**) in real space, and we defined

$$\hat{g} = \left(\hat{1} + \frac{\nabla^2\hat{\mathcal{P}}^{\mathrm{T}}}{q^2\epsilon_A}\right)^{-1}, \tag{10}$$

which, as shown below, plays the role of a metric.

Using equation (9) we write equation (6) as

$$\hat{\mathcal{W}}_M^{-1} = \frac{u}{\epsilon_A}\hat{g}_{\mathrm{aa}}(u - \hat{\mathcal{B}}\hat{g})_{\mathrm{aa}}^{-1}, \tag{11}$$

where we have taken advantage of the fact that $\hat{g}$ is unrelated to the texture of the crystal, so that it does not couple the average to fluctuating fields.

Using Bloch›s theorem [43], we consider an electric field of the form [21]

$$\mathbf{E}_{\mathbf{k}}(\mathbf{r}) = \sum_{\mathbf{G}} \mathbf{E}_{\mathbf{G}} e^{i(\mathbf{k}+\mathbf{G})\cdot\mathbf{r}}, \tag{12}$$

where **k** is a given wavevector, {**G**} is the reciprocal lattice of our crystal and the coefficients**EG** represent the field in reciprocal space. In this representation, all operators may be written as matrices with vector index pairs **G**,**G'**, besides other possible indices, such as Cartesian ones. Thus, we represent the longitudinal projector as the matrix

$$\mathcal{P}^{\mathrm{L}}_{\mathbf{GG}'} = \delta_{\mathbf{GG}'}\frac{(\mathbf{k}+\mathbf{G})}{|\mathbf{k}+\mathbf{G}|}\frac{(\mathbf{k}+\mathbf{G}')}{|\mathbf{k}+\mathbf{G}'|} \tag{13}$$

with δ**GG**, the Kronecker›s delta, so the transverse projector becomes $\mathcal{P}^{\mathrm{T}}_{\mathbf{GG}'} = \mathbf{1}\delta_{\mathbf{GG}'} - \mathcal{P}^{\mathrm{L}}_{\mathbf{GG}'}$ with **1** the Cartesian identity matrix. The Laplacian operator in reciprocal space is represented by

$$\nabla^2 \rightarrow -(\mathbf{k}+\mathbf{G})^2\delta_{\mathbf{GG}'} \tag{14}$$

and we can define the average projector as a cutoff in the reciprocal space [6]

$$(\mathcal{P}_{\mathrm{a}})_{\mathbf{GG}'} = \delta_{\mathbf{G0}}\delta_{\mathbf{G}'\mathbf{0}}, \tag{15}$$

average fields simply keep the contributions with vector **k** while all other wavevectors are filtered out.

As shown by equation (11), we only require

$$(u - \hat{\mathcal{B}}\hat{g})_{\mathrm{aa}}^{-1} = (u - \hat{\mathcal{B}}\hat{g})_{\mathbf{00}}^{-1} \tag{16}$$

to obtain the macroscopic inverse wave operator, where the subindices **0** denote the projection onto the subspace with **G** = **0**.

## RECURSIVE METHOD

The calculation of equation (16) is analogous to that of a projected Green's function [44, 45]

$$\mathcal{G}_{\mathrm{aa}}(\varepsilon) = \langle a|\hat{\mathcal{G}}(\varepsilon)|a\rangle \tag{17}$$

onto a given state |a⟩ , where

$$\hat{\mathcal{G}}(\varepsilon) = (\varepsilon - \hat{\mathcal{H}})^{-1} \tag{18}$$

is the Green's operator corresponding to some Hermitian Hamiltonian $\hat{\mathcal{H}}$ and ε is a complex energy. In [28] a similar result was obtained, where the Hamiltonian $\hat{\mathcal{H}}$ was identified with the longitudinal projection of the characteristic function $\hat{\mathcal{B}}^{LL}$, the energy ε with the spectral variable u, and |a⟩ with a slowly varying longitudinal wave, and it was shown that the corresponding projected Green's function was proportional to the inverse of the longitudinal macroscopic dielectric function. Haydock's method may be applied to obtain the projected Green's function (17) in a very efficient way [46, 47], and it has been adapted previously [28] to the calculation of the optical response of nanostructured systems in the long-wavelength local limit. In this section, we generalize the approach of [28] to arbitrary frequencies and wavevectors.

Equations (17) and (18) are similar to equation (16), although the operator product $\hat{\mathcal{H}} = \hat{\mathcal{B}}\hat{g}$ that plays the role of a Hamiltonian does not correspond to a symmetric matrix. Nevertheless, it would correspond to a Hermitian operator if we use $\hat{g}$ as a metric, that is, if we use $\langle\psi|\hat{g}|\phi\rangle$ instead of $\langle\psi|\phi\rangle$ as the scalar product of two arbitrary states $|\psi\rangle$ and $|\phi\rangle$. Then, it is easy to verify the Hermiticity condition $[\langle\psi|\hat{g}(\hat{\mathcal{H}}|\phi\rangle)]^* = \langle\phi|\hat{g}(\hat{\mathcal{H}}|\psi\rangle)$. Note, however, that $\hat{g}$ is not positive definite.

According to equation (11), we need the average $(u - \hat{\mathcal{B}}\hat{g})^{-1}_{aa} \rightarrow \langle a|(u - \hat{\mathcal{B}}\hat{g})^{-1}|a\rangle$, where we projected onto an average state $|a\rangle = b_0|0\rangle$ consisting of a plane wave with a given wavevector **k**, frequency ω and polarization **e**. Here, $|0\rangle$ is normalized according to the metric $\hat{g}$, i.e. $\langle 0|\hat{g}|0\rangle = g_0 = \pm 1$, and the coefficient $b_0$ is a real positive number chosen to guarantee that $|a\rangle$ is normalized in the usual sense $\langle a|a\rangle = 1$. Now, we define $|-1\rangle \equiv 0$ and we obtain new states by means of the recursion relation

$$|\widetilde{n+1}\rangle \equiv \hat{\mathcal{B}}\hat{g}|n\rangle = b_{n+1}|n+1\rangle + a_n|n\rangle + g_{n-1}g_n b_n|n-1\rangle, \tag{19}$$

where all the states $|n\rangle$ are orthonormalized according to the metric $\hat{g}$, that is,

$$\langle n|\hat{g}|m\rangle = g_n\delta_{nm} \tag{20}$$

with $g_n = \pm 1$ and $\delta_{nm}$ the Kronecker›s delta function. The requirement of orthonormality yields the generalized Haydock coefficients $a_n$, $b_{n+1}$ and $g_{n+1}$ given the previous coefficients $b_n$, $g_n$ and $g_{n-1}$. Thus, $a_n$ are obtained from

$$\langle n|\hat{g}|\widetilde{n+1}\rangle = a_n g_n \tag{21}$$

and $b_{n+1}$ from

$$\langle\widetilde{n+1}|\hat{g}|\widetilde{n+1}\rangle = g_{n+1}b^2_{n+1} + g_n a^2_n + g_{n-1}b^2_n, \tag{22}$$

where we choose the sign $g_{n+1} = \pm 1$ so that $b^2_{n+1}$ is positive and we may choose $b_{n+1}$ as a real positive number. In the basis $\{|n\rangle\}$, $\hat{\mathcal{B}}\hat{g}$ is represented by a tridiagonal matrix with $a_n$ along the main diagonal, $b_n$ along the subdiagonal and $g_{n-1}g_n b_n$ along the supradiagonal, so that

$$u - \hat{\mathcal{B}}\hat{g} \rightarrow \begin{pmatrix} u-a_0 & -b_1g_1g_0 & 0 & 0 & \cdots \\ -b_1 & u-a_1 & -b_2g_2g_1 & 0 & \cdots \\ 0 & -b_2 & u-a_2 & -b_3g_3g_2 & \cdots \\ \vdots & \vdots & & \vdots & \ddots \end{pmatrix}. \tag{23}$$

Note that we may represent the states $|n\rangle$ through the corresponding Cartesian field components for each reciprocal vector **G** or for each position **r** within a unit cell. Thus, the action of $\hat{g}$ is a trivial multiplication in reciprocal space, while that of $\hat{\mathcal{B}}$ is a trivial multiplication in real space, and we may repeatedly compute $\hat{\mathcal{B}}\hat{g}|n\rangle$ and Haydock›s coefficients without performing any large matrix multiplication, by alternatingly fast-Fourier transforming our representation between real and reciprocal space. More explicitly, we represent a state $|n\rangle$ through its components $n_{\mathbf{G}i} = \langle\mathbf{G},i|n\rangle$, where $|\mathbf{G},i\rangle$ corresponds to a normalized plane wave with wavevector **k** + **G** and polarization **i** along the ith Cartesian direction. Then, $\hat{g}|n\rangle$ is represented by $\phi_{\mathbf{G},i} = \langle\mathbf{G},i|\hat{g}|n\rangle = g_{ij}(\mathbf{G})n_{\mathbf{G},j}$, where, according to equation (10),

$$g_{ij}(\mathbf{G}) = \frac{\epsilon_a q^2\delta_{ij} - (k_i+G_i)(k_j+G_j)}{\epsilon_a q^2 - (\mathbf{k}+\mathbf{G})^2} \tag{24}$$

is a small Cartesian matrix for each **G**. Performing a fast Fourier transform, we obtain $\langle\mathbf{r},i|\phi\rangle = \phi_{\mathbf{r}i}$ in real space, which we then multiply by B(**r**) to obtain $|\widetilde{n+1}\rangle$, represented in real space by $(\widetilde{n+1})_{\mathbf{r},i} = B(\mathbf{r})\phi_{\mathbf{r},i}$. Fourier transforming back into reciprocal space and orthonormalizing, we obtain Haydock coefficients $a_n$, $b_{n+1}$ and $g_{n+1}$, and the new state $|n+1\rangle$ represented by $(n+1)_{\mathbf{G}i} = \langle\mathbf{G}i|n+1\rangle$. As discussed above, we iterate this procedure to obtain enough coefficients to converge the continued fraction (25).

According to equation (11), we do not require the full inverse of the matrix (23), but only the element in the first row and the first column. Following [28], we obtain that element as a continued fraction, which when substituted into equation (11) yields

$$\hat{\mathcal{W}}^{-1}_{\mathrm{M}} \rightarrow \sum_{ij} \mathrm{e}_i(\mathcal{W}^{-1}_{\mathrm{M}})_{ij}\,\mathrm{e}_j = \frac{u}{\epsilon_A}\cfrac{g_0b_0^2}{u-a_0-\cfrac{g_0g_1b_1^2}{u-a_1-\cfrac{g_1g_2b_2^2}{u-a_2-\cfrac{g_2g_3b_3^2}{\ddots}}}}. \tag{25}$$

The right-hand side of equation (25) denotes the macroscopic response projected onto a state with the given wavevector **k**, frequency ω and polarization **e**. Due to the translational invariance of the homogenized system, the inverse wave operator for a given **k** corresponds simply to a tensor with Cartesian components $(\mathcal{W}^{-1}_{\mathrm{M}})_{ij}$. Thus, so far we have calculated its inner products with **e**, as shown by the second term of equation (25). We may repeat the calculation above for different orientations of the (possibly complex) unit polarization vector**e** and solve the resulting equations for all the individual Cartesian components $(\mathcal{W}^{-1}_{\mathrm{M}})_{ij}$. Finally, we perform a matrix inversion of this macroscopic tensor and we use equations (6) and (7) to compute the macroscopic dielectric tensor

$$\epsilon^{\mathrm{M}}_{ij}(\omega,\mathbf{k}) = \frac{1}{q^2}(k^2\delta_{ij} - k_ik_j) + \mathcal{W}^{\mathrm{M}}_{ij}(\omega,\mathbf{k}). \tag{26}$$

Equations (25) and (26) constitute the main result of our formalism. Note that, in general, this dielectric tensor would depend on both ω and **k**, that is, it is a temporal and spatially dispersive response function.

## RESULTS

To test our formalism, we first calculate the local limit of the longitudinal macroscopic dielectric function ϵ

$\epsilon_L^M(\omega) \equiv \epsilon_{xx}^M(\omega, \mathbf{k} \to 0\mathbf{x})$ of a 2D system made up of a square array of empty cylindrical holes ($\epsilon B = 1$) placed with lattice parameter a within a dielectric medium with permittivity $\epsilon A = 12$. The holes are of circular cross section with radius $\rho = 0.45a$. We assume that the axes of the cylinders are parallel to the z-axis. This system has been frequently been used as a testground for calculations of photonic crystal properties [48]. In this calculation, we took the limit $\mathbf{k} \to \mathbf{0}$ along the x-direction, and we introduced the Cartesian unit vectors **x**, **y** and **z**. We remark that it is important to assign a direction to **k**even in this limit, as the transverse T and longitudinal L projection of $\mathcal{W}^M$ differ, although, for this system, the resulting $\epsilon^M(\omega) = \epsilon^M(\omega,\mathbf{k} \to \mathbf{0})$ is isotropic within the xy-plane.

In figure 1, we show $\epsilon_L^M(\omega)$calculated for two different lattice parameters a = 40 and 120 nm. Our calculations were performed using the Perl Data Language [49].5 The figure shows that our formalism yields the same results as a full straightforward matricial calculation [32] that solves Maxwell equations in a plane wave basis. Nevertheless, our calculation is much more economical in memory usage, as we do not have to store the full matrices that represent the dynamics of the system, and is at least four orders of magnitude faster as we do not perform any matrix manipulation with our scheme. For example, to obtain a 1% accuracy, we used an N × N reciprocal lattice and an N × N pixel image of the system in real space, where $N \approx 10^2$; we have verified that the accuracy of our calculation scales with 1/N. An in-plane field is described by two coordinates for each reciprocal vector. Thus, to represent a state requires $2N^2$ numbers. The matrix calculation of [32] would require calculating, storing, multiplying and solving equations with $2N^2 \times 2N^2$ matrices that have $\approx 4 \times 10^8$ elements, while our calculation only manipulates vectors with $\approx 2 \times 10^4$elements. Thus, each point in figure 1 requires several hours when parallelized and run in a cluster with 40 cores, while it took a few seconds in one CPU of a laptop.

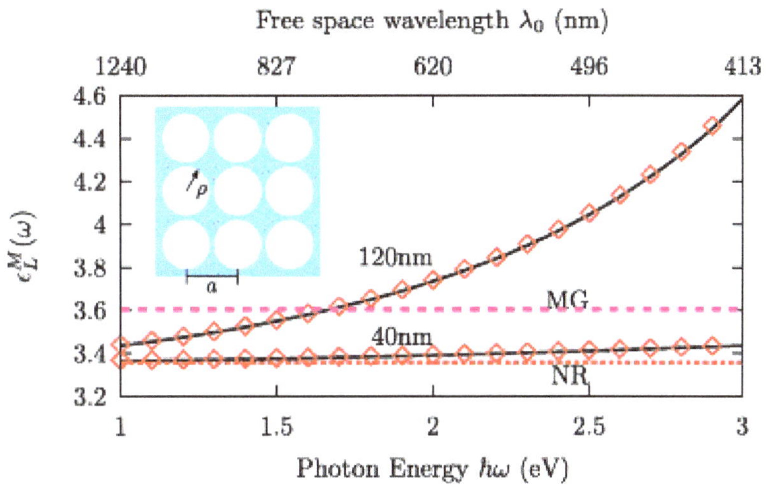


**Figure 1:** Longitudinal macroscopic dielectric function $\epsilon_L^M(\omega)$ of a 2D system (inset) made up of a square array with lattice parameter a of empty cylindrical holes ($\epsilon B = 1$) of circular cross section with radius $\rho = 0.45a$ within a dielectric medium with permittivity $\epsilon A = 12$. We show the results of using the Maxwell–Garnett (MG) formula, of a non-retarded (NR) calculation, of a numerically exact matrix calculation (diamonds) and of our theory (solid lines) for two different lattice parameters a = 40 and 120 nm.

To illustrate the effects of retardation, in the figure we also show the result of using the non-retarded (NR) calculation of [28], together with those of the well-known 2D MG effective medium formula [21]. As the dielectric functions of the components are independent of the frequency, since the constituents were taken to be non-dispersive, in this case the frequency dependence of the result arises solely from the finite ratio of the free space wavelength to the lengthscale of the system, namely, the lattice parameter a. Thus, in the $\omega \to 0$ limit our results approach those of the NR calculation. The difference between the NR and MG calculations is due to the high filling fraction $f \approx 0.64$ of the cylinders, which interact non-dipolarly with their nearest neighbors which they almost touch, while MG contains only dipolar interactions. Naturally, retardation effects become stronger as a increases. Furthermore, $\epsilon^M(\omega,\mathbf{k})$ depends only on the products qa and **k**a so that the two curves shown in figure 1 could be collapsed into one curve if appropriately plotted. We use this in figure 2, which shows the transverse response $\epsilon$ $\epsilon_T^M(\omega, \mathbf{k}_\Delta) = \epsilon_{zz}^M(\omega, \mathbf{k}_\Delta)$ corresponding to out-of-plane or transverse electric (TE) polarization as a function of the normalized frequency $qa/2\pi$. Here, we have chosen a finite in-plane wavevector $\mathbf{k}_\Delta = (\pi/2a,0,0)$, corresponding to the midpoint of the $\Delta$ line between $\Gamma$ and X in the first Brillouin zone (BZ) [43]. This figure covers a larger frequency range than figure 1 and therefore it displays a series of poles related to resonant multiple coherent reflections at the interfaces between the A and B regions.

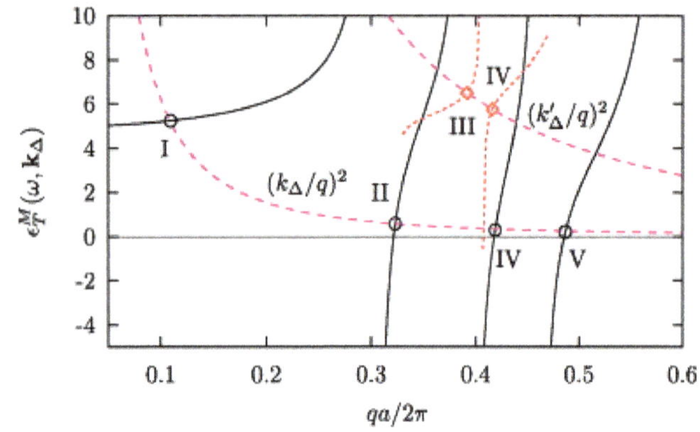


**Figure 2:** Out-of-plane macroscopic dielectric function $\epsilon_T^M(\omega, \mathbf{k}_\Delta)$ of the same system as in figure 1 for a given wavevector $\mathbf{k}_\Delta$ along the Δ line of the first BZ (solid lines). The line $(k_\Delta/q)^2$ (long dashes) intersects (circles) $\epsilon_T^M(\omega, \mathbf{k}_\Delta)$ at the frequencies $\omega\alpha(\mathbf{k}_\Delta)$ of the TE normal modes, with α = I, II, IV and V. The intersections (diamonds) of $\epsilon_T^M(\omega, \mathbf{k}'_\Delta)$ (short dashes) with $(k'_\Delta/q)^2$ (long dashes) corresponding to a shifted wavevector $\mathbf{k}'_\Delta = \mathbf{k}_\Delta + (2\pi/a)\mathbf{y}$ yield the modes α = III and IV (see the text).

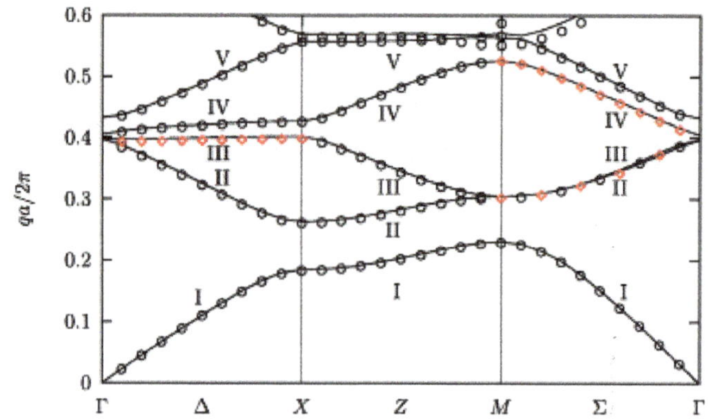


**Figure 3:** Photonic bands of the TE modes of the same system as in figure 1 obtained by solving equation (27) (circles). For comparison, we also show the photonic bands obtained from the solution of the corresponding eigenvalue problem, equation (28), as described in the text (solid lines). We also show the modes obtained by shifting the wavevector $\mathbf{k} \rightarrow \mathbf{k}' = \mathbf{k} + (2\pi/a)\mathbf{y}$ (diamonds).

The dispersion relation of transverse waves with TE polarization propagating along the xy-plane can be simply written as [50]

$$\epsilon_T^M(\omega, \mathbf{k}) = \frac{k^2}{q^2}. \tag{27}$$

Therefore, in figure 2 we also plotted the curve $(k_\Delta/q)^2$. Those points where it intersects the curves for $\epsilon_T^M$ correspond to normal TE modes of the system with frequencies $\omega\alpha(\mathbf{k} = \mathbf{k}_\Delta), \alpha$ = I, II, .... Here, we use roman numerals to number the modes in increasing frequency order.

In figure 3, we show the photonic bands (circles) of the TE modes of our system obtained by solving the dispersion relation (27) as $\mathbf{k}$ travels along the line Γ–Δ–X–Z–M–Σ–Γ within the first BZ. In order to test our theory, we also plot the TE modes obtained by solving the eigenvalue equation [51]

$$\left(\frac{1}{\sqrt{\epsilon(\mathbf{r})}}\nabla^2\frac{1}{\sqrt{\epsilon(\mathbf{r})}}\right)\left(\sqrt{\epsilon(\mathbf{r})}E_z(\mathbf{r})\right) = -\frac{\omega^2}{c^2}\left(\sqrt{\epsilon(\mathbf{r})}E_z(\mathbf{r})\right) \tag{28}$$

through an implicitly restarted Arnoldi iteration [52], applying $1/\sqrt{\epsilon(\mathbf{r})}$ in real space and $\nabla \rightarrow i(\mathbf{k} + \mathbf{G})$ in reciprocal space [42]. We further verified our results by recalculating them with the freely available MIT Photonic-Bands Package (MPB) [53] and comparing it with previous results [48] for the same system. The agreement between these calculations shows that it is feasible to use the macroscopic dielectric response of the system to obtain its photonic band structure. However, to succeed, we have to account fully for the spatial dispersion of $\epsilon^M_T(\omega,\mathbf{k})$.

We remark that although we obtained the correct photonic bands even for large wavevectors, going all the way around the first BZ, there are some regions where we failed to obtain all of the normal modes. For example, our procedure did not produce the third band (labeled III in the figure) along the Δ line, nor the fourth band (IV) along the Σ line, between M and Γ. Furthermore, it did not yield the second band (II), which is almost degenerate with the third band (III) along Σ. It is easily shown that the microscopic field corresponding to the missing band along Δ is antisymmetric with respect to the reflection $G_y \leftrightarrow -G_y$ about the Δ line. Similarly, the missing bands along the region Σ are antisymmetric with respect to the reflection $G_x \leftrightarrow G_y$ about the Σ line. Thus, the reciprocal vector **0** does not contribute to the electric field for those bands [54], i.e. the missing modes have no macroscopic field components and thus, apparently, they may not be obtained from the roots of the macroscopic dispersion relation (27).

Nevertheless, within our formalism, the wavevector **k** is actually not restricted to lie within the first BZ. Thus, we may calculate the macroscopic dielectric function for wavevectors beyond the first BZ, and obtain the modes in the extended scheme. In figure 2, we show part of the macroscopic dielectric function $\epsilon^M_T(\omega,\mathbf{k}'_\Delta)$ with a wavevector that has been shifted out of the first BZ from $\mathbf{k}_\Delta$ along the y-direction by the reciprocal vector $(2\pi/a)\mathbf{y}$, i.e. $\mathbf{k}'_\Delta = \mathbf{k}_\Delta + (2\pi/a)\mathbf{y}$. We remark that $\epsilon^M(\omega,\mathbf{k})$ is not a periodic function of **k**, as it corresponds to a specific plane wave, not to a microscopic Bloch's wave. The intersection of the curve $(k'_\Delta/q)^2$ with $\epsilon_T^M(\omega, \mathbf{k}'_\Delta)$ yields the corresponding macroscopic modes. For example, the diamonds in figure 2 illustrate two of the normal modes that could be obtained using this shifting procedure. One of these modes is identical to the mode labeled IV that was obtained previously using the unshifted response $\epsilon_T^M(\omega, \mathbf{k}_\Delta)$Nevertheless, another mode, labeled III, appears just below $qa/2\pi = 0.4$ and corresponds to one of the previously missing modes. We may shift it back into the first BZ in order to display it in the reduced zone scheme. Proceeding in this fashion, we have obtained all of the previously missing bands, shown by diamonds in figure 3. Thus, we have shown that we can calculate the full photonic band structure for the TE modes from the macroscopic non-local dielectric function of the composite system. The converse, obtaining the non-local response from the dispersion relation [38], can also be attempted, although the results would not be uniquely defined and are not expected to be valid for wavevectors and frequencies that do not lie in the dispersion relation. Our procedure has no such limitation.

A similar procedure to that discussed above can be employed also for the transverse magnetic (TM) modes (with an out of plane magnetic field) of the system, with the electric field within the xy-plane. In figure 4, we show the in-plane macroscopic dielectric function $\epsilon^{M}_{xx}(\omega, \mathbf{k}_\Delta)$ and $\epsilon^{M}_{yy}(\omega, \mathbf{k}_\Delta)$ of the system for the same wavevector $\mathbf{k}_\Delta$ as in figure 2. This wavevector and the system are symmetric under $y \leftrightarrow -y$ reflections. Therefore, $\epsilon^{M}_{xy} = 0$ and we may identify the transverse and longitudinal responses as $\epsilon^{M}_{T}(\omega, \mathbf{k}_\Delta) = \epsilon^{M}_{yy}(\omega, \mathbf{k}_\Delta)$), respectively. The normal modes are then obtained from the transverse dispersion relation (27) and the longitudinal dispersion relation [50]

$$\epsilon^{M}_{L}(\omega, \mathbf{k}) = 0. \tag{29}$$

These are identified with circles and diamonds in the figure.

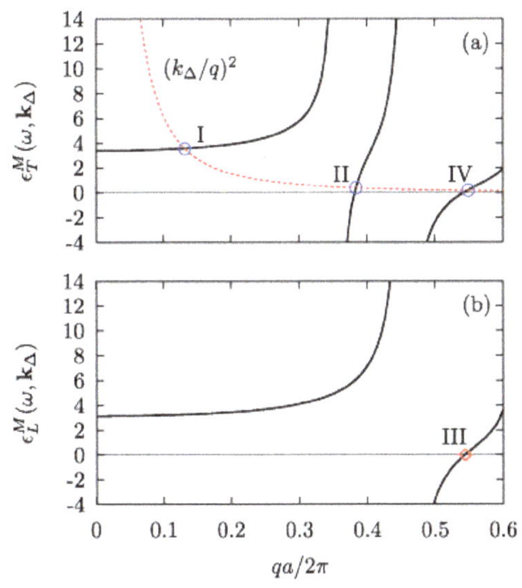


**Figure 4:** (a) In-plane macroscopic dielectric functions $\epsilon^{M}_{L}(\omega, \mathbf{k}_\Delta)$ and (b) $\epsilon^{M}_{T}(\omega, \mathbf{k}_\Delta)$ of the same system as in figure 1 for the same wavevector $\mathbf{k}_\Delta$ as in figure 2. The line $(k_\Delta/q)^2$ (dashes) intersects (circles) $\epsilon^{M}_{T}(\omega, \mathbf{k}_\Delta)$ at the frequencies $\omega\alpha(\mathbf{k}_\Delta)$ of the TM transverse normal modes, with α = I, II and IV. The zeros (diamonds) of $\epsilon^{M}_{T}(\omega, \mathbf{k}_\Delta)$ yields the longitudinal mode III (see the text).

In figure 5, we perform a similar calculation but along the Σ line, for the wavevector $\mathbf{k}_\Sigma$ = (π/2a,π/2a). Along this line, the wavevector and the system are symmetric under the reflection $x \leftrightarrow y$, and thus $\epsilon^{M}_{xx}(\omega, \mathbf{k}_\Sigma) = \epsilon^{M}_{yy}(\omega, \mathbf{k}_\Sigma)$ We may now identify the transverse and longitudinal responses as $\epsilon^{M}_{T}(\omega, \mathbf{k}_\Sigma) = \epsilon^{M}_{xx}(\omega, \mathbf{k}_\Sigma) - \epsilon^{M}_{xy}(\omega, \mathbf{k}_\Sigma)$ and $\epsilon^{M}_{L}(\omega, \mathbf{k}_\Sigma) = \epsilon^{M}_{xx}(\omega, \mathbf{k}_\Sigma) + \epsilon^{M}_{xy}(\omega, \mathbf{k}_\Sigma)$, ), respectively. In figure 5(a), we show $\epsilon^{M}_{T}(\omega, \mathbf{k}_\Sigma)$, $(k_\Sigma/q)^2$ $(k_\Sigma/q)^2$ and their intersections (circles) which yield the transverse modes, while in figure 5(b), we display $\epsilon^{M}_{L}(\omega, \mathbf{k}_\Sigma)$ and its zero (diamond) which correspond to a longitudinal mode.

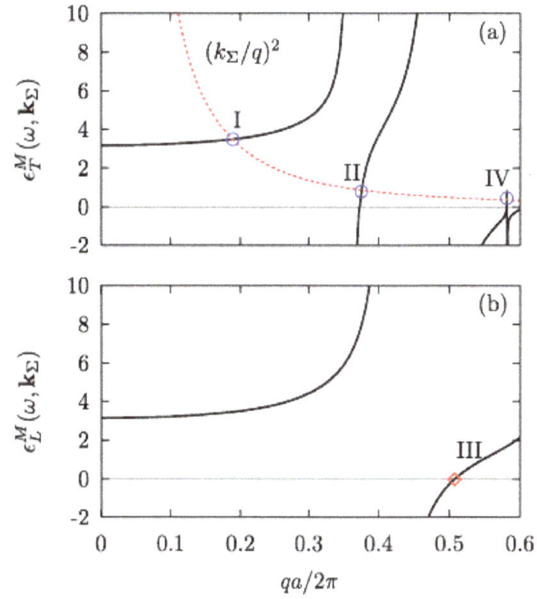


**Figure 5:** (a) In-plane macroscopic dielectric functions $\epsilon^{M}_{T}(\omega, \mathbf{k}_\Sigma)$ and (b) $\epsilon^{M}_{L}(\omega, \mathbf{k}_\Sigma)$ of the same system as in figure 1 for a given wavevector $\mathbf{k}_\Sigma$ along the Σ line. The line $(k_\Sigma/q)^2$ (dashes) intersects (circles) $\epsilon^{M}_{T}(\omega, \mathbf{k}_\Sigma)$ at the frequencies $\omega\alpha(\mathbf{k}_\Sigma)$ of the TM transverse normal modes, with α = I, II and IV. The zeros (diamonds) of $\epsilon^{M}_{T}(\omega, \mathbf{k}_\Sigma)$ yield the longitudinal mode III (see the text).

Along the Z line, from X to M, there is no crystal symmetry operation that leaves the wavevector invariant, and thus transverse and longitudinal fields mix among themselves, i.e. there are no longitudinal or transverse modes. Nevertheless, we may obtain the frequencies of the mixed modes through the singularities of the wave operator matrix

$$\det\left[\mathcal{W}^{M}(\omega, \mathbf{k})\right] = 0 \tag{30}$$

as illustrated in figure 6 for a wavevector **k**Z = (π/a,π/2a) along the Z line.

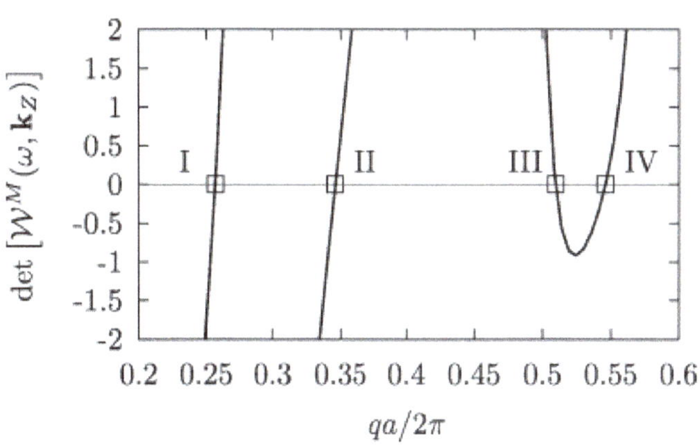


**Figure 6:** Determinant of the macroscopic wave operator matrix $\mathcal{W}^{M}(\omega, \mathbf{k}_Z)$ for a given wavevector **k**Z along the Z line. The normal modes of the system ωα(**k**Z), α = I–IV, correspond to its zeros (squares).

From calculations such as those illustrated by figures 4–6, or more generally, from the zeros of $\det\left[\mathcal{W}^{M}(\omega,\mathbf{k})\right]$ for arbitrary wavevectors **k**, we can obtain the full TM photonic band structure, as shown by figure 7. In order to test our theory, we also plot the TM modes obtained by solving the eigenvalue equation [51]

$$\nabla\cdot\frac{1}{\epsilon(\mathbf{r})}\nabla B_z(\mathbf{r}) = -\frac{\omega^2}{c^2}B_z(\mathbf{r}) \tag{31}$$

using similar methods [42, 52] as for the TE case and comparing them successfully with previous results [48] for the same system. The agreement between these calculations shows that it is also feasible to use the macroscopic dielectric response of the system for obtaining its photonic band structure in the TM case.

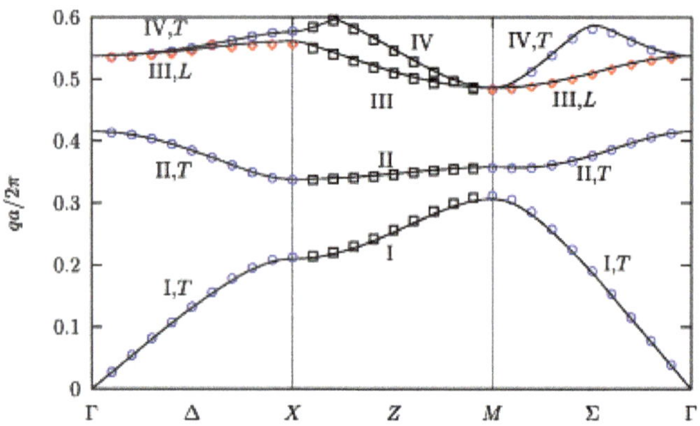


**Figure 7:** Photonic bands of the TM modes of the same system as in figure 1. The modes between Γ and X and between M and Γ may be obtained by solving equation (27) for the transverse (T) modes (circles) and equation (29) for the longitudinal (L) modes (diamonds) or equation (30) for both of them. The modes from X → M are obtained from equation (30) and have a mixed polarization (squares). For comparison, we also show the photonic bands obtained from the solution of the corresponding eigenvalue problem, equation (31), as described in the text (solid lines).

An approximate band structure could be obtained for transverse waves in the region of small wavevectors $\mathbf{k}\to\mathbf{0}$ by expanding the lhs of equation (27) up to second order in k and solving for $k^2$. The result is a local dispersion relation

$$k^2 = q^2\epsilon^{M}(\omega)\mu(\omega), \tag{32}$$

where the non-locality of $\epsilon^{M}_{T}(\omega,\mathbf{k})$ is partially accounted for through an effective magnetic permeability [25],

$$\mu(\omega) = \left.\frac{1}{1-\frac{q^2}{2}\frac{\partial^2}{\partial k^2}\epsilon^{M}_{T}(\omega,k\mathbf{x})}\right|_{k=0}. \tag{33}$$

In figure 8(a), we show the resulting approximate bands for the TE case. We can see that the acoustic band within the local approximation is indistinguishable from the exact result for small wavevectors. This local acoustic band extends up to the frequency $qa/2\pi \approx 0.2$ for which μ has a pole, see figure 8(c). The local approximation also reproduces the upper band of the exact calculation (labeled V in figure 3), but only in the immediate vicinity of Γ, close to the zero $qa/2\pi \approx 0.43$ of $\epsilon^{M}(\omega)$. For larger wavevectors it acquires a larger positive dispersion. As discussed above, the band labeled III in figure 3 does not couple to macroscopic fields for **k** along the Δ line, while the band labeled IV does not couple along the Σ line. Thus, neither band couples to macroscopic fields at Γ, and as a consequence, these bands are not reproduced at all by the local approximation (32).

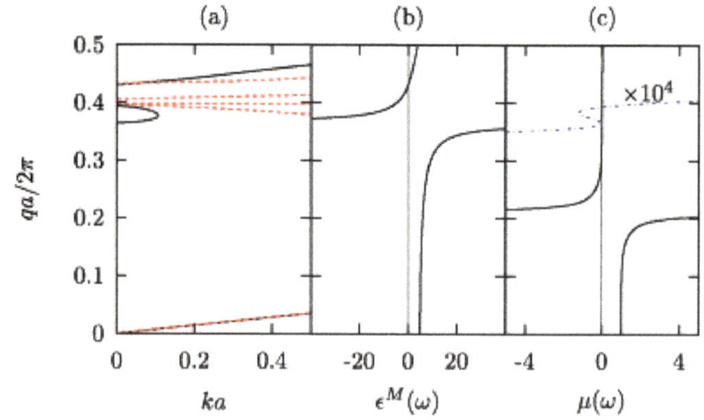


**Figure 8:** (a) Local dispersion relation for the same system as in figure 1 for the TE case (solid line), obtained from equation (32). (b) Local dielectric function $\epsilon^{M}(\omega)$. (c) Local magnetic permeability $\mu(\omega)$ (solid line) and $10^4\mu(\omega)$ (dash-dotted line). For comparison purposes, we also show the exact dispersion relation of (a) (dashed line).

In the region $qa/2\pi \approx 0.36$–$0.40$, the local approximation predicts a curious band that displays backbending. Its lower part begins at the pole of $\epsilon^{M}(\omega)$, which corresponds to a double zero of $\mu(\omega)$ [42]. This part of the local dispersion relation is spurious, as the second-order Taylor expansion of the dielectric function that yields equations (33) and (32) starting from equation (27) would be meaningless at a pole. Consequently, this part of the local band does not correspond to any band within the exact calculation. On the other hand, the upper part of the backbending band starts at a simple zero of μ, which arises from a different kind of singular behavior, consisting of a pole in $\epsilon^{M}_{T}(\omega,\mathbf{k})$ at $qa/2\pi \approx 0.40$, which is suppressed at $k = 0$ as its weight is approximately proportional to $k^2$. This part of the local dispersion relation agrees with band II of figure 3 at Γ and displays a negative dispersion as does band II. Furthermore, for this band, both $\epsilon^{M}$ and μ are negative. Thus, our photonic crystal, made up of holes within a dispersionless dielectric, behaves for some frequencies like a left-handed metamaterial [55, 56]. Nevertheless, caution should be exercised when using equations (33) and (32) close to singularities of $\epsilon^{M}_{T}(\omega,\mathbf{k})$

## CONCLUSIONS

We have shown that the macroscopic inverse wave operator of composite systems with arbitrary spatial fluctuations is simply given by the average projection of the corresponding microscopic operator. These operators may be related to the macroscopic and microscopic dielectric functions respectively, thus yielding a general procedure for incorporating the effects of spatial fluctuations into the calculation of all the components of the macroscopic dielectric tensor of the system. We have extended Haydock's recursive scheme in order to calculate very efficiently the response of periodic two-component systems in terms of a continued fraction whose coefficients may be obtained without recourse to operations with the large matrices that characterize the microscopic response and the field equations. We illustrated and tested our results through the calculation of the wavevector and frequency-dependent macroscopic dielectric tensor of a 2D dielectric system with non-dispersive non-dissipative components. We identified the transverse and longitudinal components of the response for wavevectors along symmetry lines and obtained all the components of the tensorial response for the case of reduced symmetry. The macroscopic response has a series of poles, related to resonant multiple coherent reflections, and by accounting for its spatial dispersion we obtained the full photonic band structure of the system from the dispersion relations corresponding to the homogenized material. Besides yielding the correct bands from a macroscopic approach, our scheme allowed us to classify the polarization of each mode as either transverse, longitudinal or mixed. Even though the macroscopic approach might fail to yield some modes, namely, those which have no coupling to the macroscopic field due to symmetry, they may be recovered by extending the notion of macroscopic state, allowing its wavevector to lie beyond the first BZ. The non-locality of the dielectric response may be partially accounted for through a magnetic permeability which can then be employed in the calculation of the optical properties of the system. We compared the band structure obtained through this local approximation to the exact results and we obtained partial agreement close to Γ for those bands that do couple to long wavelength fields. In particular, we showed that this approach can yield a negative dispersion in frequency regions where both the permittivity and permeability are negative. Nevertheless, we discussed some shortcomings of the local approach. Although here we presented results for dispersionless transparent dielectrics, our calculation does not require that the materials that make up the system be non-dispersive or non-dissipative, so that calculations for real dielectrics and metals may be performed with the same low computational costs.

## ACKNOWLEDGMENTS

JSPH is grateful for a scholarship awarded by CONACYT. This work was partially supported by ANPCyT grant no. PICT-PRH-135-2008 (GPO), CONACyT grant no. 153930 (BSM) and DGAPA-UNAM grant no. IN108413 (WLM).

## REFERENCES

1. Bredov M, Rumiantsev V and Toptiguin L 1985 ´ Classical Electrodynamics (Moscow: Mir)
2. Notomi M 2000 Theory of light propagation in strongly modulated photonic crystals: refractionlike behavior in the vicinity of the photonic band gap Phys. Rev. B 62 10696–705
3. Barrera R G, Reyes-Coronado A and Garc´ıa-Valenzuela A 2007 Nonlocal nature of the electrodynamic response of colloidal systems Phys. Rev. B 75 184202
4. Yeh P, Yariv A and Hong C-S 1977 Electromagnetic propagation in periodic stratified media: I. General theory J. Opt. Soc. Am. 67 423–38
5. Luis Mochan W and Barrera R G 1985 Electromagnetic response of systems with spatial fluctuations: I. ´ General formalism Phys. Rev. B 32 4984–8
6. Luis Mochan W and Barrera R G 1985 Electromagnetic response of systems with spatial fluctuations: II. ´ Applications Phys. Rev. B 32 4989–5001
7. Kosaka H, Kawashima T, Tomita A, Notomi M, Tamamura T, Sato T and Kawakami S 1998 Superprism phenomena in photonic crystals Phys. Rev. B 58 R10096–9
8. Stroud D and Pan F P 1978 Self-consistent approach to electromagnetic wave propagation in composite media: application to model granular metals Phys. Rev. B 17 1602–10
9. Ortiz G P and Mochan W L 2003 Scaling of light scattered from fractal aggregates at resonance ´ Phys. Rev. B 67 184204
10. Ortiz G P and Mochan W L 2003 Scaling condition for multiple scattering in fractal aggregates ´ Physica B 338 54
11. Hilke M 2009 Seeing Anderson localization Phys. Rev. A 80 063820
12. Vasconcelos M S and Albuquerque E L 1998 Plasmon–polariton fractal spectra in quasiperiodic multilayers Phys. Rev. B 57 2826–33
13. Albuquerque E L and Cottam M G 1993 Superlattice plasmon–polaritons Phys. Rep. 233 67–135
14. Chen J et al 2011 Observation of the inverse doppler effect in negative-index materials at optical frequencies Nature Photon. 5 239–45
15. Zhang B, Luo Y, Liu X and Barbastathis G 2011 Macroscopic invisibility cloak for visible light Phys. Rev. Lett. 106 033901
16. Lezec H J, Dionne J A and Atwater H A 2007 Negative refraction at visible frequencies Science 316 430–2
17. Urzhumov Y A and Shvets G 2008 Optical magnetism and negative refraction in plasmonic metamaterials Solid State Commun. 146 208–20
18. Milton G W, McPhedran R C and McKenzie D R 1981 Transport properties of arrays of intersecting cylinders Appl. Phys. A 25 23–30
19. Bergman D J and Dunn K-J 1992 Bulk effective dielectric constant of a composite with a periodic microgeometry Phys. Rev. B 45 13262–71
20. Tao R, Chen Z and Sheng P 1990 First-principles Fourier approach for the calculation of the effective dielectric constant of periodic composites Phys. Rev. B 41 2417–20
21. Datta S, Chan C T, Ho K M and Soukoulis C M 1993 Effective dielectric constant of periodic composite structures Phys. Rev. B 48 14936–43
22. Alexopoulos A 2010 Effective-medium theory of surfaces and metasurfaces containing two-dimensional binary inclusions Phys. Rev. E 81 046607
23. Doyle W T 1989 Optical properties of a suspension of metal spheres Phys. Rev. B 39 9852–8
24. Ludwig A and Webb K J 2010 Accuracy of effective medium parameter extraction procedures for optical metamaterials Phys. Rev. B 81 113103
25. Costa J T, Silveirinha M G and Maslovski S I 2009 Finite-

difference frequency-domain method for the extraction of effective parameters of metamaterials Phys. Rev. B 80 235124

26. Ortiz G P, Lopez-Bastidas C, Maytorena J A and Moch ´ an W L 2003 Bulk response of composite from finite ´ samples Physica B 338 103
27. Silveirinha M G 2006 Nonlocal homogenization model for a periodic array of epsilon-negative rods Phys. Rev. E 73 046612
28. Mochan L W, Ortiz G P and Mendoza B S 2010 Efficient homogenization procedure for the calculation of optical properties of 3D nanostructured composites Opt. Express 18 22119–27
29. Myroshnychenko V and Brosseau C 2010 Analysis of the effective permittivity in percolative composites using finite element calculations Proc. 8th Int. Conf. on Electrical Transport and Optical Properties of Inhomogeneous Media, ETOPIM-8; Physica B 405 3046–9
30. Guenneau S, Zolla F and Nicolet A 2007 Homogenization of 3D finite photonic crystals with heterogeneous permittivity and permeability Waves Random Media 17 653–97
31. Cabuz A I, Nicolet A, Zolla F, Felbacq D and Bouchitt ˘ e G 2011 Homogenization of nonlocal wire ´ metamaterial via a renormalization approach J. Opt. Soc. Am. B 28 1275–82
32. Ortiz G P, Mart´ınez-Zerega B E, Mendoza B S and Moch ´ an L W 2009 Effective optical response of ´ metamaterials Phys. Rev. B 79 245132
33. Raghunathan S B and Budko N V 2010 Effective permittivity of finite inhomogeneous objects Phys. Rev. B 81 054206
34. Halevi P, Krokhin A A and Arriaga J 1999 Photonic crystal optics and homogenization of 2D periodic composites Phys. Rev. Lett. 82 719–22
35. Reyes-Arendano J A ˜ et al 2011 From photonic crystals to metamaterials: the bianisotropic response New J. Phys. 13 073041
36. Simovski C R 2011 On electromagnetic characterization and homogenization of nanostructured metamaterials J. Opt. 13 013001
37. Menzel C, Paul T, Rockstuhl C, Pertsch T, Tretyakov S and Lederer F 2010 Validity of effective material parameters for optical fishnet metamaterials Phys. Rev. B 81 035320
38. Elser J, Podolskiy V A, Salakhutdinov I and Avrutsky I 2007 Nonlocal effects in effective-medium response of nanolayered metamaterials Appl. Phys. Lett. 90 191109
39. Barrera R G, Reyes-Coronado A and Garc´ıa-Valenzuela A 2005 Nonlocal effective medium for the electromagnetic response of colloidal systems: a t-matrix approach Progress in Electromagnetic Research Symp. (Hangzhou, China, 2005) pp 646–9
40. Jackson J D 1998 Classical Electrodynamics (New York: Wiley)
41. Mochan L W 2005 Plasmons ´ Encyclopedia of Condensed Matter Physics ed G Bassani, G Liedl and P Wyder p 310–17
42. Perez-Huerta J S, Mendoza B S, Ortiz G and Luis Moch ´ an W in preparation ´
43. Sakoda K 2004 Optical Properties of Photonic Crystals (Springer Series in Optical Sciences) (Berlin: Springer)
44. Sutton A P 2004 Electronic Structure of Materials (Oxford: Oxford University Press)
45. Datta S 1997 Electronic Transport in Mesoscopic Systems (Cambridge Studies in Semiconductor Physics and Microelectronic Engineering) (Cambridge: Cambridge University Press)
46. Haydock R, Heine V and Kelly M J 1972 Electronic structure based on the local atomic environment for tight-binding bands J. Phys. C: Solid State Phys. 5 2845
47. Haydock R, Heine V and Kelly M J 1975 Electronic structure based on the local atomic environment for tight-binding bands: II J. Phys. C: Solid State Phys. 8 2591
48. Busch K, von Freymann G, Linden S, Mingaleev S F, Tkeshelashvili L and Wegener M 2007 Periodic nanostructures for photonics Phys. Rep. 444 101–202
49. Glazebrook K and Economou F 1997 PDL: the Perl data language Dr. Dobb's J. 9719 45
50. Landau L D and Lifshits E M 1984 Electrodynamics of continuous media Course of Theoretical Physics (Oxford: Pergamon)
51. Joannopoulos J D, Johnson S G, Winn J N and Meade R D 2008 Photonic Crystals: Molding the Flow of Light 2nd edn (Princeton, NJ: Princeton University Press)
52. Saad Y 2011 Numerical Methods for Large Eigenvalues Problems (Philadelphia: SIAM)
53. Johnson S G and Joannopoulos J D 2001 Block-iterative frequency-domain methods for Maxwell's equations in a planewave basis Opt. Express 8 173–90
54. Robertson W M and Arjavalingam G 1992 Measurement of photonic band structure in a two-dimensional periodic dielectric array Phys. Rev. Lett. 68 2023
55. Peng L, Ran L, Chen H, Zhang H, Kong J A and Grzegorczyk T M 2007 Experimental observation of lefthanded behavior in an array of standard dielectric resonators Phys. Rev. Lett. 98 157403
56. Vynck K, Felbacq D, Centeno E, Cabuz A I, Cassagne D and Guizal B 2009 All-dielectric rod-type ˘ metamaterials at optical frequencies Phys. Rev. Lett. 102 133901

CHAPTER 7

# FIBER GUIDING AT THE DIRAC FREQUENCY BEYOND PHOTONIC BANDGAPS

Kang Xie[1], Wei Zhang[1], Allan D Boardman[2], Haiming Jiang[1], Zhijia Hu[1,3], Yong Liu[4], Ming Xie[2], Qiuping Mao[1], Lei Hu[1], Qian Li[1], Tianyu Yang[1], Fei Wen[1] and Erlei Wang[1]

[1]School of Instrument Science and Opto-electronic Engineering, Hefei University of Technology, Hefei, Anhui 230009, China

[2]Department of Physics, University of Salford, Salford, Manchester M5 4WT, UK

[3]Department of Polymer Science and Engineering, University of Science and Technology of China, Hefei, Anhui 230026, China

[4]School of Optoelectronic Information, University of Electronic Science and Technology of China, Chengdu, Sichuan 610054, China

## ABSTRACT

**Light trapping within waveguides is a key practice of modern optics, both scientifically and technologically. Photonic crystal fibers traditionally rely on total internal reflection (index-guiding fibers) or a photonic bandgap (photonic-bandgap fibers) to achieve field confinement. Here, we report the discovery of a new light trapping within fibers by the so-called Dirac point of photonic band structures. Our analysis reveals that the Dirac point can establish suppression of radiation losses and consequently a novel guided mode for propagation in photonic crystal fibers. What is known as the Dirac point is a conical singularity of a photonic band structure where wave motion obeys the famous Dirac equation. We find the unexpected phenomenon of wave localization at this point beyond photonic bandgaps. This guiding relies on the Dirac point rather than total internal reflection or photonic bandgaps, thus providing a sort of advancement in conceptual understanding over the traditional fiber guiding. The result presented here demonstrates the discovery of a new type of photonic crystal fibers, with unique characteristics that could lead to new applications in fiber sensors and lasers. The Dirac equation is a special symbol of relativistic quantum mechanics. Because of the similarity between band structures of a solid and a photonic crystal, the discovery of the Dirac-point-induced wave trapping in photonic crystals could provide novel insights into many relativistic quantum effects of the transport phenomena of photons, phonons, and electrons.**

## INTRODUCTION

Optical resonators and waveguides are widely used in modern optics[1], including optical couplers/filters[2,3,4], microlasers[5,6], and mainstream optical fibers[7,8,9,10]. Any dramatic change to the building blocks will generate a completely new range of optical devices. The study of the fundamental properties of resonators and waveguides thus attracts much of the current research[1]. Field confinement in optical fibers is traditionally based on total internal reflection. A high-index core surrounded by a cladding with a lower refractive index has been the most basic requirement[7]. Recently, a novel class of optical fibers that permit the guidance of light in a low-index core region has emerged[8,9,10]. These so-called photonic-bandgap fibers operate through bandgap effects of photonic crystals[11,12], which occur because of periodic microstructuring of the dielectric in the cladding region. Photonic crystals have been studied extensively for their bandgap[1] and special in-band dispersion effects such as negative refraction[13]. However, another really interesting feature of photonic crystals is the Dirac points that

**Citation:** Kang Xie, Wei Zhang, Allan D Boardman, Haiming Jiang, Zhijia Hu, Yong Liu, Ming Xie, Qiuping Mao, Lei Hu, Qian Li, Tianyu Yang, Fei Wen and Erlei Wang,Fiber guiding at the Dirac frequency beyond photonic bandgaps, doi:10.1038/lsa.2015.77.

appear at corners of the Brillouin zone[14,15,16,17]. A Dirac point, as illustrated in Figure 1, is a conical singularity surrounded by a region of linear dispersion in the band structure of triangular, hexagonal, or kagome lattices. At this special point, the Maxwell equations can be replaced by the two-dimensional massless Dirac equation for relativistic particles: $-iv_D(\sigma_x\partial_x+\sigma_y\partial_y)\Psi=(\omega-\omega_D)\Psi$, where $v_D$ is the group velocity, $\omega_D/2\pi$ is the Dirac frequency, $\sigma_x$ and $\sigma_y$ are Pauli matrices, and $\Psi=\begin{bmatrix}\psi_1\\ \psi_2\end{bmatrix}$ represents wave functions of two degenerate Bloch states. Thus, analogies with the electronic energy band structure of graphene can be sustained[18]. Pseudo-diffusive transmission in such lattices has been observed[15,16]. A localized mode has also recently been identified at the Dirac frequency[17]. However, the localized mode observed thus far is non-propagating because the solution addressed in Ref. 17 is uniform in the longitudinal direction. The most common waveguides based on triangular lattices are photonic crystal fibers[8,9,10]. In the study of the fiber problem, the mode is confined in the transverse plane while carrying electromagnetic energy in the longitudinal direction; thus, the propagating effect along the fiber axis is essential for these devices to operate. The fiber problem cannot be considered two-dimensional by fixing $k_z$, the wave vector in the longitudinal direction, to zero. In this study, we explore the quasi-two-dimensional system. We demonstrate the discovery of a new type of light trapping by the Dirac point of photonic crystal structures for out-of-plane propagation ($k_z\neq0$). A new type of photonic crystal fibers can be created based on it. This finding provides a sort of advancement in conceptual understanding over the conventional photonic crystal fibers that rely on either total internal reflection or bandgaps.

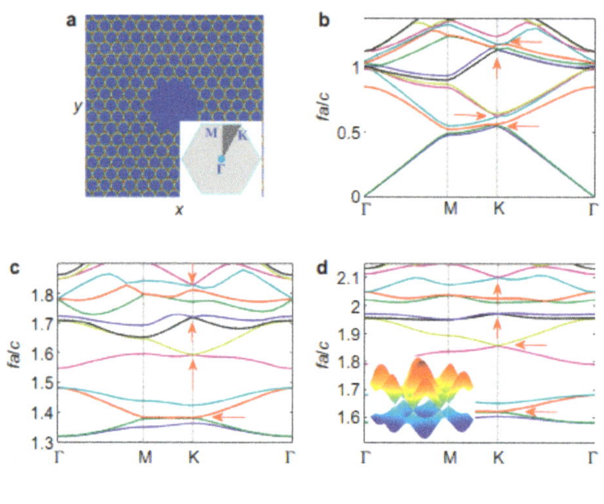


**Figure 1:** Cross-section and band structures of photonic crystal fiber. (**a**) A typical cross-section of photonic crystal fibers, where the lattice constant is a=2.21 μm, the hole radius is $r_a$=0.47a, and the central hollow defect radius is R=1.9a. The fiber is made of either silica glass($\varepsilon_r=1.45^2$), germania glass ($\varepsilon_r=1.59^2$), or SF6 glass ($\varepsilon_r=1.8^2$). The inset in **a** is the first Brillouin zone (shaded light gray) of the reciprocal lattice for the triangular lattice of air holes, revealing the high-symmetry points Γ, M, K at the corners of the irreducible Brillouin zone(shaded dark gray). (**b–d**) Band structures of the fiber cladding made of germania glass, at $k_za/2\pi=0.001$ (**b**), $k_za/2\pi=1.581$ (**c**), and$k_za/2\pi=1.910$ (**d**). The four lowest order Dirac points are indicated by red arrows. A well-isolated, wide-open Dirac spectrum is shown in **d** in the band structure for $k_za/2\pi=1.910$. The inset to **d** shows an enlarged three-dimensional view around the second Dirac point, where the two bands touch as a pair of linear Dirac cones at the six Brillouin zone corners.

# MATERIALS AND METHODS

## Calculations of Band Structures by the Plane Wave Expansion Method

A typical cross-section of the photonic crystal fiber is shown in Figure 1a, where the first Brillouin zone of the corresponding reciprocal lattice is shown as an inset. The fiber is formed from either silica glass ($\varepsilon_r=1.45^2$), germania glass ($\varepsilon_r=1.59^2$), or SF6 glass ($\varepsilon_r=1.8^2$). The cladding consists of a triangular lattice of air holes ($r_a$=0.47a), where a=2.21 μm is the center-to-center distance between the two nearest holes. A central hollow defect (R=1.9a) serves as the core of the fiber. The associated band structures of the triangular lattice for three different $k_z$ are calculated by the plane wave expansion (PWE) method[1], and the results are presented in Figure 1b–1d. Variations of the Dirac point versus $k_z$ are plotted in Figure 2a–2c for the three different fibers. Guided modes of the fibers are also studied by the PWE method, where a 14 × 16 supercell is used for the calculation to include the fiber core as a defect of the photonic crystal. The results for the fundamental guided modes are presented in Figure 2a–2c as red diamonds. Some details of the mode profiles are given in Figure 3. As a very basic property of photonic crystal fibers, the fundamental mode consists of an approximately degenerate pair of orthogonal, quasi-linearly polarized states[9].

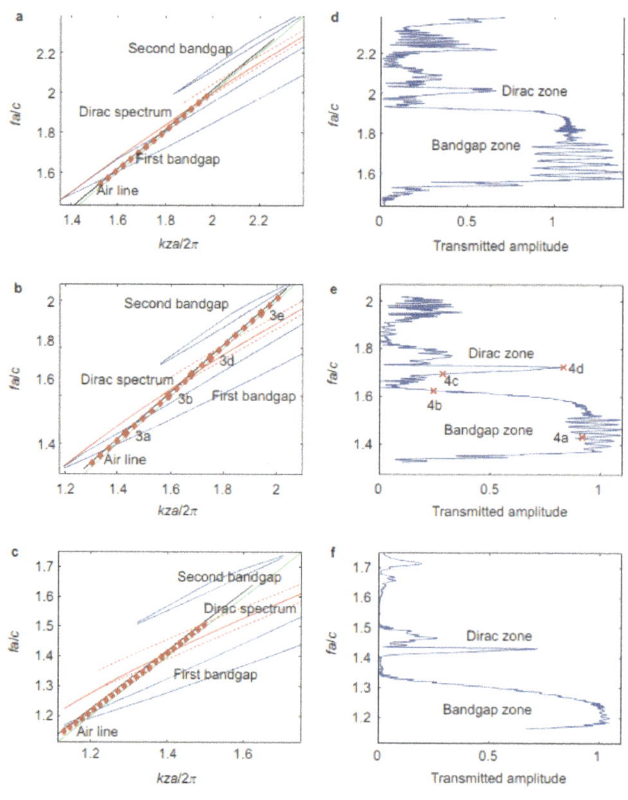


**Figure 2:** Propagation diagrams and transmission spectra of photonic crystal fibers. All the fibers have the same cross-section as that shown inFigure 1a, where a=2.21 μm, $r_a$=0.47a, and R=1.9a. (**a–c**) Propagation diagrams for the silica glass fiber (**a**, $\varepsilon_r=1.45^2$), germania glass fiber(**b**, $\varepsilon_r=1.59^2$), and SF6 glass fiber (**c**, $\varepsilon_r=1.8^2$). The solid red line represents the Dirac frequency of the fiber cladding. The dashed red lines mark the upper and lower edges of the Dirac spectrum. The black line is the theoretically calculated dispersion curve of the fundamental fiber mode obtained using the analytical model (1). The red diamonds are the numerically calculated data points of the fundamen-

tal fiber mode obtained using the PWE method. Field distributions of the modes at points 3a–3e, indicated by larger diamonds in **b**, are presented in Figure 3a–3e. The blue curves mark the edges of the photonic bandgaps. The green line represents the light line $k_z=\omega/c$ of air. (**d–f**) Numerically simulated (by FDTD)transmission spectra of narrow pulses passing through a section (1200μm in length) of the silica glass fiber (**d**), germania glass fiber (**e**), and SF6 glass fiber (**f**). Propagation scenarios of modes at the four points 4a–4d, marked by red crosses in **e**, are presented in Figure 4a–4d.

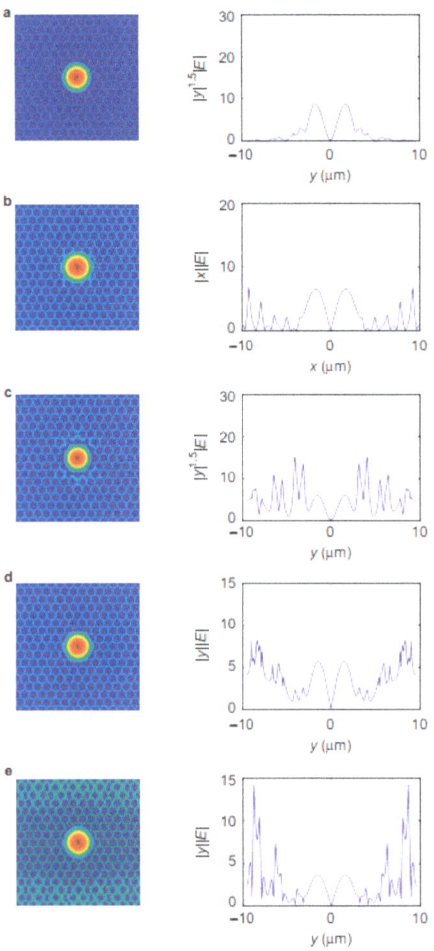


**Figure 3:** Guided modes in a triangular lattice photonic crystal fiber of germania glass. The fiber structure is shown in Figure 1a, wh ere a=2.21 μm,$r_a$=0.47a, R=1.9a, and $\varepsilon_r=1.59^2$. The modes were numerically calculated using the PWE method. The light-blue circles in the dark blue background give the positions of the hole–glass interfaces. (**a–e**)correspond to the red diamonds labeled 3a–3e in Figure 2b: (**a**) a gap mode ($\omega a/2\pi c$= 1.432), (**b**) a mode at a frequency between the photonic band gap and the lower edge of the Dirac spectrum($\omega a/2\pi c$=1.601), (**c**) the Dirac mode ($\omega a/2\pi c$=1.689), (**d**) a mode in the Dirac spectrum ($\omega a/2\pi c$=1.761), and (**e**) a mode at a frequency higher than the upper edge of the Dirac spectrum ($\omega a/2\pi c$=1.950). The electric-field pattern |E| is shown on the left, and the product $r^{3/2}$|E| orr|E| on the x- or y-axis is shown on the right.

## Finite-Difference time-Domain (FDTD) Numerical Simulations

In nonmagnetic media, a complete description of electromagnetic waves in time and space is given by the Maxwell equations $\nabla\times E=-\mu_0\frac{\partial H}{\partial t}, \nabla\times H=\varepsilon_0\varepsilon_r\frac{\partial E}{\partial t}$. The Maxwell equations can be solved directly by the well-known FDTD method[19]without invoking the functional form $\exp(i\omega t-ik_z z)$. A three-dimensional FDTD model was established with a perfectly matched layer boundary condition on all the edges. The structure is uniform in the longitudinal direction, extending from z=0 μm to z=1200 μm. On the transverse plane, the structure has a profile of the refractive index given by Figure 1a, with an area of 66.3 μm × 65.07 μm to include 30 layers and 34 layers in the x and y directions, respectively. The step sizes in space are$\Delta x=\Delta z=a/20=0.1105$ μm and $\Delta y=a\times\sqrt{3}/36\approx0.1063$ μm. The step size in time is$\Delta t=0.1995$ fs, which satisfies the Courant–Friedrichs–Ley condition $c\Delta t\leq(1/\Delta x^2+1/\Delta y^2+1/\Delta z^2)^{-1/2}$, where c is the velocity of light. For the study of transmission spectrum, a broadband Gaussian pulse is input from the front end of the fiber. The incident pulse has a Gaussian amplitude profile in the transverse plane, with a waist radius equal to the central hollow core radius of the fiber. The incident pulse has a finite duration in time, with a Gaussian spectrum centered at the Dirac frequency $\omega_D/2\pi$ and a frequency span of $0.5\omega_D/2\pi$. This wave is simulated to evolve in the 1200-μm-long germania glass fiber for 8000 fs, which is long enough for the waves to pass through the structure in all directions. The full-time response at the rear end of the fiber is recorded. The transmission spectra are obtained from the time-domain response by Fourier transform. The outcomes presented in Figure 2d–2f reveal peaks at frequencies inside the bandgap and at the Dirac frequency. To investigate evolutions of the different frequency components in detail, waves at specific frequencies are picked out for further simulations. For this purpose, the incident field of a continuous Gaussian beam (single harmonic) is used. The waist radius of the beam in the transverse plane is again set equal to the radius of the fiber core. The wave is simulated to evolve for long enough time to reach a steady state. The field at the end of the simulation (t=8000 fs) is recorded by a monitor in the yz cut plane, and the results are presented in Figure 4 for four different frequency components. Leakage-free propagation of a guided mode at the Dirac frequency is visualized, confirming the transmission spectrum of Figure 2e.

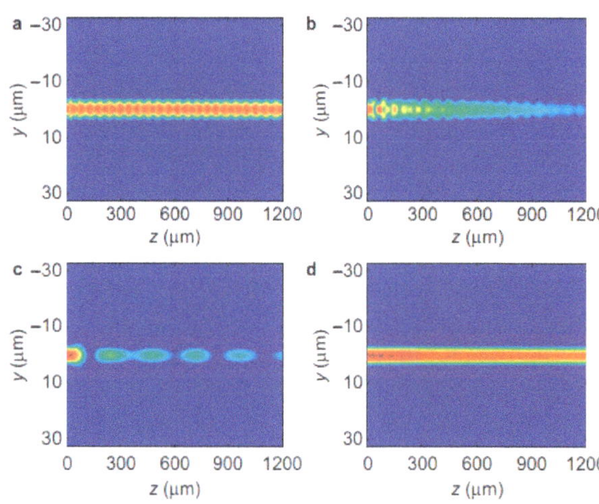


**Figure 4:** The steady-state electric fields in the germania fiber at the end of the FDTD simulations. The fiber has the cross-section structure shown inFigure 1a, where a=2.21 μm, $r_a$=0.47a, R=1.9a, and $\varepsilon_r=1.59^2$. The |E|fields are recorded at t=8000 fs and displayed by a monitor in the yzcut plane that runs directly through the center axis of the fiber at x=0, ranging from z=0 to z=1200 μm in the z-direction and from y=–32.535 to y=32.535 μm in the y-direction. (**a–d**) corre-

spond to the four red crosses labeled 4a–4d in Figure 2e: (**a**) |E| field for $\omega a/2\pi c$= 1.432, (**b**)|E| field for $\omega a/2\pi c$=1.624, (**c**) |E| field for $\omega a/2\pi c$=1.694, and (**d**) |E|field for $\omega a/2\pi c$=1.724.

## RESULTS AND DISCUSSIONS

### Dirac Cone for Fiber-Like Propagation

The fiber problem is different from the two-dimensional problem addressed in Ref.17 mainly for two reasons. First, the propagation direction in the fiber is perpendicular to the lattice direction. The restriction on the longitudinal propagation wave vector $k_z$ must be lifted. Second, the refractive index contrast of a photonic crystal fiber is relatively small. Here, the refractive index contrast is not a free parameter to select but is dictated by the constituent materials of the fiber, which is in turn dependent on the available technology. Silica glass and germania glass are the most common materials for fiber fabrication. Other types of glasses such as SF6 can also be used to extrude a specially designed fiber[20]. A well-developed Dirac cone has been shown to exist in the in-plane ($k_z$=0) band structure of a photonic crystal consisting of a triangular lattice of air columns in a GaAs substrate (refractive index contrast of 4−1)[17]. However, the Dirac cone degrades as the refractive index contrast decreases, indicating that a small refractive index contrast is a disadvantage. We begin our analysis by revealing the existence of well-developed Dirac cones in structures with refractive index contrasts as small as that of glass and air for out-of-plane propagation, which is the actual situation for fiber-like propagation. The band structure of an air-germania triangular lattice (index contrast of 1.59-1) is studied using the PWE method for various $k_z$. Figure 1bdisplays the band structure of the lattice for $k_z a/2\pi$=0.001. The lowest four Dirac points are shown clustered two by two, with none of them well isolated from other bands. The situation improves as $k_z$ increases. The band structure for $k_z a/2\pi$=1.581 is displayed in Figure 1c. The second Dirac point, the one that involves the fifth and sixth bands, has moved away from the others. This trend continues, as is evident by the band structure shown in Figure 1d for $k_z a/2\pi$=1.910, where the second Dirac point stands out and evolves into a pair of cones. A three-dimensional plot of the second Dirac point is presented as an inset to Figure 1d, revealing the existence of well-developed Dirac cones with an outstretched Dirac spectrum for propagation out of the plane. The Dirac spectrum is defined as the frequency spectrum of linear dispersion around the Dirac point in the band structure. This example illustrates how a dwarf Dirac cone evolves into a well-developed Dirac cone as $k_z$ increases. This phenomenon is possible because the longitudinal wave vector $k_z$ strongly flattens the bands, thus leaving space for the Dirac cones. Consequently, the air-glass index contrast is sufficient to open the Dirac cone in these flatter bands with an appreciable frequency range of linear dispersion.

To better visualize the influence of $k_z$ on shape of the Dirac cone and to determine the threshold for the opening of the Dirac spectrum, let's concentrate on the second Dirac point and examine its process of transition with the change of $k_z$. Starting from Figure 1d for $k_z a/2\pi$ =1.910, it is apparently visible in the band structure that a linear double cone spectrum is formed by the fifth and sixth bands. However, as$k_z a/2\pi$ decreases, ends of the fifth band start to bend upwards in relative to the Dirac point, and thus, the frequency range of the linear spectrum around the Dirac point is reduced. When $k_z a/2\pi$ =1.581, height of the magenta curve at the M point reaches that of the Dirac point, as shown in Figure 1c. In this situation, the frequency of the Dirac point overlaps with the frequency of the lower branch of the Dirac cone at the M point, eliminating the special effect of the conical singularity (vanishing density of states) altogether; thus, in effect the frequency range of the linear spectrum in the vicinity of the Dirac frequency disappears completely. As$k_z a/2\pi$ decreases further, the fifth band bends up to a higher frequency than the Dirac frequency. The Dirac frequency remains buried in frequencies of the lower branch of the Dirac cone, and the Dirac spectrum is not clearly defined, as forFigure 1b. The value $k_z a/2\pi$ =1.581 at the turning point is taken as the threshold for the opening of the Dirac spectrum.

The above discussion suggests that well-developed Dirac cones are obtainable for photonic crystal fibers at suitable $k_z$. Among the four Dirac points indicated by the red arrows in Figure 1, only the second Dirac point develops a wide open Dirac spectrum. Width ofthe Dirac spectrum remains zero for the other three Dirac points. Since these dwarf Dirac cones cannot support localized mode, they are ignored in the subsequent discussions. The dependence of the second Dirac spectrum as $k_z$ varies is plotted in Figure 2a–2c for fibers of silica glass, germania glass, and SF6 glass, respectively. A conspicuous Dirac spectrum appears in the band structure as $k_z$ increases above a threshold value, where the solid red line intersects one of the dashed red lines. The threshold value of $k_z$ for the opening of the Dirac spectrum varies from $k_z a/2\pi \approx 1.878$ for the silica/air structure (Figure 2a) to $k_z a/2\pi \approx 1.222$ for the SF6/air structure (Figure 2c), depending critically on the refractive index contrast of the fiber cladding. The Dirac spectrum is also dependent on the air hole radius. For the best performance, we require $0.45a<r_a<0.49a$.

### Wave Localization at the DIRAC Frequency

The Dirac spectrum of the fiber cladding enables a new type of guiding for waves at the Dirac frequency. Guided modes of the fibers are studied by the PWE method on a supercell. The eigen frequency of the fundamental guided mode versus $k_z/2\pi$ is plotted in Figure 2a–2c as red diamonds for the silica glass fiber, germania glass fiber, and SF6 glass fiber. In Figure 2d–2f, the FDTD simulated amplitude spectra of narrow pulses transmitted through a section of these fibers are plotted. For a hollow-core fiber, the mode energy is largely contained in the center air hole region;therefore, the effective index of the guided mode is always smaller than unity. For this reason, the guided modes in Figure 2a–2c cannot appear below the air line and instead follow the air line closely as the frequency varies. For the photonic crystal fiber made of silica glass, the refractive index contrast is so small that the opening of the Dirac spectrum occurs at large wave vector $k_z/2\pi$. It is apparent from Figure 2a that the Dirac spectrum starts so close to the air line that vast majority of the Dirac spectrum appears below the air line. The Dirac cone above the air line in this fiber is yet immature. It may not function well to sustain the Dirac mode. Indeed, the mode profile calculated at the Dirac frequency for this fiber reveals a large residual field outside the core. This opinion is supported by the results from the FDTD simulations. The FDTD simulated amplitude spectrum in Figure 2d exhibits a dwarf peak of transmission in the region of the 'Dirac zone'.

Both results suggest that a photonic crystal fiber made of silica glass is not suitable for Dirac mode propagation. The Dirac spectrum opens up at smaller wave vector $k_z/2\pi$ in the photonic crystal fibers made of germania glass or SF6 glass. As shown in Figure 2b–2c, the Dirac spectrum of such a higher index contrast fiber is shifted to the left such that a substantial portion appears above the air line. Propagation of the Dirac mode is therefore well supported in both of the fibers. Comparing the SF6 glass fiber with the germania glass fiber, the curve of the Dirac frequency intersects the dispersion curve of the guided mode at a larger angle in Figure 2c than in Figure 2b. This result implies that the bandwidth of the Dirac mode in the SF6 glass fiber is smaller than that in the germania glass fiber. This observation is confirmed inFigure 2e–2f by the narrower transmission peak in the 'Dirac zone' of the SF6 glass fiber than that for the germania glass fiber. It suggests that a broadband propagation of the Dirac mode is feasible by proper selection of the fiber material and careful design of the structure.

The mode that falls within the bandgap is a conventional gap mode. An example of the gap mode at a point marked 3a in Figure 2b (represented by a larger diamond) is plotted in Figure 3a. Outside the bandgap, the mode becomes a guided resonance that is leaky but retains its maximum within the air core for a relatively broad spectral range. Examples of the guided resonances at the points marked 3b and 3e in Figure 2b are plotted in Figure 3b and 3e, respectively. The fields of these modes are not well confined to the core and will leak away into the continuum of radiation states in the cladding and thus have finite propagation distances. At the Dirac frequency $\omega_D a/2\pi c = 1.689$, marked 3c in Figure 2b, the density of radiation states is zero[21]; therefore, the propagation distance of the resonance goes to infinity, and light becomes perfectly guided again in the fiber. The mode at this point, which is plotted in Figure 3c for its field distribution, is the new type of guided mode (Dirac mode) identified in this work. This state is no longer a leaky resonance but an eigen mode that does not decay in propagation. Compared with the gap mode in Figure 3a, one noticeable feature of the Dirac mode is the relatively large peripheral field surrounding the core. The field profile of the Dirac mode can be leveled up by multiplying it by $r^{3/2}$, as shown in Figure 3c by a slice cut on the y-axis. The almost constant level of oscillation in the envelope of the product as the tail extends outwards demonstrates that the envelope function of the new state decays algebraically as $1/r^{3/2}$, rather than exponentially, with distance away from its center. It is demonstrated in Figure 3a that the field profile of a conventional gap mode cannot be leveled up by $r^{3/2}$. As the mode frequency deviates from the Dirac frequency, the density of radiation states increases linearly in the vicinity of the Dirac point and abruptly as the mode frequency goes beyond the Dirac spectrum. The mode becomes a guided resonance similar to that observed in the neighborhood of a bandgap. An example of the mode at the point marked 3d in Figure 2b is plotted in Figure 3d. All the leaky modes (Figure 3b, 3d and 3e) have tails that decay slower than 1/r, indicating that their total energies are not convergent.

The existence of the Dirac mode is reconfirmed by peaks in the transmission spectrum of the fiber. As shown in Figure 2e–2f, two transparent zones are apparent, as marked by 'Bandgap zone' and 'Dirac zone', respectively. The 'Bandgap zone' corresponds to the position of a bandgap, whereas the high transmission peak in the 'Dirac zone' is associated with the leakage-free propagation of the Dirac mode. By cross-checking Figure 2e–2f with Figure 2b–2c, a very good match of the two independent set of calculations is obtained. In Figure 2e–2f, the sharp peaks appear at $\omega a/2\pi c = 1.724$ and 1.429 in FDTD simulations for the germania glass fiber and the SF6 glass fiber, respectively. In Figure 2b–2c, which emerge from the PWE calculations, however, the intersection point of the curves of the Dirac frequency and the dispersion is at $\omega a/2\pi c = 1.689$ for the germania glass fiber and at 1.415 for the SF6 glass fiber. The locations of the intersection points in Figure 2b–2c for the onset of the Dirac modes agree well with the positions of the sharp transmission peaks in Figure 2e–2f for the leakage-free propagation, leaving discrepancies of approximately 2% and 1%, respectively. At a frequency near the intersection point, all the bandgaps in Figure 2b–2c are far away from the air line; thus, they are not reachable by a core-guided mode. The Dirac mode is therefore the only localized mode that can be confined to the air core in this frequency range. This situation guarantees pure Dirac-mode operation of the fiber in the 'Dirac zone', where the fiber selectively supports only the Dirac mode for leakage-free propagation.

## Physical Mechanism of the Light Guiding at the Dirac Frequency

To mathematically show that an optical bound state is feasible at the Dirac frequency, we examine the governing equation of the waves in such a photonic crystal. Close to the Dirac frequency, photon transport in the transverse plane of the lattice is describable by the massless Dirac equation[14,15]: $-iv_D(\sigma_x\partial_x+\sigma_y\partial_y)\Psi=(\omega-\omega_D)\Psi$. It is well known that the Dirac equation substantiates a half-bound solution of the form $\psi_{1,2}\propto e^{\pm i\theta}/r$ at the Dirac frequency. This type of mode has been previously studied for charge carriers in graphene[18], where the wave function is the probability of finding the spinors in space. This relativistic quantum mechanics model can be adopted to optics by a slight modification to the interpretation of the modal function. In photonic crystals, the wave function $\psi_{1,2}\propto e^{\pm i\theta}/r$ represents the envelope amplitude of a basis function (Bloch state)[14], and the basis wave function intrinsic to a cylindrical geometry is a cylindrical wave function[17] $\Phi \propto J_0(Kr)$, where$K=4\pi/(3a)$. Therefore, at the Dirac frequency, the physical quantities $E_z$ and $H_z$ in the photonic crystal are characterized by $J_0(Kr)e^{\pm i\theta}/r$. Because the amplitude of the Bessel function decays as $1/\sqrt{r}$, the overall spatial dependence of the corresponding optical state asymptotically approaches $r^{-3/2}$ at large distances. This result coincides precisely with the asymptotic algebraic spatial profile exhibited by Figure 3c, thus verifying the mathematical model of the Dirac equation. The $r^{-3/2}$ spatial profile ensures square integrability of the physical quantities and confirms that the Dirac mode in photonic crystal fibers is a well-behaved, true bound state.

As outlined in Supplementary Information, using $J_0(Kr)e^{\pm i\theta}/r$ for the wave solution of Maxwell equations in the photonic crystal cladding, the eigenvalue equation of the fundamental guided mode of the fiber is derived as

$$\left(W_1(k_1R)-W_0(KR)+\frac{1}{K^2R^2}\right)\left(W_1(k_1R)-\varepsilon_{eff}W_0(KR)+\varepsilon_{eff}\frac{1}{K^2R^2}\right)$$

$$= \frac{k_z^2}{k_0^2}\left(\frac{1}{k_1^2R^2} - \frac{1}{K^2R^2}\right)^2, \quad (1)$$

where $k_0=\omega/c,\ k_1^2=k_0^2-k_z^2,\ \varepsilon_{eff}=(K^2+k_z^2)/k_0^2,\ W_N(x)=\frac{J_N'(x)}{xJ_N(x)}$. Upon solving Equation (1) for the eigen frequency $k_0c/2\pi$ as a function of the wave vector$k_z/2\pi$, the dispersion relation of the guided mode is obtained. The results are plotted in Figure 2a–2c as black lines for the three different fibers. Good agreement with numerical results is observed. The analytical model (1) is derived at the Dirac frequency $\omega_D/2\pi$. When applied to a frequency $\omega/2\pi$ other than $\omega_D/2\pi$, we quantitatively require $\delta\omega(\equiv\omega-\omega_D)<<\omega_D$ for the result to be valid. In practice, we typically find it to be accurate for frequency within the Dirac spectrum, and often qualitatively extensible for even larger frequency range, as shown in Figure 2a–2c. For the case of germania glass fiber (Figure 2b), the black line intersects the red line at $\omega a/2\pi c=1.668$. This theoretical result compares well with the value$\omega a/2\pi c=1.689$ from the PWE calculation for the onset of the Dirac mode, leaving only a discrepancy of approximately 1.2%. The electric field pattern of the mode at the Dirac frequency obtained using this analytical model has a single lobe monopole pattern inside the hollow core and a power-law profile in the surrounding cladding, which compares well with the numerically calculated field distribution presented inFigure 3c. These basic similarities between the theoretical and numerical patterns further validate the theory and numerical findings.

The trapping of light at the Dirac frequency is more intuitively recognized by examining the density of states of the waves propagating in the surrounding medium. Right at the Dirac frequency, the density of radiation states in a triangular lattice photonic crystal is precisely zero[21]; thus, outgoing waves are forbidden in the surrounding medium, and field confinement to the center hollow core becomes possible. Because the zero density of states at the Dirac frequency is a special feature of the band structure, and the band structure itself is a consequence of Bragg scattering of light by the photonic crystal lattice, at a deeper level of understanding, the localization phenomenon observed at the Dirac frequency is essentially due to Bragg scattering, which determines the overall band structure. Rather than leading to a complete gap, however, we find an isolated conical singularity where the band gap is vanishingly small. The key feature of this band structure is that the two bands touch as a pair of cones at corners of the Brillouin zone in a linear fashion, with linear dispersion and a linearly vanishing density of states in its vicinity. One would expect to see a true localized mode precisely at the point where the density of states is zero and a guided resonance with finite lifetime away from it. The fact that there is not a band of true localized modes around the Dirac frequency gives the mode special properties such as an algebraic profile and sharp frequency response. This light-trapping phenomenon occurs at a single frequency and should be distinguished from a recent observation of an 'embedded eigenvalue'[22,23] that also occurs at a single frequency. The existence of the Dirac mode is due to disappearance of radiation states at the Dirac frequency, whereas the trapped light in Refs. 22 and 23 occurs in the presence of radiation states;however, somehow, mutual coupling of the trapped light and radiation states is suppressed by destructive interference. In light of this discussion, the two phenomena are different in how they break up coupling of the trapped light and radiation states.

## Numerical Confirmation of the Guided Mode at the Dirac Frequency

To confirm the existence of a trapped state at the Dirac frequency, numerical tools[24] based on the FDTD method[19] were adopted to simulate the evolutions of waves passing through a section (1200 μm long) of the germania glass fiber. InFigure 4a, a typical gap mode propagating over this distance is presented. The oscillation in amplitude is due to a mismatch between the input wave and the eigenmode of the fiber. When the frequency goes beyond the bandgap, the mode still concentrates much of its field energy in the core. Nevertheless, the mode is no longer a true-bound state but a guided resonance that resembles the so-called ARROW mode[25,26]. Like the ARROW mode of a cylindrical waveguide, this mode has a $|E|\propto 1/\sqrt{r}$ asymptotic profile in space. It is a radiative mode itself ($\int|E|^2 r dr d\theta\rightarrow\infty$) and is therefore inherently leaky. An example of the guided resonance is shown in Figure 4b. Further away from the bandgap, the resonance is less concentrated within the center core until the resonance broadens away and melts into the continuum. As the localized mode penetrates into the region of the Dirac spectrum, the density of radiation states starts to decrease because of the presence of the Dirac point. Guided resonance similar to that observed in the neighborhood of a bandgap makes an appearance in the Dirac spectrum. The mode, such as that shown in Figure 4c, for example, remains leaky because it is situated inside a band of a continuum of states and mutual coupling is inevitable. When the frequency reaches the Dirac frequency, the density of radiation states vanishes; thus, coupling of the Dirac mode and the radiation states disappears completely. The profile of the mode thus approaches $|E|\propto r^{-3/2}$, which is square integrable, indicating that the mode becomes trapped again. Mode propagation at the Dirac frequency is shown inFigure 4d, an almost leakage-free propagation is achieved. The leakage-free operational window of the fiber can be best visualized by a transmission spectrum of the fiber itself, as shown in Figure 2d–2f, where the Dirac frequency corresponds to a sharp transmission peak in each case. The simulations presented in Figure 4a–4d correspond to the four points 4a–4d marked by red crosses in Figure 2e. From the observation of the low and high transmissivity and the corresponding locations on the transmission spectrum, Figure 4a–4d is consistent with behaviors expected at the points 4a–4d; therefore, the propagation scenarios of these modes are understood. This finding represents an important discovery, based on which a new type of fiber guiding is established at the Dirac frequency beyond a bandgap.

If the lattice extends to infinity and material absorption is ignored, the Dirac mode in Figure 4d and the gap mode in Figure 4a are both lossless. In practice, however, a real lattice has a finite size, which results in a loss that is associated with field penetration across the boundary into the surrounding medium. Because the tail of the gap mode decays exponentially, whereas the tail of the Dirac mode decays algebraically with distance, for the same lattice extent, the residual field of the Dirac mode on the boundary is stronger than that of the gap mode. Consequently, the Dirac mode is more lossy than a gap mode in general. Because of the leakage to the surrounding medium, propagation at a transmission peak is not truly leakage-free in a fiber of finite cross section, but, as shown in Figure 4a and 4d, declines slowly with distance following an exponential law $P=P_0e^{-\alpha z}$ for its power. For the germania glass fiber, the attenuation constant α is estimated to be $1.5 \times 10^{-4}$ μm$^{-1}$ for the Dirac mode and ~0 for the gap mode. Fortunately, the loss due to the finite lattice size can be made as small as desired by increasing the number of surrounding crystal periods.

## Practical Fiber Design, Experimental Validation, and Applications

The design concern for the new proposed fiber is quite similar to the design concern of an ordinary photonic crystal fiber. The complexity of the photonic crystal structure denies an analytical solution; thus, a lot of computer software (see Ref. 24for some examples) has been developed to facilitate the design of practical photonic crystal fibers. This commercially available software can be directly applied to the design activity of the new proposed fiber because it still belongs to the group of photonic crystal fibers. The only difference is a shift of attention from the bandgap to the Dirac point. To be specific, the first concern of the design is whether a Dirac mode can be supported by a fiber with given parameters. The band structure of a photonic crystal fiber can be calculated using one of the available commercial software products or using a simple MATLAB code based on the PWE method[1]. Our studies demonstrate that the Dirac mode is not well supported in a photonic crystal fiber if the refractive index contrast is as small as that of a silica/air (1.45−1) system. We confirm that photonic crystal fibers made of SF6 glass ($n_r$=1.8) or germania glass ($n_r$=1.59) are suitable for the Dirac mode propagations. From these studies, the constituent materials of the fiber can be selected, and the air hole radius can be optimized. The second concern for the design is the precise position of the Dirac frequency. As shown in Figure 2b–2c, the curves of the Dirac frequency and the dispersion intersect at $\omega a/2\pi c \approx 1.689$ for the germania glass fiber and at 1.415 for the SF6 glass fiber. The frequency at the intersection point can be used as a rough estimate of the operational frequency for leakage-free propagation of the Dirac mode. For the lattice constant a=2.21 μm, these frequencies correspond to wavelengths of approximately 1308 and 1562 nm, respectively. The selected fiber materials, germania and SF6, are transparent at these wavelengths. Note that the Dirac frequency can be scaled to any desired value by adjusting the lattice period a.

Experimental validation of the Dirac mode can be established by the identification of the Dirac transmission peak on spectra measured on the specially designed photonic crystal fibers. From numerical simulations the Dirac peak shows up nicely in transmission spectra of photonic crystal fibers made of SF6 glass or germania glass. The above two fibers are therefore potential media for Dirac peak transmission measurements. In addition, polymer fiber is easy to handle and widely available. The refractive index of polystyrene, which is 1.590, is very close to that of germania glass, so polystyrene fiber is a convenient alternative for the Dirac transmission peak measurement. Experimental verification of the theory presented here is handled smoothly, and this will be the subject of a later paper.

Photonic crystal fibers were traditionally divided into two broad classes, namely, index-guiding fibers and photonic-bandgap fibers, depending on whether they used index guiding or photonic bandgaps for optical confinement. Our finding of Dirac point guiding enriches the varieties of photonic crystal fibers. Bandgap confinement is attractive because it allows light to be guided within a hollow core. This capability not only minimizes the effects of material absorptions, undesired nonlinearities, and any other unwanted properties of the bulk materials but also facilitates applications in biological sensors or organic light-emitting devices, where enhancement of laser interactions with dilute or gaseous media is required. Our new proposed fiber also allows light to be guided within a hollow core; thus, any benefit a photonic-bandgap fiber can offer is also available in the Dirac fiber. If photonic-bandgap fibers already exist; what is the purpose of exploring a new type of guiding phenomenon? The first reason is simple curiosity: the idea of using a Dirac point to trap light in a hollow core is intriguing. Second, the Dirac-point-induced guiding does have unique features. The Dirac mode is different from the gap mode in terms of the resulting frequency characteristics and spatial profile features. Our new proposed fiber can therefore be regarded as a supplement to existing photonic-bandgap fibers. For example, the unique power-law profile of the Dirac mode results in loose lateral confinement, thus enabling long-range interaction between multiple cores. This feature could lead to the creation of new coupling and sensing characteristics for use in optical couplers, laser arrays, and fiber sensors. The sharp frequency response of the Dirac mode can take on the role of a resonant cavity with the establishment of frequency selectivity in applications of coherent random fiber lasers, where resonant cavities are not available but where frequency selectivity is desired[27].

## CONCLUSIONS

A new type of light trapping was demonstrated in fibers at the Dirac frequency in the absence of a bandgap and total internal reflection. This discovery will extend the possibilities of fiber optics and will lead to many new applications in fiber lasing/sensing systems. We believe that the effects discovered in the present study for electromagnetic waves in photonic crystals could occur in other wave phenomenon systems, including acoustic waves in phononic crystals and matter waves in semiconductors. A further search for different systems with the extraordinary localization of photons, as well as phonons and electrons, can expand the applicability of our results.

## REFERENCES

1. Skorobogatiy M, Yang J. Fundamentals of Photonic Crystal Guiding. Cambridge: Cambridge University Press, 2009.
2. Liu H, Zhu S. Coupled magnetic resonator optical waveguides. Laser & Photon Rev 2013; 7: 882–900.
3. Meade RD, Devenyi A, Joannopoulos JD, Alerhand OL, Smith DA et al. Novel applications of photonic band gap materials: low-loss bends and high Qcavities. J Appl Phys 1994; 75: 4753–4755.
4. Fan S, Villeneuve PR, Joannopoulos JD, Haus HA. Channel drop tunneling through localized states. Phys Rev Lett 1998; 80: 960–963.
5. Shakoor A, Savio RL, Cardile P, Portalupi SL, Gerace D et al. Room temperature all–silicon photonic crystal nanocavity light emitting diode at sub–bandgap

wavelengths. Laser & Photon Rev 2013; 7: 114–121.

6. Diao Z, Bonzon C, Scalari G, Beck M, Faist J et al. Continuous-wave vertically emitting photonic crystal terahertz laser. Laser & Photon Rev 2013; 7: L45–L50.

7. Kao KC, Hockham GA. Dielectric–fibre surface waveguides for optical frequencies. IEEE Proc J Optoele 1986; 133: 191–198.

8. Knight JC, Broeng J, Birks TA, Russell PS. Photonic band gap guidance in optical fibers. Science 1998; 282: 1476–1478.

9. Steel MJ, White TP, de Sterke CM, McPhedran RC, Botten LC. Symmetry and degeneracy in microstructured optical fibers. Opt Lett 2001; 26: 488–490. |

10. Knight JC. Photonic crystal fibres. Nature 2003; 424: 847–851.

11. Yablonovitch E. Inhibited spontaneous emission in solid-state physics and electronics. Phys Rev Lett 1987; 58: 2059–2062.

12. John S. Strong localization of photons in certain disordered dielectric superlattices. Phys Rev Lett 1987; 58: 2486–2489.

13. Gajić R, Meisels R, Kuchar F, Hingerl K. All–angle left–handed negative refraction in Kagomé and honeycomb lattice photonic crystals. Phys Rev B2006; 73: 165310.

14. Raghu S, Haldane FD. Analogs of quantum–Hall–effect edge states in photonic crystals. Phys Rev A 2008; 78: 033834. |

15. Sepkhanov RA, Bazaliy YB, Beenakker CW. Extremal transmission at the Dirac point of a photonic band structure. Phys Rev A 2007; 75: 063813. |

16. Zandbergen SR, de Dood MJ. Experimental observation of strong edge effects on the pseudodiffusive transport of light in photonic graphene. Phys Rev Lett2010; 104: 043903. |

17. Xie K, Jiang HM, Boardman AD, Liu Y, Wu ZH et al. Trapped photons at a Dirac point: a new horizon for photonic crystals. Laser & Photon Rev 2014; 8: 583–589.

18. Castro Neto AH, Guinea F, Peres NM, Novoselov KS, Geim AK. The electronic properties of graphene. Rev Mod Phys 2009; 81: 109–162.

19. Taflove A, Hagness SC. Computational Electrodynamics: The Finite–difference Time–domain Method. Boston: Artech House, 2005.

20. Ravi Kanth Kumar VV, George AK, Reeves WH, Knight JC, Russell PS et al. Extruded soft glass photonic crystal fiber for ultrabroad supercontinuum generation. Opt Express 2002; 10: 1520–1525.

21. Plihal M, Maradudin AA. Photonic band structure of two–dimensional systems: the triangular lattice. Phys Rev B 1991; 44: 8565–8571. |

22. Hsu CW, Zhen B, Lee J, Chua SL, Johnson SG et al. Observation of trapped light within the radiation continuum. Nature 2013; 499: 188–191.

23. Hsu CW, Zhen B, Chua SL, Johnson SG, Joannopoulos JD et al. Bloch surface eigenstates within the radiation continuum. Light Sci & App 2013; 2: e84;doi:10.1038/lsa2013.40.

24. The open source MIT photonic bands software package at http://ab-initio.mit.edu/wiki/index.php/Main_Page ; Lumerical FDTD Solutions athttps://www.lumerical.com/; OptiFDTD at http://optiwave.com/; The RSoft Photonics Suite from RSoft Design Group athttp://optics.synopsys.com/rsoft/.

25. Rowland KJ, Afshar S, Monro TM. Bandgaps and antiresonances in integrated-ARROWs and Bragg fibers; a simple model. Opt Exp 2008; 16: 17935–17951.

26. Langbein U, Trutschel U, Unger A, Duguay M. Rigorous mode solver for multilayer cylindrical waveguide structures using constraints optimization.Opt Quant Electron 2009; 41: 223–233.

27. Hu ZJ, Zhang Q, Miao B, Fu Q, Zou G et al. Coherent Random fiber laser based on nanoparticles scattering in the extremely weakly scattering regime.Phys Rev Lett 2012; 109: 253901.

CHAPTER 8

# BREAKDOWN OF BOSE-EINSTEIN DISTRIBUTION IN PHOTONIC CRYSTALS

Ping-Yuan Lo, Heng-Na Xiong & Wei-Min Zhang

Department of Physics and Center for Quantum Information Science, National Cheng Kung University, Tainan 70101, Taiwan, Republic of China.

## ABSTRACT

**In the last two decades, considerable advances have been made in the investigation of nano-photonics in photonic crystals. Previous theoretical investigations of photon dynamics were carried out at zero temperature. Here, we investigate micro/nano cavity photonics in photonic crystals at finite temperature. Due to photonic-band-gap-induced localized long-lived photon dynamics, we discover that cavity photons in photonic crystals do not obey Bose-Einstein statistical distribution. Within the photonic band gap and in the vicinity of the band edge, cavity photons combine the long-lived non-Markovain dynamics with thermal fluctuations together to form photon states that memorize the initial cavity state information. As a result, Bose-Einstein distribution is completely broken down in these regimes, even if the thermal energy is larger or much larger than the cavity detuning energy. In this investigation, a crossover phenomenon from equilibrium to nonequilibrium steady states is also revealed.**

## INTRODUCTION

Photonic band gap (PBG) structures in photonic crystals (PCs) together with the characteristic dispersion properties have stimulated considerable interest in the study of fundamental photonic science and also in the development of new photonic technology[1,2,3]. The most significant new features induced by the PBG are the inhibition of atom spontaneous emission and the localization of light[4,5,6]. As a result, ultrahigh quality-factor cavity has been realized on-chip with PBG structures[7]. This provides the opportunity to control and manipulate light for photonic information technology[8]. In the past two decades, quantum optics with a few-level atom placed inside PCs have been extensively explored[9], and cavity QED with the features of atomic population trapping and atom-photon bound states in the vicinity of the photonic band edge (PBE) has been examined[10,11,12,13,14]. These features are obtained mainly at zero temperature by solving the Schrödinger equation in which the PCs contain only one single photon emitted from an atom which is initially in the excited state. On the other hand, when the number of photons increases, light propagating in PCs was understood using the classical Maxwell equations, which is also defined at zero temperature[1,3]. Indeed, photonic quantum dynamics, even for a pure micro/nano cavity embedded in PCs, has not yet been solved at finite temperature. Practically, understanding photonic quantum dynamics at finite temperature is important for the development of all-optical circuits incorporating cavities and PBG waveguides embedded in PCs in the microwave regime.

**Citation:** Ping-Yuan Lo, Heng-Na Xiong & Wei-Min Zhang, Break-down of Bose-Einstein Distribution in Photonic Crystals, doi:10.1038/srep09423

All-optical circuits for networks on-chips consist of micro/nano cavities and waveguides[8]. Micro/nano cavities in PCs are created by point defects, and PBG waveguides can be made with coupled defect arrays. Frequencies of cavities and waveguides can be easily tuned by changing the size and/or the shape of defects. In this article, we investigate micro/nano cavity photonics in PCs at finite temperature. Due to PBG-induced localized long-lived non-Markovian photon dynamics, we find that cavity photons in PCs do not obey Bose-Einstein statistical distribution. Within the PBG region and also in the vicinity of the PBE, cavity photons combine the nontrivial non-Markovian dissipations with thermal fluctuations together to form photon states that can memorize the initial cavity state information. As a result, Bose-Einstein statistical distribution for photons is completely broken down in these regimes, even though the photonic thermal energy is larger or much larger than the cavity detuning energy. Also, a crossover phenomenon from equilibrium to nonequilibrium steady states is revealed in low-dimensional photonic crytslas.

## RESULTS

### Dissipation and Dissipationless Photon Dynamics in Photonic Crystals

The methodology for the study of cavity photon dynamics in photonic crystals at finite temperature was developed in our previous work[15,16,17,18], which is summarized in the Methods Section in the end of the article. To investigate the evolution of cavity photon states in PCs, we shall solve first the dynamics of photon dissipation and fluctuations, determined by Eqs. (19)–(20) given in Methods. Different spectral density J(ω) of the PCs, which requires the knowledge of DOS of the PCs, will provide different photon dissipation and fluctuations. In standard quantum optics, the spectral density J(ω) is usually treated as a constant in the weak-coupling limit, where the Weisskopf-Winger approximation or the Markovian master equation is valid so that the photon damping rate is time-independent (i.e. memoryless). As a result, all cavity photon modes have a finite lifetime and the cavity photon states will ultimately evolve into thermal equilibrium with the environment after decoherence takes place[19], and photons inside the cavity must obey Bose-Einstein distribution[20,21,22]. However, this well-known result is no longer satisfied for micro/nano cavities in PCs, due to the presence of the PBG, as we will show below.

In principle, the DOS for different PCs, denoted by $\varrho_{PC}(\omega)$, should be calculated by solving photon eigenfrequencies and eigenfunctions from the Maxwell's equations for different photonic crystal structures[2,23,24]. In the literature, different photonic DOS have been introduced for different PCs[9,10,11,12,14,25,26,27,28,29]. For examples, for 1D PCs, the corresponding DOS is given by $\varrho_{PC}(\omega) \propto \frac{1}{\sqrt{\omega-\omega_e}}\Theta(\omega-\omega_e)$, where $\Theta(\omega-\omega_e)$ is the Heaviside step function and $\omega_e$ is the frequency at the PBE. This DOS is also often used for isotropic 3D PCs, which could predict qualitatively correct behaviors of the non-Weisskopf-Winger decay and the photon-atom bound state in PCs[9,10,11,12,14], but it may overestimate the spontaneous emission rate of atoms due to the absence of the singularity of DOS in reality. Thus, for 3D PCs, the DOS near the PBE may be modeled by an anisotropic DOS[9,10,11]: $\varrho_{PC}(\omega) \propto \sqrt{\omega-\omega_e}\Theta(\omega-\omega_e)$, which could be further modified when the vectorial property of EM field is taken into account[25,26]. For 2D PCs with van Hove saddle point singularity[27,28,29], the photonic DOS exhibits a logarithmic divergence near the PBE and can be approximated by $\varrho_{PC}(\omega) \propto -\left[\ln\left|(\omega-\omega_0)/\omega_0\right|-1\right]\Theta(\omega-\omega_e)$, where $\omega_0$ is the center of the logarithmic peak. The spectral density J(ω) is microscopically defined as the multiplication of the DOS of the PCs with the photon tunneling amplitude V(ω) between the cavity and the PCs,

$$J(\omega) = \varrho(\omega)|V(\omega)|^2. \tag{1}$$

Assume that the cavity mode equally couples to all possible modes near the PBE of PCs, i.e. treating V(ω) as a constant, then the corresponding spectral densities J(ω) for different dimensional PCs are fully determined by the corresponding DOS, which are summarized in Table 1 and are also plotted in Fig. 1(a). We take the unit $\hbar = 1$ hereafter so that both the frequency ω and the spectral density J(ω) have the same unit as the energy.

**Table 1:** Characters of different photonic crystal structures. For different DOS of different dimensional PCs, the corresponding different spectral densities J(*ω*) are listed and the reservoir-induced photon self-energy corrections Σ(*ω*) are calculated, which completely determine the dissipative photon dynamics in PCs. The parameters C, *η* **and** *χ* **are coupling strengths (with a proper rescaling) between the cavity and PCs for 1D, 2D and 3D PCs, respectively. A smooth high-frequency cutoff $\Omega_C$ is introduced for avoiding the divergence of DOS in 3D PCs, and a sharp high-frequency cutoff at $\Omega_d$ is set up for maintaining the positivity of DOS in 2D PCs. Here $Li_2$** (x) is a dilogarithm function and erfc (x) is a complementary error function

| Photonic Crystals (PCs) | Spectral density J(ω) for different DOS | Reservoir-induced self-energy correction Σ(ω) |
|---|---|---|
| 1D | $\frac{C}{\pi}\frac{1}{\sqrt{\omega-\omega_e}}\Theta(\omega-\omega_e)$ | $-\frac{C}{\sqrt{\omega_e-\omega}}$ |
| 2D | $-\eta\left[\ln\left\|\frac{\omega-\omega_0}{\omega_0}\right\|-1\right]\Theta(\omega-\omega_e)\Theta(\Omega_d-\omega)$ | $\eta\left[\mathrm{Li}_2\left(\frac{\Omega_d-\omega_0}{\omega-\omega_0}\right)-\mathrm{Li}_2\left(\frac{\omega_0-\omega_e}{\omega_0-\omega}\right)-\ln\frac{\omega_0-\omega_e}{\Omega_d-\omega_0}\ln\frac{\omega_e-\omega}{\omega_0-\omega}\right]$ |
| 3D | $\chi\sqrt{\frac{\omega-\omega_e}{\Omega_C}}\exp\left(-\frac{\omega-\omega_e}{\Omega_C}\right)\Theta(\omega-\omega_e)$ | $\chi\left[\pi\sqrt{\frac{\omega_e-\omega}{\Omega_C}}\exp\left(-\frac{\omega-\omega_e}{\Omega_C}\right)\mathrm{erfc}\sqrt{\frac{\omega_e-\omega}{\Omega_C}}-\sqrt{\pi}\right]$ |

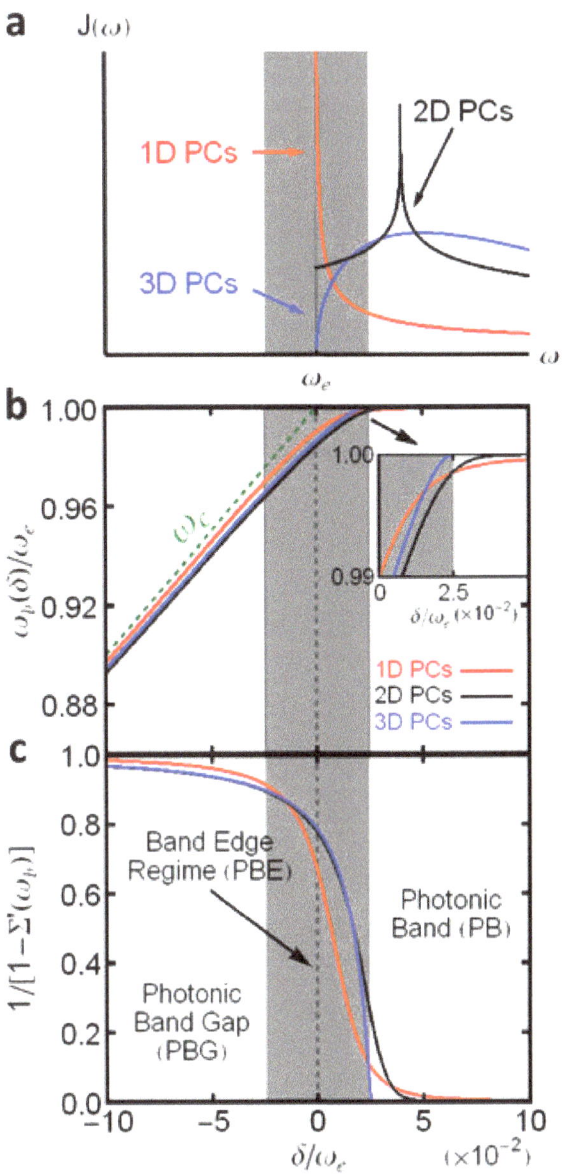


**Figure 1:** Band structures of photonic crystals and localized photon modes.

(a) Spectral densities listed in Table 1 for different DOS of 1D, 2D and 3D PCs (with different colors) are plotted respectively in the vicinity of photonic band edge $\omega_e$; (b) The corresponding localized photon mode frequency $\omega_b$ as a function of the detuning $\delta = \omega_c - \omega_e$; and (c) The corresponding localized photon mode amplitudes, given in Eq. (2). The localized photon mode shows a crossover for 1D and 2D PCs, and a critical transition for 3D PCs when the cavity frequency $\omega_c$ changes from the PBG region to the PB region. The parameters given in Table 1 take as follows: the coupling strengths for 1D, 2D and 3D PCs are $C^{2/3} = 0.01\omega_e$, $\eta = 0.001\omega_e$ and $\chi = 0.014\omega_e$, respectively. Experimentally, the typical photonic band edge frequency $\omega_e$ ranges from a few GHz to a few tens GHz for most of 1D, 2D and 3D PCs30·31, then the corresponding coupling strengths used in our calculations are ranged from a few MHz to a few hundreds MHz. The cutoff parameters $\Omega_d \approx 3.87\omega_e$ and $\Omega_C = 0.1\omega_e$; and the logarithmic divergence center $\omega_0 = 1.04\omega_e$.

Consider the cavity frequency $\omega_c$ lies not too far away from the PBE. The dissipative photon dynamics is described by the cavity field propagating Green function $u(t, t_0)$ in photonic crystals through the relation $\langle a(t)\rangle = u(t, t_0)\langle a(t_0)\rangle$, and is determined by the dissipative integro-differential equation (19). The general solution of Eq. (19) is given in Ref. 15,

$$u(t,t_0)=\frac{1}{1-\Sigma'(\omega_b)}e^{-i\omega_b(t-t_0)}+\int_{\omega_e}^{\infty}d\omega\frac{J(\omega)e^{-i\omega(t-t_0)}}{[\omega-\omega_c-\Delta(\omega)]^2+\pi^2J^2(\omega)}, \tag{2}$$

where $\Sigma'(\omega_b)=[\partial\Sigma(\omega)/\partial\omega]_{\omega=\omega_b}$ and $\Sigma(\omega)$ is the reservoir-induced cavity photon self-energy correction,

$$\Sigma(\omega)=\int_{\omega_e}^{\infty}d\omega'\frac{J(\omega')}{\omega-\omega'}. \tag{3}$$

The explicit solution of cavity photon self-energy corrections for different spectral densities is also presented in Table 1. The frequency $\omega_b$ in Eq. (2) is the localized photon mode frequency located inside the PBG ($0 < \omega_b < \omega_e$) but not too far away from the PBE, and it is determined by the pole condition: $\omega_b - \omega_c - \Delta(\omega_b) = 0$, where $\Delta(\omega)=\mathcal{P}\left[\int d\omega'\frac{J(\omega')}{\omega-\omega'}\right]$ is a principal-value integral.

Equation (2) provides a general solution of the non-Markovian dissipative photon dynamics for micro/nano cavities in various photonic crystal structures. It shows that the cavity photon dynamics in PCs always contains two parts15: a localized photon mode (the first term) plus a non-exponential photon damping (the second term). The localized photon mode is a long-lived non-Markovian effect (dissipationless), induced by the PBG structure in PCs. The corresponding frequency $\omega_b$ lies always within the PBG. The non-exponential photon damping comes from the non-analyticity of the photon self-energy correction, which is determined by the DOS profile of the PCs. This non-exponential damping is a short-time non-Markovian memory effect, and it will become an exponential (Markov) decay in the photonic band (PB) region, as we will show later. The contributions of both the localized photon mode and the non-exponential photon damping strongly rely on the detuning $\delta = \omega_c - \omega_e$.

In Fig. 1b, we plot the localized photon mode frequency as a function of the detuning $\delta$ for 1D, 2D and 3D PCs, respectively. The values of the localized mode frequencies for three different dimensional PCs are indeed very close to each other. However, a carefully check (see the inset in Fig. 1b) shows that for 3D PCs with an anisotropic DOS, the localized photon mode exists only when $\delta < \sqrt{\pi}\chi$, namely, the cavity frequency $\omega_c$ must be tuned into the PBG or near the PBE. Here $\chi$ is a coupling strength between the cavity and the PCs (see Table 1). For 1D and 2D PCs, the localized photon mode exists for any location of the cavity frequency $\omega_c$. This is because for 1D and 2D PCs with the DOS given in Table 1, the self-energy corrections are negatively divergent when $\omega \to \omega_e - 0^+$ so that the pole condition is always satisfied for the localized photon mode. However, this does not ensure that the localized photon mode in 1D and 2D PCs must have a significant contribution to the photon dynamics for the whole range of the cavity frequency.

The importance of the localized photon mode is determined by the localized photon mode amplitude, $1/[1-\Sigma'(\omega_b)]$ given in Eq. (2), which is plotted in Fig. 1c as a function of the detuning δ for all the three different dimensional PCs. It shows that for 3D PCs with an anisotropic DOS, the localized photon mode amplitude vanishes for $\delta > \sqrt{\pi}\chi$, indicating a critical transition for the occurrence of the localized mode. For 1D and 2D PCs, the localized photon mode amplitude decays to almost zero when $\omega_c$ is tuned into the PB region, as a crossover phenomenon. However, the overall effect of these localized photon modes on cavity photon dynamics has the very similar behavior for different PCs given in Table 1, as shown in Fig. 1.

The results presented in Fig. 1b–c provide indeed the full steady-state information of the cavity photon dynamics. This is because the non-exponential photon damping [i.e. the second term in Eq. (2)] will decay to zero so that only the localized photon mode contributes to the steady-state cavity photon field. In other words, the steady-state cavity field amplitude quantifies the contribution of the localized photon mode as a dissipationless effect, given by

$$|u(t\to\infty,t_0)| = 1/[1-\Sigma'(\omega_b)]. \quad (4)$$

On the other hand, the non-exponential damping term in Eq. (2) characterizes the time-dependent cavity photon damping rate κ(t) in the intermediate dissipation time region through the relation[15,16]

$$\kappa(t) = -\mathrm{Re}[\dot{u}(t,t_0)/u(t,t_0)]. \quad (5)$$

The detailed photon dynamics in terms of the cavity field amplitude $|u(t,t_0)|$ is provided in Fig. 2(a) for 1D, 2D and 3D PCs, respectively, with several different detuning δ taking from the PBG region to the PB region. The corresponding damping rate κ(t) is given in Fig. 2b, except for the case with $\delta = 0.1\omega_e$ which will be discussed later. The results show that the cavity dynamics changes dramatically when $\omega_c$ crosses over from the PBG region to the PB region. Since the range of $u(t, t_0)$ is given by $1 \geq |u(t, t_0)| \geq 0$, we define the crossover region with the condition $0.9 \gtrsim |u(t\to\infty,t_0)| \gtrsim 0.1$ which corresponds to $-0.025\omega_e \lesssim \delta \lesssim 0.025\omega_e$, as shown in Fig. 1. The results in Fig. 2b show that the damping rate κ(t) will rapidly approach to zero within the PBG ($\delta < -0.025\omega_e$) and in the vicinity of the PBE ($-0.025\omega_e \lesssim \delta \lesssim 0.025\omega_e$). In other words, the photon damping rate will approach to zero after some time, due to the existence of the localized photon dynamics. In the PB region ($\delta > 0.025\omega_e$) where either the localized mode vanishes (for 3D PCs) or becomes negligible (for 1D and 2D PCs), then the cavity photons undergo a full dissipation process, and can be approximately characterized by an exponential damping for $\delta \gg 0.025\omega_e$, as shown by the black curves in Fig. 2a for $\delta = 0.1\omega_e$. In this case, the damping rateκ(t) will oscillate rapidly in time when $u(t, t_0)$ decays to zero, see Eq. (5). This rapidly oscillating damping rate, originated from $u(t, t_0)$ approaching zero, has no physical consequence because the photon dissipation is almost completed after this point of time. Thus, alternatively we determine the damping rate κ by fitting the dashed-dot black curves in Fig. 2a with an exponential function ~exp(−κt). The resulting cavity photon damping rate for $\delta = 0.1\omega_e$ is given by $\kappa \simeq 3.16^{-3}\omega_e$, $1.24 \times 10^{-2}\omega_e$, and $1.72 \times 10^{-2}\omega_e$ for 1D, 2D and 3D PCs, respectively. Experimentally, the typical photonic band edge frequency$\omega_e$ ranges from a few GHz to a few tens GHz for most of 1D, 2D and 3D PCs fabricated in the microwave regime[30,31], this corresponds to the damping rate κ ranging from a few tens MHz to hundreds MHz when the cavity frequency $\omega_c$ lies inside the PB region.

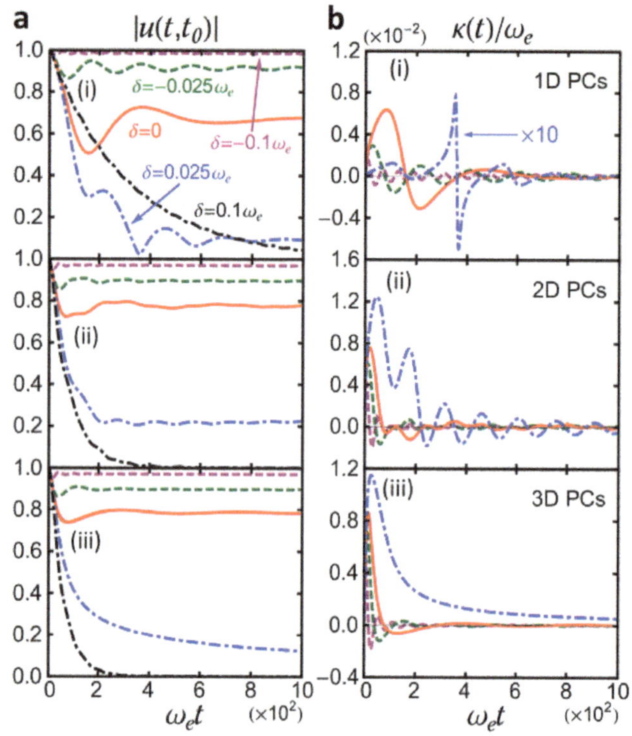


**Figure 2:** Dissipation and dissipationless photon dynamics.

(a) Dissipation and dissiptionless photon dynamics in terms of the cavity field, $\langle a(t)\rangle = u(t,t_0)\ \langle a(t_0)\rangle$ , and (b) the corresponding decay rate κ(t), are plotted for (i) 1D PCs; (ii) 2D PCs and (iii) 3D PCs, with several different detuning δ. It shows how the dissipative cavity photons becomes dissipationless when the cavity frequency moves from the PB region into the PBG region for different dimensional PCs. For $\delta = 0.1\omega_e$, the photon dynamics exhibits almost a perfect exponential decay, and by fitting the damping with an exponential function, the resulting damping rate is $\kappa = 3.16 \times 10^{-3}\omega_e$, $1.24 \times 10^{-2}\omega_e$ and $1.72 \times 10^{-2}\omega_e$ for 1D, 2D and 3D photonic crystals, respectively. Here $t_0 = 0$, and other parameters are the same as given in Fig. 1.

Combining the localized photon mode together with the photon dissipation dynamics, as shown in Fig. 1 and Fig. 2, we can see that when the cavity mode is tuned away from the PBE with $\delta > 0.025\omega_e$, the localized mode has a negligible effect on cavity photon dynamics for 1D and 2D PCs, and vanishes completely for 3D PCs. The cavity photon field decays almost exponentially for $\delta \gg 0{,}025\omega_e$, which is a Markov decay, as shown in Fig. 2a by the dashed-dot black curves with $\delta = 0.1\omega_e$. In the crossover region $-0.025\omega_e \lesssim \delta \lesssim 0.025\omega_e$, i.e. the cavity frequency is tuned into the vicinity of the PBE, the contribution of the localized mode to the photon dynamics increases rapidly, while the contribution of the nonexponential photon damping goes in an opposite way, namely the damping rate κ(t) is decreased to zero after some time (see Fig. 2b). When the cavity frequency lies inside the PBG ($\delta < -0.025\omega_e$), the cavity field has almost no damping (the damping rate κ(t) quickly decays to zero), and the photon dynamics is dominated by the localized photon mode. Thus light can be confined in the defect of the photonic crystal, providing a high-Q micro/nano cavity. In fact, this nontrivial dissipative cavity photon dynamics reproduces the same result with regard to atomic population trapping and atom-photon bound states in the vicinity of the PBG where an atom is placed in photonic crystals9·10·11·12.

The importance of the above results is that the localized photon mode provides dissipationless cavity photon dynamics when the cavity frequency $\omega_c$ lies inside the PBG or in the vicinity of the PBE. This dissipationless localized-mode contribution to the steady-state cavity photon field is universal for different dimensional PCs with different DOS given in Table 1, because it only relies on the presence of PBG. The localized photon mode decreases rather quickly but smoothly in the crossover region ($-0.025\omega_e \lesssim \delta \lesssim 0.025\omega_e$), and becomes negligible for $\delta > 0.025\omega_e$ for 1D and 2D PCs, where the cavity photon dynamics becomes dissipative. For the 3D PCs, the cavity photon dynamics shows the same dissipationless behavior before reaching the critical point $\delta_c = \sqrt{\pi}\chi \simeq 0.025\omega_e$ , where the localized photon mode dominates the photon dynamics inside the PBG ($\delta < -0.025\omega_e$) and decays rapidly near the PBE ($-0.025\omega_e \lesssim \delta \lesssim 0.025\omega_e$). When $\delta > \delta_c$, the localized photon mode vanishes for 3D PCs, and the cavity photon dynamics becomes fully dissipative. With these results, a general picture for dissipation and dissipationless photon dynamics in photonic crystals is provided.

## Thermal Photon Fluctuations

The above exact solution of dissipation and dissipationless cavity photon dynamics in photonic crystals can be used to describe thermal photon fluctuations through the fluctuated photon correlation functionv(t, t), which is fully determined by the generalized non-equilibrium fluctuation-dissipation theorem15,

$$v(t,t) = \int_{t_0}^{t} dt_1 \int_{t_0}^{t} dt_2 u^*(t_1,t_0)\tilde{g}(t_1,t_2)u(t_2,t_0). \tag{6}$$

In Eq. (6), the two-time correlation function $\tilde{g}(t_1,t_2) = \int d\omega J(\omega)\bar{n}(\omega,T)e^{-i\omega(t-t')}$ depicts photon fluctuations induced by the thermal photonic crystal, where $\bar{n}(\omega,T) = 1/[e^{\hbar\omega/k_BT} - 1]$ is the initial photon distribution in PCs at temperature T. When the cavity approaches the steady state, photon fluctuations are simply determined by the modified steady-state fluctuation-dissipation theorem, $v(t,t\rightarrow\infty) = \int_{\omega_e}^{\infty} d\omega \mathcal{V}(\omega)$ with\

$$v(t,t\rightarrow\infty) = \int_{\omega_e}^{\infty} d\omega \mathcal{V}(\omega) \text{ with}$$

$$\mathcal{V}(\omega) = \bar{n}(\omega,T)[\mathcal{D}_l(\omega) + \mathcal{D}_d(\omega)], \tag{7}$$

where $\mathcal{D}_l(\omega) = \frac{J(\omega)}{(\omega-\omega_b)^2[1-\Sigma'(\omega_b)]^2}$ and $\mathcal{D}_d = \frac{J(\omega)}{[\omega-\omega_c-\Delta(\omega)]^2 + \pi^2 J^2(\omega)}$. The first term is the localized photon mode contribution that modifies the conventional equilibrium fluctuation-dissipation theorem. This additional contribution is negligible when the cavity frequency is tuned into the PB region, i.e. $\delta > 0.025\omega_e$, where $\omega_b \rightarrow \omega_e$ so that $\mathcal{D}_l(\omega) \rightarrow 0$ (seeFig. 1b–c). Consequently, the solution (7) is reduced to the standard equilibrium fluctuation-dissipation theorem: $\mathcal{V}(\omega) = \bar{n}(\omega,T)\mathcal{D}_d(\omega)$. In the high temperature limit, it reproduces Einstein's fluctuation-dissipation relation.

Photon fluctuations are presented in Fig. 3a(i)–(iii) respectively for 1D, 2D and 3D PCs at a given temperature. It shows that photon fluctuations evolve in a similar way for different PCs. When $\omega_c$ is tuned into the PB region ($\delta > 0.025\omega_e$), photons continuously flow into the cavity from the photonic crystal until the cavity reaches its steady state. In this case, the cavity thermally equilibrates with the photonic crystal. When $\omega_c$ lies inside the PBE region ($-0.025\omega_e \lesssim \delta \lesssim 0.025\omega_e$), photons also flow into the cavity at the beginning, but some of photons are then transmitted back into the photonic crystal, due to the environment-induced memory effects and the effect induced by the localized photon mode. When $\omega_c$ is tuned into the PBG region ($\delta < -0.025\omega_e$), photons hardly flow from the photonic crystal into the cavity, because the localized photon mode dominates the cavity photon dynamics in the PBG region where the reservoir-induced dissipation is negligible, even though the thermal energy, $k_BT = \hbar\omega_e$, is much larger than the detuning energy. Thus, thermal fluctuations are suppressed significantly by the photon localization in the PBG region and also near the PBE.

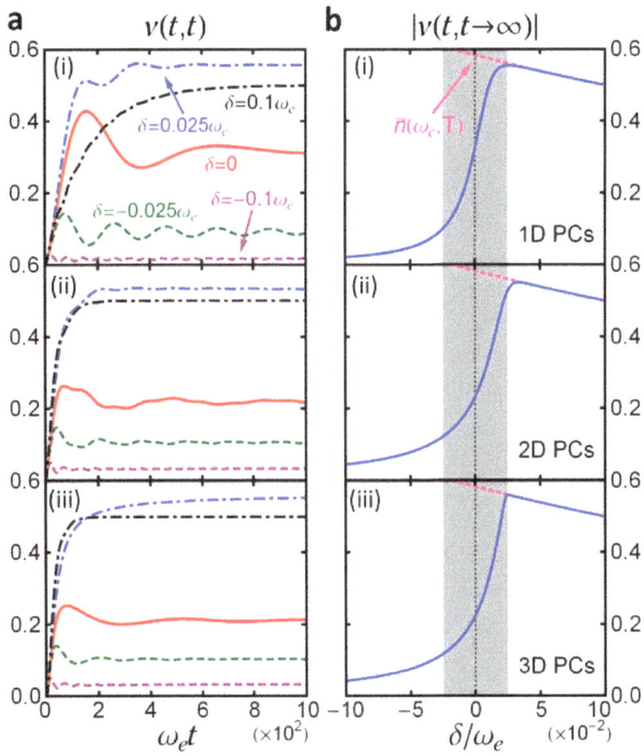


**Figure 3:** Fluctuated photon dynamics.

(a) Time-evolution of cavity photon fluctuations, in terms of the photon correlation functionv(t, t), are plotted for (i) 1D PCs, (ii) 2D PCs and (iii) 3D PCs. Different curves correspond to different detuning $\delta = \omega_c - \omega_e$, as also shown in Fig. 2a, with the photonic crystal temperature $k_BT = \hbar\omega_e$; (b) The corresponding steady-state values of photon fluctuations, given by the solid-blue curve, as a function of the detuning. The pink-dashed curve is the result of the Bose-Einstein distribution $\bar{n}(\omega_c,T)$.

The physical picture of the above photon fluctuations can be seen clearly by connecting the photon correlations with the measurable average cavity photon number (i.e. field intensity)[16,18,32]: $n(t) = \langle a^\dagger(t)a(t)\rangle = |u(t, t_0)|^2 n(t_0) + v(t, t)$, where $n(t_0)$ is the initial cavity intensity. It shows that the fluctuated photon correlation function $v(t, t)$ is just the thermal-fluctuation-induced average photon number in the cavity. In Fig. 3b, we present the steady-state photon fluctuations at a given temperature, see the solid-blue curve. The dashed-pink curve is the thermal photon distribution in PCs at the same temperature: $\bar{n}(\omega_c,T)=1/\left[e^{\hbar\omega_c/k_BT}-1\right]$. It shows that when $\delta > 0.025\omega_e$, the thermal-fluctuation-induced steady-state average photon number in the cavity is identical to the thermal photon distribution: $n(t\to\infty)=v(t,t\to\infty)=\bar{n}(\omega_c,T)$. In this regime, the initial cavity photons are totally lost into PCs as $u(t, t_0) \to 0$ in the steady-state limit, as a consequence of the Weisskopf-Winger decay. The steady-state cavity photons all come from thermal fluctuations of the reservoir. As a result, the cavity photons approach equilibrium with the reservoir and obey Bose-Einstein distribution.

However, near the PBE ($-0.025\omega_e \lesssim \delta \lesssim 0.025\omega_e$), the thermal-fluctuation-induced photon number deviates significantly from the thermal photon distribution, as shown in Fig. 3b. In other words, Bose-Einstein distribution no longer works for cavity photons in the PBE region, due to the localized photon effect. When the cavity frequency $\omega_c$ lies inside the PBG ($\delta < -0.025\omega_e$), the thermal fluctuations approach to zero, and thus the cavity is fully determined by the localized photons, even though the cavity mode equally couples to all possible modes in PCs and the thermal energy is much larger than the detuning, $k_BT \gg \delta$. Consequently, Bose-Einstein distribution is broken down. This provides a nontrivial phenomenon of the thermal fluctuation dynamics from equilibrium to nonequilibrium steady states. Here nonequilibrium steady states are defined by these steady states which can memorize the initial state information of the system so that the equilibrium hypothesis cannot be satisfied. Such steady states will be given explicitly in the next subsections.

It is recently shown that non-Markovian dynamics in open systems can induce a critical transition from equilibrium to nonequilibrium steady states, when a localized mode occurs[15,22,33]. Such a critical transition appears for the 3D PCs with an anisotropic DOS, where there is a critical detuning for the occurrence of localized photon mode, $\delta_c = \sqrt{\pi}\chi$. However, for the 1D and 2D PCs, our result shows that there is no such a critical transition due to the fact that the localized photon mode always exists. The photon fluctuations then undergoes a crossover from equilibrium to nonequilibrium steady states through the change of the detuning $\delta$. No matter it is a crossover or a critical transition, the results in Fig. 3b shows that thermal photon fluctuations cannot make the cavities embedded in PCs to approach equilibrium with the PCs when the cavity frequency lies inside the PBG or near the PBE. This is a general feature for cavity photons in PCs.

## Breakdown of Bose-Einstein distribution through the time-evolution of photonic Fock States

Based on the exact cavity photon dissipation and fluctuation dynamics in PCs, we shall explicitly examine the breakdown of Bose-Einstein distribution through the evolution of cavity photonic states, and also the crossover from equilibrium to nonequilibrium steady states. This must be done by solving the exact master equation (15) given in the Methods. To be more specific, we consider first the cavity to be initially in a Fock state with an arbitrary photon number $n_0$, i.e. $\rho(t_0) = |n_0\rangle\langle n_0|$, which may be prepared experimentally through the real-time quantum feedback control[34]. By solving the master equation (15), the cavity state at arbitrary time t is given by

$$\rho(t) = \sum_{n=0}^{\infty} \mathcal{P}_n^{(n_0)}(t)|n\rangle\langle n|, \tag{8}$$

$$\mathcal{P}_n^{(n_0)}(t) = \frac{[v(t,t)]^n}{[1+v(t,t)]^{n+1}}[1-\Omega(t)]^{n_0} \times \sum_{k=0}^{\min\{n_0,n\}} \binom{n_0}{k}\binom{n}{k}\left[\frac{1}{v(t,t)}\frac{\Omega(t)}{1-\Omega(t)}\right]^k, \tag{9}$$

where $\Omega(t) = \frac{|u(t,t_0)|^2}{1+v(t,t)}$. This result shows that an initial photon Fock state will evolve into a mixed state of different Fock states, and the probability in each Fock state $|n\rangle$ is $\mathcal{P}_n^{(n_0)}(t)$.

As we have shown, the breakdown of Bose-Einstein distribution relies on the dissipationless photon dynamics. And different dimensional PCs show almost the same dissipationless photon dynamics, determined by the localized photon mode due to the existence of the PBG. To be specific, we will present in the following the numerical solution of the state evolution for 1D PCs. The 2D and 3D PCs must provide a similar solution, based on the universal dissipation and fluctuation photon dynamics provided by Eqs. (2) and (6), and also the explicit numerical results presented in Fig. 2 and Fig. 3. The time-evolution of the cavity photon distribution $\mathcal{P}_n^{(n_0)}(t)$ for the initial state $|n_0 = 5\rangle$ is given in Fig. 4. The steady-state limit, $\mathcal{P}_n^{(n_0)}(t\to\infty)$, is shown in Fig. 5, where several different initial states have been considered to demonstrate the initial-state dependence of the steady photon states.

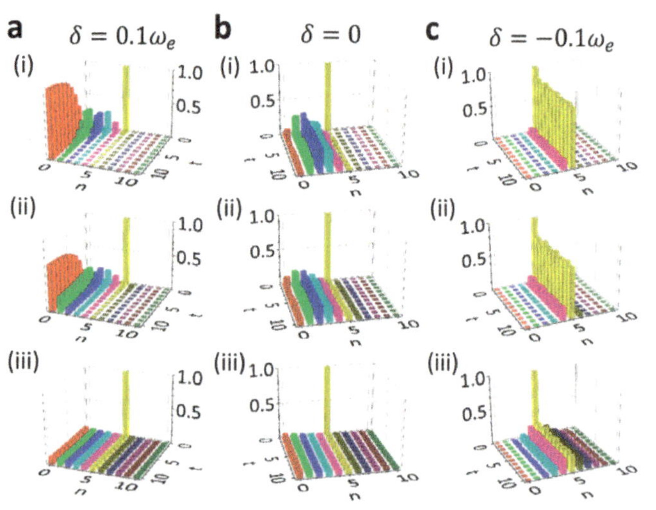


**Figure 4:** Time-evolution of Fock states.

The time-evolution of the probability distribution, $\mathcal{P}_n^{(n_0)}(t)$ with different detuning $\delta$: (a) $\delta = -0.1\omega_e$; (b) $\delta = 0$; and (c) $\delta = 0.1\omega_e$, for photonic crystals at different temperature T: (i) $k_BT=0.2\hbar\omega_e$; (ii) $k_BT=\hbar\omega_e$; and (iii) $k_BT=10\hbar\omega_e$. The cavity is initially prepared in the Fock state $|n_0 = 5\rangle$. The time t is scaled by $10^2\omega_e$.

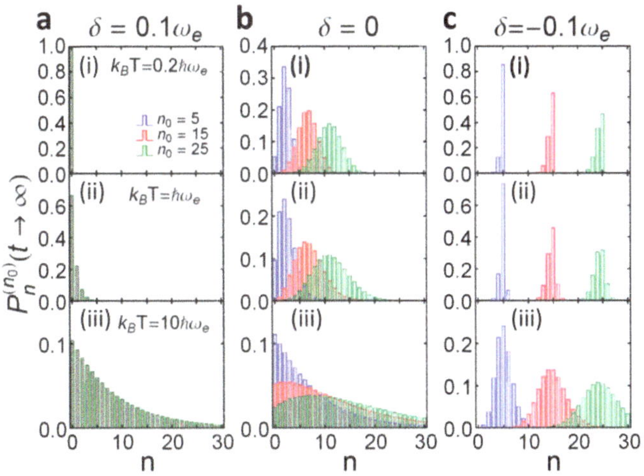


**Figure 5:** Steady-state photon distributions I.

The steady-state cavity photon distribution, $P_n^{(n_0)}(t\to\infty)$ for different initial Fock states $|n_0\rangle$, $n_0 = 5$, 15 and 25 (in terms of different colors); with different detuning δ: (a) $\delta = 0.1\omega_e$, (b) δ = 0, and (c) $\delta = -0.1\omega_e$; and different temperatures T of the photonic crystals: (i) $k_BT = 0.2\hbar\omega_e$, (ii) $k_BT = \hbar\omega_e$, and (iii) $k_BT = 10\hbar\omega_e$, as given in the figure.

The results show that if we tune $\omega_c$ into the PB region (e.g. $\delta = 0.1\omega_e$), then $u(t, t_0)$ will gradually decay to zero. This indicates that photons in the cavity will gradually be damped into the photonic crystal, and photons in the photonic crystal are transferred into the cavity through thermal fluctuations. In this case, the contribution of the localized photon mode is negligible. The cavity state ultimately reaches thermal equilibrium with the photonic crystal, and Bose-Einstein statistical distribution is produced,

$$\mathcal{P}_n^{(n_0)}(t\to\infty) = \frac{[\bar{n}(\omega_c,T)]^n}{[1+\bar{n}(\omega_c,T)]^{n+1}}, \tag{10}$$

where the initial state information is completely washed out, and the cavity photon steady state is independent of the initial states, as shown in Fig. 4a and Fig. 5a.

Near the PBE (e.g. δ = 0), the cavity still loses photons into the photonic crystal and gains photons from the photonic crystal through thermal fluctuations. At a low temperature ($k_BT = 0.2\hbar\omega_e$), the photon loss and photon gain make the cavity become a mixed state of several Fock states $|n\rangle$ only for $n < n_0$, and mainly distributed around $n = n_0/2$, seeFig. 4b–(i) and also the steady-state limit by Fig. 5b–(i) where the initial state dependence is manifested. The photon distribution deviates significantly from the standard Bose-Einstein distribution. At a relatively high temperature ($k_BT = \hbar\omega_e$), the cavity state also becomes a mixed state of several Fock states $|n\rangle$, distributed mainly among $n \leq n_0/2$, but the distribution is broader, see Fig. 4b–(ii) and also the steady-state limit in Fig. 5b–(ii). In this case, the initial-state dependence of the steady state is still shown up and the derivation of the photon distribution from the standard Bose-Einstein distribution is obvious. As the temperature becomes rather high ($k_BT = 10\hbar\omega_e$), the time-evolution of the cavity state behaves quite differently because thermal fluctuations play a more important role now, thus the structure of the initial state is quickly destroyed, as shown in Fig. 4b–(iii). However, the cavity does not really approach thermal equilibrium with the photonic crystal, because $\Omega(t) = \frac{|u(t,t_0)|^2}{1+v(t,t)}$ is small but not negligible so that the distribution given by Eq. (9) is still different from the thermal state distribution. The cavity steady states for different initial states do not differ from each other as much as these in the low temperature cases, but the initial-state dependence is still observable, as shown inFig. 5b–(iii).

When the cavity frequency $\omega_c$ lies inside the PBG, e.g. $\delta = -0.1\omega_e$ given inFig. 4c and Fig. 5c, the steady-state cavity photon distribution strongly depends on the initial state. In particular, if the photonic crystal temperature is not too high ( $k_BT = 0.2\hbar\omega_e$ and $k_BT = \hbar\omega_e$), the cavity remains in its initial Fock state as a photon localization state. It only has a small chance to decay to the Fock state $|n\rangle$ with $n < n_0$ [see Fig. 4c(i) with $k_BT = 0.2\hbar\omega_e$], and an even smaller probability to be in the Fock state $|n\rangle$ with $n > n_0$ [see Fig. 4c(ii) when $k_BT$ increases to $\hbar\omega_e$]. The steady-state cavity photons are distributed mainly in the regime $n < n_0$with the maximum peak at $n = n_0$, see Fig. 5c(i)–(ii). However, at a higher temperature $k_BT = 10\hbar\omega_e$, the thermal fluctuation becomes strong such that some of the initial state information will be lost and the cavity evolves into a mixed state covering several Fock states around the initial one $|n_0\rangle$ [see Fig. 4c(iii)]. The steady-state cavity photons are distributed more broadly, but are still centered around the initial photon number $n_0$, as shown in Fig. 5c(iii). This is because the localized photon mode dominates the photon dynamics over thermal photon fluctuations. The overall initial-state dependence of the steady states shown in Fig. 5b–c indicates that the equilibrium hypothesis in statistical mechanics is no longer obeyed and Bose-Einstein statistical distribution is completely broken down, even though the initial thermal energy of the photonic crystals is larger or much larger than the detuning energy.

## Breakdown of Bose-Einstein Distribution Through the Time-Evolution of Coherent States

Because Fock sates are highly non-classical and may not be easy to prepare in experiments, here we examine the case the cavity is initially in a coherent state $|\alpha_0\rangle$. By solving the master equation (15), the cavity state at an arbitrary later time t is given by

$$\rho(t) = \mathcal{D}[\alpha(t)]\rho_T[v(t,t)]\mathcal{D}^{-1}[\alpha(t)], \tag{11}$$

where $\mathcal{D}[\alpha(t)] = \exp\{\alpha(t)a^\dagger - \alpha^*(t)a\}$ is a displacement operator with $\alpha(t) = u(t, t_0)\alpha_0$, and

$$\rho_T[v(t,t)] = \sum_{n=0}^{\infty} \frac{[v(t,t)^n]}{[1+v(t,t)]^{n+1}}|n\rangle\langle n| \tag{12}$$

is a thermal-like state with average particle number v(t, t). Equation (11) shows that the initial cavity state will evolve into a displaced thermal-like state35·36 which is the mixture of displaced number states $\mathcal{D}[\alpha(t)]|n\rangle$37. In the photon number representation, Eq. (11) can be written as

$$\langle m|\rho(t)|n\rangle = e^{-\Omega(t)|\alpha_0|^2} \frac{[\alpha(t)]^m[\alpha^*(t)]^n}{[1+v(t,t)]^{m+n+1}} \times \sum_{k=0}^{\min\{m,n\}} \frac{\sqrt{m!n!}}{(m-k)!(n-k)!k!}\left[\frac{v(t,t)}{\Omega(t)|\alpha_0|^2}\right]^k, \quad (13)$$

where the probability of the cavity containing n photons (i.e. cavity photon distribution) is given by the diagonal terms of Eq. (13).

The cavity photon distribution in the steady-state limit is presented inFig. 6. It shows again that when the cavity frequency $\omega_c$ lies inside the PB region ($\delta = 0.1\omega_e$), the cavity state (11) will evolve into the thermal state $\rho_T[\bar{n}(\omega_c,T)]$, which indicates that the cavity photon number obeys Bose-Einstein distribution, see Fig. 6a. If the cavity mode is tuned at the PBE ($\delta = 0$), the cavity steady state will depend on its initial state, as shown in Fig. 6b. For low temperature (e.g. $k_BT=0.2\hbar\omega_e$), the thermal fluctuation is suppressed ($v(t, t) \to 0$), and the cavity will evolve into the coherent state $|\alpha(t)\rangle$, see Eq. (11). In this case, the cavity photon distribution in the steady state is Poissonian-like, with average photon number $\langle n\rangle|_{t\to\infty} = |\alpha(t)|^2 \propto |\alpha_0|^2$, see Fig. 6b–(i). As the temperature goes higher ($k_BT=\hbar\omega_e$), the cavity photon distribution is deviated away from the Poisson distribution due to thermal fluctuations, see Fig. 6b–(ii). At high temperature ($k_BT=10\hbar\omega_e$), since thermal fluctuation plays a more important role, the initial-state dependence is not as significant as that in the low temperature, and the cavity photon distribution is not Poissonian-like, but still deviates from Bose-Einstein distribution, seeFig. 6b–(iii). When the cavity frequency lies inside the PBG ($\delta = -0.1\omega_e$), the cavity steady state strongly depends on its initial state. The photon distribution in the cavity remains almost the same as in its initial state when the temperature is not very high ($k_BT=0.2\hbar\omega_e$ and $k_BT=\hbar\omega_e$), see the first two graphs in Fig. 6c. The initial-state dependence of the photon distribution is still strong when the temperature gets higher ($k_BT=10\hbar\omega_e$), see Fig. 6c–(iii). The photon distribution in this case is slightly broadened from the initial states due to thermal fluctuations at high temperature, but still centered around $n \approx |\alpha_0|^2$, as a manifestation of photon localization. These results obtained for different initial coherent states at different temperature behave indeed in the same way as the results obtained for different initial Fock states under the same temperature.

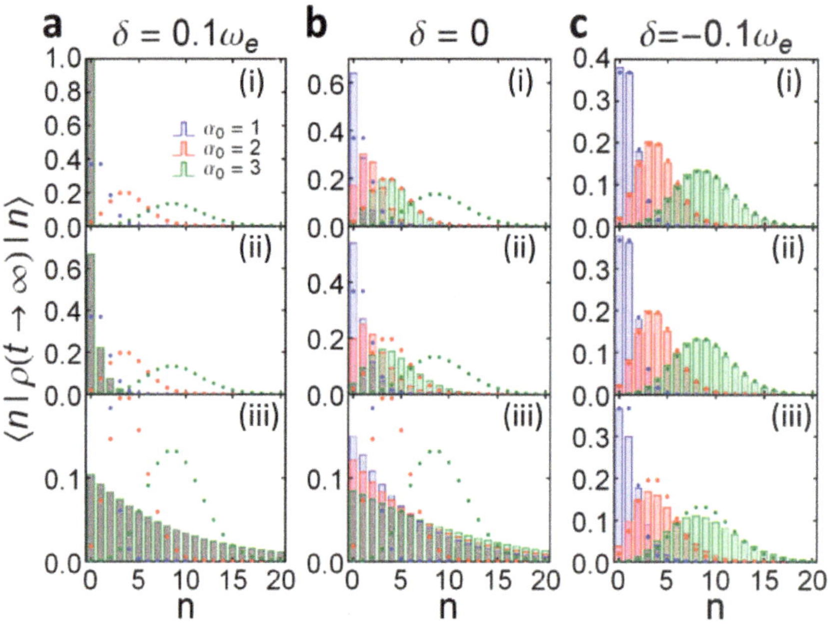


**Figure 6:** Steady-state photon distributions II.

The steady-state cavity photon distribution (colored bars), $\langle n|\,\rho\,(t \to \infty)\,|n\rangle$ for different initial coherent states $|\alpha_0\rangle$ (shown by the dotted curves with different colors); with different detuning δ: (a) $\delta = 0.1\omega_e$, (b) $\delta = 0$, and (c) $\delta = -0.1\omega_e$ at different temperatures of photonic crystals: (i) $k_BT=0.2\hbar\omega_e$, (ii) $k_BT=\hbar\omega_e$, and (iii) $k_BT=10\hbar\omega_e$

## DISCUSSION

The breakdown of Bose-Einstein distribution in PCs, explored in details through the time-evolution of various cavity photon states for 1D PCs, is also valid for 2D and 3D PCs with different DOS listed in Table 1. This is because the time evolution of the cavity photon states, solved analytically by Eqs. (8) and (11), is fully determined by the photon dissipation and fluctuation dynamics of Eqs. (2) and (6). The steady-state behaviors of Eqs. (2) and (6) for different PCs are almost the same, as shown in Fig. 2 and Fig. 3. There is only a slight difference on the photon dissipation and fluctuation dynamics between 1D and 2D PCs with 3D PCs. For 1D and 2D PCs, varying the detuning leads to a crossover from equilibrium to nonequilibrium steady states for the photon dynamics, while for 3D PCs with an anisotropic DOS, it gives a critical transition, rather than a crossover, due to the existence of a critical detuning for the existence of the localized mode, see explicitly the results presented in Fig. 1 to Fig. 3. Such a critical transition is also found recently in other open systems22·33 but not for the crossover phenomenon. Thus, photon dissipation and fluctuation dynamics investigated in this work reveal a new nontrivial property, i.e. a crossover phenomenon for photon dynamics in low-dimensional photonic crystals.

In conclusion, we show that when the cavity frequency lies inside the PBG or near the PBE in photonic crystals, Bose-Einstein statistical distribution is broken down for cavity photons. This conclusion is generally valid for various photonic band gap structures in PCs. For the 1D and 2D PCs, the breakdown of Bose-Einstein distribution leads to a crossover from equilibrium to nonequilibrium cavity steady states, while for 3D PCs with an anisotropic DOS, the breakdown of Bose-Einstein distribution corresponds to a critical transition rather than a crossover. No matter whether it is a crossover or a critical transition, the breakdown of Bose-Einstein distribution is a consequence of localization photons due to the presence of PBG structures in PCs. Therefore the conclusion is also valid for other nanomaterials with band gap structures. It could provide a hitherto unexplored challenge on photon statistics. Furthermore, this non-trivial photon dynamics can be examined via quantum non-demolition measurement[38,39], by sending sequences of circular Rydberg atoms through the photonic crystal micro/nano cavity, which carry information without destroying the cavity photon state. In particular, the cavity photon state can be measured using an experimental setup similar to that given in Ref. 38. Such experiments could be done with microcavities in low-dimensional photonic crystals in the microwave regime.

## METHODS

To investigate photon dynamics of micro/nano cavities which are coupled each other through waveguides embedded in PCs at finite temperature, we treat both the PCs and waveguides as reservoirs of the cavities. Thus the entire system of the micro/nano cavities (defects) embedded in photonic crystals can be described by the Fano-Anderson model (a model of impurity electrons coupled with continuous states introduced by Anderson[40] in solid-state physics, and discrete states embedded in a continuum proposed by Fano[41] in atomic spectra). The corresponding Fano-Anderson Hamiltonian is given by Refs.[16,40,41,42,43]:

$$H = \sum \omega_{ci} a_i^\dagger a_i + \sum_k \omega_k b_k^\dagger b_k + \sum_{ik} \left( V_{ik} a^\dagger b_k + V_{ik}^* b_k^\dagger a_i \right), \tag{14}$$

where $a_i \left(a_i^\dagger\right)$ is the annihilation (creation) operator of the micro/nano cavity modes (defects), and $b_k \left(b_k^\dagger\right)$ the annihilation (creation) operator of the photonic modes of photonic crystals (continuum). The coefficients $V_{ik}$ are tunneling amplitudes of photons between the micro/nano cavities and the photonic crystals. The Hamiltonian (14) has also the same form as the one for cavity photon loss in open space[20], by replacing the photonic crystals with an open space.

Consider initially the photonic crystals are in equilibrium state. Integrating out completely the reservoir degrees of freedom of photonic crystals via the influence functional[44,45] in the coherent state representation[46], arbitrary photon states for micro/nano cavities in PCs are then governed by the following exact master equation[15,16,17,18,32,43,47,48]

$$\begin{aligned} \dot{\rho}(t) = & -i\left[H_c'(t), \rho(t)\right] + \sum_{ij} \Big\{ \kappa_{ij}(t) \Big[ 2a_j \rho(t) a_i^\dagger \\ & - a_i^\dagger a_j \rho(t) - \rho(t) a_i^\dagger a_j \Big] + \tilde{\kappa}_{ij}(t) \Big[ a_i^\dagger \rho(t) a_j \\ & + a_j \rho(t) a_i^\dagger - a_i^\dagger a_j \rho(t) - \rho(t) a_j a_i^\dagger \Big] \Big\}. \end{aligned} \tag{15}$$

Here ρ (t) is the reduced density matrix for cavity states, $H'_c(t) = \sum_{i,j} \omega'_{cij}(t) a_i^\dagger a_j$ is the renormalized Hamiltonian of cavities with the environment-modified cavity frequencies $\omega'_{cii}(t) \equiv \omega'_{ci}(t)$ and the environment-induced couplings between different cavities $\omega'_{cij}(t)$, after the environmental degrees of freedom are completely integrated out. The coefficients $\kappa_{ij}$ (t) and $\tilde{\kappa}_{ij}(t)$ characterize photon dissipations and fluctuations in PCs at finite temperature. The renormalized frequency, $\omega'_{cij}$, and the time-dependent dissipation and fluctuation coefficients, $\kappa_{ij}$(t) and $\tilde{\kappa}_{ij}(t)$, can be exactly and non-perturbatively determined by the dissipation-fluctuation Dyson equation, as shown explicitly in our early works[16,17], and will also be given later.

It might be worth pointing out that the first master equation derived from the original Feynman-Vernon influence functional[44] was obtained by Caldeira and Leggett for the Brownian motion with a high temperature environment[45] which is also called the Caldeira-Leggett master equation in the literature. For the Brownian motion, the system-reservoir coupling Hamiltonian is given by $H_I = \sum_k \lambda_k x q_k$, where x and $q_k$ are the positions of the principal harmonic oscillator (as the Brownian particle) and all other harmonic oscillators in the reservoir, respectively. In terms of the second quantization, this system-reservoir coupling Hamiltonian can be rewritten as $H_I = \sum_k V_k \left(a^\dagger b_k + b_k^\dagger a + a^\dagger b_k^\dagger + a b_k\right)$ which shows that the amplitudes for particle tunneling processes between the system and the reservoir (the first two terms) is equal to the amplitudes of particle pair production and annihilation processes (the last two terms), both are given by $V_k$. However, in quantum optics, it is well-known that the photon tunneling amplitudes cannot be the same as the amplitudes of two-photon pair production and annihilation processes. As a result, the Caldeira-Leggett model is not applicable to photonics.

To make the cavity photon dynamics in thermal photonic crystals more specific, we consider a single-mode micro/nano cavity in photonic crystals. The photon dissipation and fluctuations, characterized by the dissipation and fluctuation coefficients κ (t) and $\tilde{\kappa}(t)$ (all the sub-indices (i, j) in Eq. (15) can be dropped now), are determined non-perturbatively and exactly by the nonequilibrium Green functions[49,50] through the relations[15,16,17,18]:

$$\omega'_c(t) = -\mathrm{Im}\left[\dot{u}(t,t_0)/u(t,t_0)\right], \tag{16}$$

$$\kappa(t) = -\mathrm{Re}\left[\dot{u}(t,t_0)/u(t,t_0)\right], \tag{17}$$

$$\tilde{\kappa}(t) = \dot{v}(t,t) + 2v(t,t)\kappa(t).) \tag{18}$$

Here u (t, $t_0$) is the cavity photon field propagating Green's function describing the photon field relaxation, and v (t, t) characterizes the reservoir-induced photon thermal fluctuations. These two Green functions, u (t, $t_0$) and v (t, t), are determined by the following integrodifferential dissipation equation and the nonequilibrium fluctuation-dissipation theorem, respectively15·16,

$$\dot{u}(t,t_0) = -i\omega_c u(t,t_0) - \int_{t_0}^{t} dt' g(t-t') u(t',t_0), \quad (19)$$

$$v(t,t) = \int_{t_0}^{t} dt_1 \int_{t_0}^{t} dt_2 u^*(t_1,t_0) \tilde{g}(t_1-t_2) u(t_2,t_0), \quad (20)$$

where $\omega_c$ is the original cavity frequency. The integral kernels in Eqs. (19)–(20) characterize all the back-actions between the cavity and photonic crystals, and can be determined uniquely by the spectral density J (ε) of photonic crystals through the relations: $g(t-t') = \int d\omega J(\omega) e^{-i\omega(t-t')}$, and $\tilde{g}(t-t') = \int d\omega J(\omega) \bar{n}(\omega,T) e^{-i\omega(t-t')}$, where $\bar{n}(\omega,T) = 1/[e^{\hbar\omega/k_B T} - 1]$ is the initial photon distribution in PCs at temperature T. The spectral density J(ω) is microscopically defined as a multiplication of the density of states (DOS) $\varrho(\omega)$ of PCs with the photon scattering amplitudes $V_k$ between the cavity and PCs,

$$J(\omega) = \sum_k |V_k|^2 \delta(\omega - \omega_k) = \varrho(\omega) |V(\omega)|^2. \quad (21)$$

In the second equality we have taken the continuous photonic modes of PCs so that $V_k \rightarrow V(\omega)$, and the index i of $V_{ik}$ in Eq. (14) has also been dropped for single-mode cavity. For arbitrary spectral density J(ω), the general solution of Eq. (19) can be obtained exactly15, which is given by Eq. (2). By solving Eqs. (16)–(18) and (15) through Eqs. (19)–(20), the complete solution of photon dissipation and fluctuations can be obtained, as we presented in details in the paper.

## REFERENCES


1. Joannopoulos, J. D., Johnson, S. G., Winn, J. N. & Meade, R. D. Photonic Crystals: Modeling the Flow of Light (Princeton, New York, 2008).
2. Busch, K., von Freymann, G., Linden, S., Mingaleev, S. F., Tkeshelashvili, L. & Wegener, M. Periodic nanostructures for photonics. Phys. Rep. 444, 101 (2007).
3. Notomi, M. Manipulating light with strongly modulated photonic crystals. Rep. Prog. Phys. 73, 096501 (2010).
4. Yablonovitch, E. Inhibited Spontaneous Emission in Solid-State Physics and Electronics. Phys. Rev. Lett. 58, 2059 (1987).
5. John, S. Strong localization of photons in certain disordered dielectric superlattices. Phys. Rev. Lett. 58, 2486 (1987).
6. Yablonovitch, E., Gmitter, T. J. & Leung, K. M. Photonic band structure: The facecentered-cubic case employing nonspherical atoms. Phys. Rev. Lett. 67, 2295 (1991).
7. Armani, D. K., Kippenberg, T. J., Spillane, S. M. & Vahala, K. J. Ultra-high-Q toroid microcavity on a chip. Nature 421, 925 (2003).
8. Joannopoulos, J. D., Villeneuve, P. R. & Fan, S. Photnic crystals: putting a new twist on light. Nature 386, 143 (1997).
9. Lambropoulos, P., Nikolopoulos, G. M., Nielsen, T. R. & Bay, S. Fundamental quantum optics in structured reservoirs. Rep. Prog. Phys. 63, 455 (2000).
10. John, S. & Wang, J. Quantum Electrodynamics near a Photonic Band Gap: Photon Bound States and Dressed Atoms. Phys. Rev. Lett. 64, 2418 (1990).
11. John, S. & Wang, J. Quantum optics of localized light in a photonic band gap. Phys. Rev. B 43, 12772 (1991).
12. Kofman, A. G., Kurizki, G. & Sherman, B. Spontaneous and Induced Atomic Decay in Photonic Band Structures. J. Mod. Opt. 41, 353 (1994).
13. Zhu, S. Y., Yang, Y., Chen, H., Zheng, H. & Zubairy, M. S. Spontaneous Radiation and Lamb Shift in Three-Dimensional Photonic Crystals. Phys. Rev. Lett. 84, 2136 (2000).
14. Wang, X. H., Gu, B. Y., Wang, R. & Xu, H. Q. Decay kinetic properties of atoms in photonic crystals with absolute gaps. Phys. Rev. Lett. 91, 113904 (2003).
15. Zhang, W. M., Lo, P. Y., Xiong, H. N., Tu, M. W. Y. & Nori, F. General NonMarkovian Dynamics of Open Quantum Systems. Phys. Rev. Lett. 109, 170402 (2012).
16. Lei, C. U. & Zhang, W. M. A quantum photonic dissipative transport theory. Ann. Phys. 327, 1408 (2012).
17. Tu, M. W. Y. & Zhang, W. M. Non-Markovian decoherence theory for a doubledot charge qubit. Phys. Rev. B 78, 235311 (2008).
18. Jin, J., Tu, M. W. Y., Zhang, W. M. & Yan, Y. J. Non-equilibrium quantum theory for nanodevices based on the Feynman-Vernon influence functional. New. J. Phys. 12, 083013 (2010).
19. Brune, M.et al. Process Tomography of Field Damping and Measurement of Fock State Lifetimes by Quantum Nondemolition Photon Counting in a Cavity. Phys. Rev. Lett. 101, 240402 (2008).
20. Carmichael, H. J. An Open Systems Approach to Quantum Optics, Lecture Notes in Physics Vol. m18, (Springer-Verlag, Berlin, 1993).
21. Scully, M. O. & Zubairy, M. S. Quantum Optics (Cambridge University press, 1997).
22. Xiong, H. N., Lo, P. Y., Zhang, W. M., Nori, F. & Feng, D. H. Non-Markovian Dynamics Impact on the Foundations of Statistical Mechanics. arXiv: 1311.1282 (2013) (unpublished).
23. Noda, S. & Baba, T. Roadmap on photonic crystals (Dordrecht: Kluwer Academic Publishers Groups, 2003).
24. Sakoda, K. Optical properties of photonic crystals (Berlin, Springer, 2005).
25. Li, Z. Y., Lin, L. L. & Zhang, Z. Q. Spontaneous Emission from Photonic Crystals: Full Vectorial Calculations. Phys. Rev. Lett. 84, 4341 (2000).
26. Li, Z. Y. & Xia, Y. Full vectorial model for quantum optics in three-dimensional photonic crystals. Phys. Rev. A 63, 043817 (2001).
27. McPhedran, R. C., Botten, L. C., McOrist, J., Asatryan,

A. A., de Sterke, C. M. & Nicorovici, N. A. Density of states functions for photonic crystals. Phys. Rev. E 69, 016609 (2004).

28. Birner, A., Wehrspohn, R. B., Go¨sele, U. M. & Busch, K. Silicon-Based Photonic Crystals. Adv. Mater. 13, 377 (2001).
29. Ibanescu, M., Reed, E. J. & Joannopoulos, J. D. Enhanced Photonic Band-Gap Confinement via Van Hove Saddle Point Singularities. Phys. Rev. Lett. 96, 033904 (2006).
30. Smith, D. R., McCall, S. L., Platzman, P. M., Dalichaoich, R., Kroll, N. & Schultz, S. Photonic band structure and defects in one and two dimensions. J. Opt. Soc. Am. B 10, 314 (1993).
31. O¨ zbay, E. et al. Measurement of a three-dimensional photonic band gap in a crystal structure made of dielectric rods. Phys. Rev. B 50, 1945 (1994).
32. Xiong, H. N., Zhang, W. M., Wang, X. & Wu, M. H. Exact non-Markovian cavity dynamics strongly coupled to a reservoir. Phys. Rev. A 82, 012105 (2010).
33. Cai, C. Y., Yang, L. P. & Sun, C. P. Threshold for nonthermal stabilization of open quantum systems. Phys. Rev. A 89, 012128 (2014).
34. Sayrin, C. et al. Real-time quantum feedback prepares and stabilizes photon number states. Nature 477, 73 (2011).
35. Lachs, G. Theoretical Aspects of Mixtures of Thermal and Coherent Radiation. Phys. Rev. 138, B1012 (1965).
36. Marian, P. & Marian, T. A. Squeezed states with thermal noise. I. Photon-number statistics. Phys. Rev. A 47, 4474 (1993).
37. de Oliveira, F. A. M., kim, M. S., Knight, P. L. & Buzˇek, V. Properties of displaced number states. Phys. Rev. A 41, 2645 (1990).
38. Gleyzes, S.et al. Quantum jumps of light recording the birth and death of a photon in a cavity. Nature 446, 297 (2007).
39. Guerlin, C. et al. Progressive field-state collapse and quantum non-demolition photon counting. Nature 448, 889 (2007).
40. Anderson, P. W. Localized magnetic states in metals. Phys. Rev. 124, 41 (1961).
41. Fano, U. Effects of Configuration interaction on intensities and phase shift. Phys. Rev. 124, 1866 (1961).
42. Mahan, G. D. Many-Body Physics 3rd Ed. (Kluwer Academic/Plenum Publishers, New Yoek, 2000).
43. 43.Wu, M. H., Lei, C. U., Zhang, W. M. & Xiong, H. N. Non-Markovian dynamics of a microcavity coupled to a waveguide in photonic crystals. Opt. Express 18, 18407 (2010).
44. Feynman, R. P. & Vernon, F. L. The theory of a general quantum system interacting with a linear dissipative system. Ann. Phys. 24, 118 (1963).
45. Caldeira, A. O. & Leggett, A. J. Path integral approach to quantum Brownian motion. Physica 121A, 587 (1983).
46. Zhang, W. M., Feng, D. H. & Gilmore, R. Coherent states: theory and some applications. Rev. Mod. Phys. 62, 867 (1990).
47. Tan, H. T. & Zhang, W. M. Non-Markovian dynamics of an open quantum system with initial system-reservoir correlations: A nanocavity coupled to a coupledresonator optical waveguide. Phys. Rev. A 83, 032102 (2011).
48. Tang, L. P., Cai, C. Y., Xu, D. Z., Zhang, W. M. & Sun, C. P. Master equation and dispersive probing of a non-Markovian process. Phys. Rev. A 87, 012110 (2013).
49. Schwinger, J. Brownian Motion of a Quantum Oscillator. J. Math. Phys. 2, 407 (1961).
50. Keldysh, L. V. Diagram Technique for Nonequilibrium Processes. Sov. Phys. JETP 20, 1018 (1965).

CHAPTER 9

# PHASE DIAGRAM FOR THE TRANSITION FROM PHOTONIC CRYSTALS TO DIELECTRIC METAMATERIALS

Mikhail V. Rybin[1,2], Dmitry S. Filonov[2], Kirill B. Samusev[1,2], Pavel A. Belov[2], Yuri S. Kivshar[2,3] & Mikhail F. Limonov[1,2]

[1] Ioffe Institute, St Petersburg 194021, Russia

[2]Department of Nanophotonics and Metamaterials, ITMO University, St Petersburg 197101, Russia

[3]Nonlinear Physics Center and the ARC Center of Excellence CUDOS, Australian National University, Canberra Australian Capital Territory 0200, Australia

## ABSTRACT

**Photonic crystals and dielectric metamaterials represent two different classes of artificial media but are often composed of similar structural elements. The question is how to distinguish these two types of periodic structures when their parameters, such as permittivity and lattice constant, vary continuously. Here we discuss transition between photonic crystals and dielectric metamaterials and introduce the concept of a phase diagram, based on the physics of Mie and Bragg resonances. We show that a periodic photonic structure transforms into a metamaterial when the Mie gap opens up below the lowest Bragg bandgap where the homogenization approach can be justified and the effective permeability becomes negative. Our theoretical approach is confirmed by microwave experiments for a metacrystal composed of tubes filled with heated water. This analysis yields deep insight into the properties of periodic structures, and provides a useful tool for designing different classes of electromagnetic materials with variable parameters.**

## INTRODUCTION

The concepts of phase transitions and phase diagrams are among the most fundamental concepts in physics. Here we reveal and analyse a rather unusual type of phase transitions between two classes of artificial electromagnetic structures: photonic crystals (PhCs)[1, 2] and metamaterials (MMs)[3, 4]. Rapidly growing attention to the study of PhCs and MMs is due to their unique electromagnetic properties and the exciting new physics these structures may offer for novel applications in photonics[5, 6, 7, 8, 9]. The resonant properties of PhCs are driven by the familiar Bragg scattering from a periodic structure when the lattice constant becomes comparable with the wavelength of light. MMs are regarded as effective media with averaged parameters which in turn can be engineered by changing the properties of their subwavelength elements modifying the propagation of electromagnetic waves with the wavelengths much larger than the lattice constant.

Many properties of conventional MMs are usually defined by the optically induced magnetic response of metallic wires or split-ring resonators, which has been demonstrated experimentally in many studies for both microwave and near-infrared spectral regions[10, 11]. However, the scaling of such MMs to higher frequencies encounters many problems, including strong absorption of electromagnetic radiation in metals. Recently suggested electromagnetic periodic structures composed of high-permittivity dielectric elements[12, 13, 14] can avoid these problems, while still delivering an effective magnetic response. The main physics underlying the properties

**Citation:** Mikhail V. Rybin,Dmitry S. Filonov,Kirill B. Samusev,Pavel A. Belov, Yuri S. Kivshar& Mikhail F. Limonov, Phase diagram for the transition from photonic crystals to dielectric metamaterials, doi:10.1038/ncomms10102.

of such all-dielectric MMs is based on the resonant modes of the Mie scattering[15] supported by dielectric particles[16, 17]. Mie resonances provide a mechanism for the generation of an optically induced resonant magnetic response based on the displacement current, and they offer a simpler and more versatile route towards the fabrication of isotropic MMs and metasurfaces operating at optical frequencies.

PhCs and all-dielectric MMs can be composed of the similar structural elements arranged in the similar geometries. Such structures offer a unique opportunity to study a transition from PhCs to MMs when their parameters, such as dielectric permittivity and lattice period, vary continuously. The transition we describe here can be mapped onto a phase diagram; however, it is quite different from phase transitions well known in other fields such as thermodynamics. The thermodynamical phase transitions appear as transformations of a system from one state or 'phase' to another with changing external variables such as pressure, temperature, chemical potential and so on. Here we reveal and analyse a rather unusual type of phase transitions between two regimes of light propagation, that is, the regime where the propagation of light is mainly influenced by the interference between multiple Bragg scattered (PhC 'phase'), and the regime where the propagation of light is mainly determined by the properties of each single element of the materials (MM 'phase').

First, we identify two different scenarios for the transition of a two-dimensional (2D) square lattice of dielectric rods from PhC to MM (Fig. 1). The first scenario that has never been discussed previously occurs when the lattice constant a decreases in comparison with the fixed radius of rods r and dielectric permittivity ε. This transformation leads to an increase of the filling ratio r/a, and it makes possible the homogenization[18] of the periodic dielectric structure with the negative effective permeability (μ<0) at higher values of ε (ref. 19, 20, 21, 22, 23).

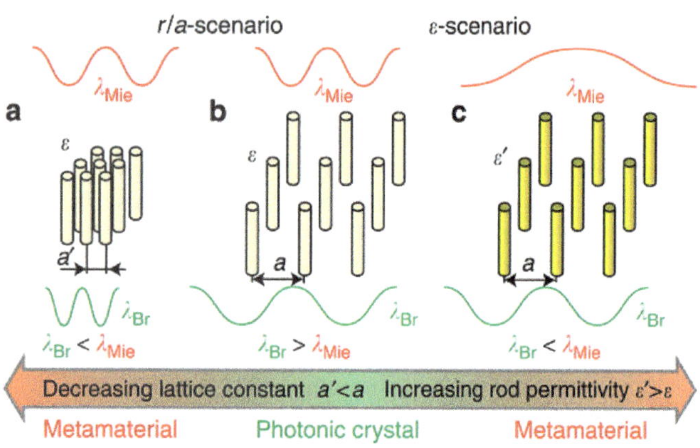


**Figure 1:** Transition from photonic crystals to metamaterials.

Schematic illustrates two possible transformations of the PhC (**b**) to MM (**a,c**) through the competition between the Bragg and Mie scattering wavelengths. To satisfy the classical homogenization condition in the low-frequency long-wavelength limit ($\lambda_{Bragg}<\lambda_{Mie}$), one may decrease the lattice constant and increase the filling ratio r/a (r/a-scenario) or, alternatively, increase the rod permittivity ε (ε-scenario).

The second scenario occurs when the dielectric permittivity of rods ε varies in a fixed square lattice. This transformation leads to an increase of the Mie scattering wavelength, and it can also provide the conditions for the homogenization approach with negative μ. While photonic structures with varying ε were studied earlier theoretically[24, 25], the transition between PhCs and MMs has never been analysed previously to our best knowledge, especially in terms of the phase diagram, as discussed below.

We notice that the aim of both scenarios is the same: to tune the wavelength of the Mie mode $\lambda_{Mie}$ higher than the Bragg wavelength $\lambda_{Br}$ that can be achieved by either decreasing the Bragg wavelength or by increasing the Mie wavelength. Decreasing the Bragg wavelength means decreasing the lattice constant (building block size) a, which will be smaller than the operating wavelength for the MM behaviour, $\lambda_{Mie}$.

Here we integrate both scenarios and introduce a novel concept of a phase transition between all-dielectric PhC and MM and suggest the corresponding phase diagram. The phase diagram PhC–MM is represented on the plane 'filling ratio versus dielectric permittivity' for an example of 2D structure composed of dielectric circular rods. Our theory is confirmed by very careful experimental design and analysis.

## RESULTS

### An Interplay Between Bragg AND Mie Resonances

To analyse the PhC–MM phase transition, we investigate theoretically electromagnetic properties of the dielectric periodic medium depending on the dielectric contrast, filling ratio and incident wave polarization. As an illustrative example, we consider here 2D square lattices of parallel dielectric circular rods infinitely long in the z direction with the radius r and real frequency independent permittivity ε. The rods are embedded in air ($\varepsilon_{air}=1$) and the arrays can be characterized by the filling ratio r/a. We have derived three key sets of spectroscopic data, namely: (i) spectra of the Mie scattering by an isolated rod; (ii) photonic band structure of an infinite 2D square lattice composed of rods; and (iii) transmittance of a 2D square lattice of 10 layers in length. All data sets have been calculated for a wide range of the rod permittivity $1 \leq \varepsilon \leq 100$ with the step of $\Delta\varepsilon=1$ (ε-scenario) and in the range of filling ratio $0 \leq r/a \leq 0.5$ with the step of 0.01 (r/a-scenario). In the r/a-scenario cylinder radius is supposed to be as a fixed parameter. Note that at r/a=0.5 all rods touch each other and at r/a>0.5 the rods penetrate into each other and the structure appears as inverted one of air holes in a dielectric matrix.

First, we consider the Mie scattering by an isolated rod. The far-field scattering can be expanded into orthogonal electromagnetic dipolar and multipolar modes, described by circular Lorenz–Mie resonant coefficients $a_n$ (transverse-electric (TE) polarization) and $b_n$ (transverse-magnetic (TM) polarization) corresponding to electric and magnetic moments, respectively[15]. Figure 2 presents the numerically calculated spectra of the Mie scattering efficiency $Q_{sca,n} = \frac{2}{x}|a_n|^2$ (ref. 26) for the dipole $TE_0$ and multipole $TE_n$ $(n \geq 1)$ modes in the range of dimensionless frequencies a/λ from 0 to 0.8. Each mode consists of an infinite set of quasi-equidistant modes $TE_{nk}$ (ref. 27). The Mie modes of the rod are denoted as $TE_{nk}$, where n is integer related to the symmetry of the mode, and $k \geq 1$ identifies resonance number for the modes with fixed symmetry n. The calculations reveal a strong decrease of the Mie eigenfrequencies and narrowing of the resonant Mie bands with ε increasing, as can be clearly seen for the lowest $TE_{01}$ magnetic Mie resonance in Fig. 2.

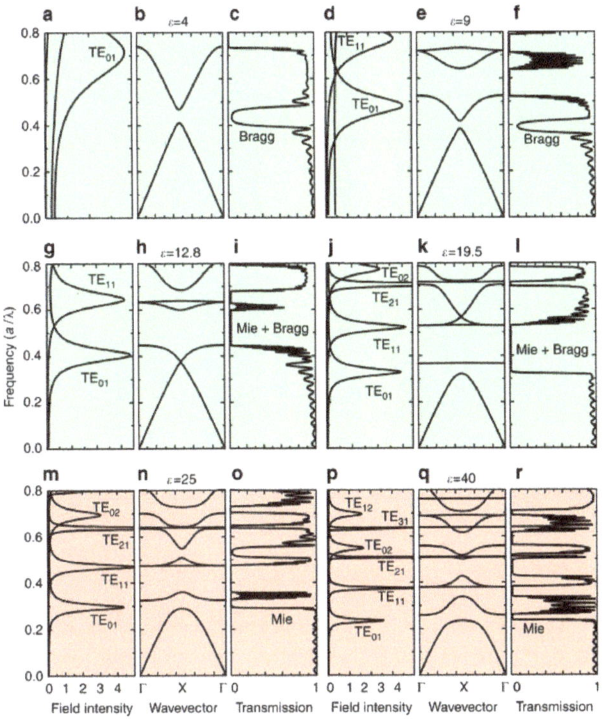


**Figure 2:** Numerical calculations.

(a), (d), (g), (j), (m), (p) Calculated Mie scattering efficiency $Q_{sca,n}$ for an isolated dielectric circular rod for $TE_n$ (n≥0) modes. (b), (e), (h), (k), (n), (q) The band structure for 2D square lattice of rods with r=0.25a in air ($\varepsilon_{air}$=1) for the TE polarization. The band structure is shown between the Γ point (wave vector k=0) and the X point (|k|=π/a along the x direction). (c), (f), (i), (l), (o), (r) The transmittance calculated for 10 lattice layers of the 2D square structure of rods in air for the TE polarization. The frequency and wave vector are plotted in dimensionless units a/λ, where λ is vacuum wavelength and a denotes the lattice constant. (a)–(c) ε=4, (d)–(f) ε=9, (g)–(i) ε=12.8, (j)–(l) ε=19.5, (m)–(o) ε=25, (p)–(r) ε=40.

The photonic band structures for the 2D square lattice of circular rods were computed numerically using the plane wave expansion method[28]. The square 2D lattice has a square Brillouin zone, with the three special points Γ, X and M corresponding to **k**=0, $\mathbf{k} = \frac{\pi}{a}\hat{x}$, and $\mathbf{k} = \frac{\pi}{a}\hat{x} + \frac{\pi}{a}\hat{y}$, respectively. We restrict here our discussion to the case for which the wave vector **k** of the eigenmodes is oriented in the Γ →X direction of the first Brillouin zone displaying the lowest Bragg frequencies in comparison with the Γ→M direction (Fig. 2). The photonic band structures were obtained with 16,384 (128 × 128) plane waves.

The transmission spectra were calculated using the CST Microwave Studio software for the wave vector of the incident beam parallel to the Γ→X direction. We considered the 2D square lattice with different thicknesses along x axis that leads to creation of the photonic bandgaps, including thicknesses of 10 lattice layers as shown in Fig. 2.

The calculated data allow us to analyse precisely the evolution of the low-frequency region of the photonic band structure formed by a complicated mixture of the dispersion curves originating from Mie and Bragg resonances and propagating modes. Since our goal is to discuss the PhC–MM transition, we concentrate on the TE polarization where the lowest $TE_{01}$ magnetic dipole resonance of the rods results in a negative magnetic permeability in the Mie gap spectral region[19].

The physics of the PhC–MM transition is displayed by the gap map (Fig. 3) that relates the Bragg and Mie gaps positions to the rod permittivity ε for given r/a=0.25. These illustrate essential features such as strong decreasing of Mie resonance frequencies with ε increase followed by intersection of the Mie and Bragg bands. The Mie resonances of individual rods form both non-dispersive flat bands in the photonic structure and Mie photonic gaps in the energy spectrum (Figs 2 and 3). For low-contrast PhCs, all Mie resonances are located is a frequency range higher than the lower Bragg gap. With increasing rod permittivity ε, the Mie resonances of single rods and correspondingly Mie bands of the structure demonstrate a shift from the higher to low frequencies, crossing the strongly dispersive Bragg bands. The creation of the lowest Mie gap driving the artificial magnetism is clearly seen both in the band structure and in the transmission spectra (Figs 2 and 3).

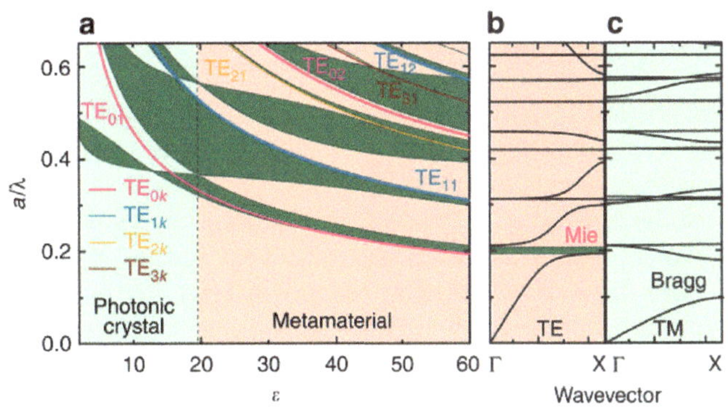


**Figure 3:** Bandgap diagram for the 2D square lattice of circular rods.

(**a**) The bandgap diagram obtained from the photonic band structures calculated for $1 \leq \varepsilon \leq 60$ with steps of Δε=1. The rods are embedded in air, r/a=0.25, and TE polarization is considered. The Bragg and Mie gaps obtained from the band structure calculations are marked by dark green. (**b**,**c**) The photonic band structure for the 2D square lattice of rods with ε=60 for TE and TM polarization, respectively, for the Γ →X scan of the wave vector **k**.

## Two Crossover Scenarios for a Metacrystal

Armed with the results of our calculations, we verify our concept experimentally by engineering a metacrystal composed of plastic circular tubes filled with water and forming a 2D square lattice with variable lattice constant a (Fig. 4). The structure of the metacrystal enables us to perform two types of experiments according to the r/a- and ε-scenarios. The r/a-scenario was realized using a special construction of the metacrystal that makes it possible to change the distance between tubes synchronously in x and y directions keeping the

2D square symmetry of the lattice (Fig. 4). For the ε-scenario, we employ the advantages of strong temperature dependence of the dielectric permittivity of water that is characterized by ε=80 at 20 °C and ε=50 at 90 °C in the microwave frequencies range from 1 to 6 GHz (refs 29, 30). On the basis of the calculated results and the parameters of the metacrystal, we chose intervals of most interest for each scenario, namely $60 \le \varepsilon \le 80$ at $0.7 \le r/a \le 0.8$ for the ε-scenario and $0.65 \le r/a \le 1.9$ at ε=60 for the r/a-scenario.

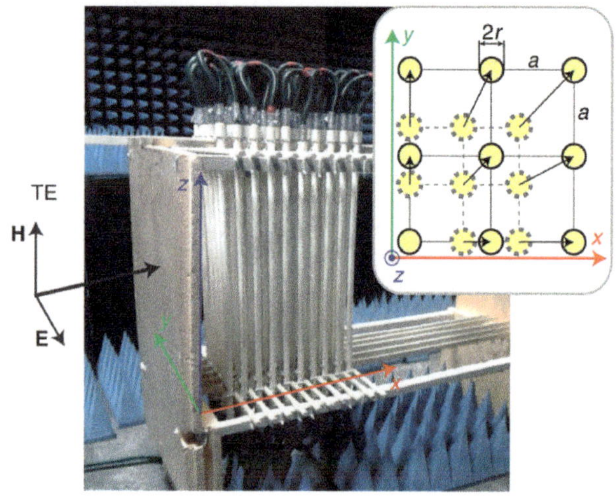


**Figure 4:** Experimental set-up.

An anechoic chamber with a metacrystal composed of 10 tubes in length (x axis) and 5 tubes in width (yaxis), 50 tubes in total. Each tube has a length of 1 m (along z axis), an outer radius of 2 cm and an inner radius of 1.7 cm. All tubes are connected in a closed series circuit together with a water heater and a temperature stabilizer. Insert: schematic illustration of the transformation of the 2D square lattice composed of dielectric rods as the lattice constant a increase. The scheme shows the square lattice from above. The rods with dielectric permittivity ε and fixed radius r are embedded in air ($\varepsilon_{air}$=1). The material is homogeneous along the z direction and periodic along x and y.

The transmission spectra of the metacrystal installed in an anechoic chamber were measured in the 1–3-GHz range. A rectangular horn antenna (TRIM 0.75–18 GHz; DR) connected to a transmitting port of the vector network analyser Agilent E8362C was used to approximate plane wave excitation. A similar horn antenna was employed as a receiver. For the electromagnetic wave incident perpendicular to the tube axis z, the two transverse polarizations appear decoupled and the vector wave problem can be reduced to two independent scalar equations. As a result, the modes of 2D PhCs can be classified as either TE modes ($E_x$, $E_y$ and $H_z$) with the electric field confined to the x–y plane and the magnetic field polarized along the axes of the rods (Fig. 4) or TM modes ($H_x$, $H_y$ and $E_z$) for which the magnetic field is oriented perpendicular to the rod axis z. For the TM polarization, such structures can not show any magnetic activity[31], so we only consider TE polarization.

Figure 5 demonstrates excellent agreement between the measured and calculated transmission spectra both for r/a and ε dependencies. For the r/a-scenario, we started our experiment from the PhC phase (r/a=0.08) and compressed the structure by decreasing the lattice constant a (Fig. 5a). The lowest Bragg gap demonstrates remarkable shift towards high-frequency region, then nearly almost disappears at r/a≈0.09 when it approaches the unchanging $TE_{01}$ Mie gap from the low-frequency side, and finally appears again at r/a≈0.095 on the high-frequency side of the Mie gap. As a result, the lowest Bragg band and the $TE_{01}$ Mie band change their positions in the energy scale inaugurating the transition into MM phase. In contrast, for the ε-scenario, the Bragg gap position changes very slowly as opposed to the strong shift of the $TE_{01}$ Mie band that finally leads to the same general result (Fig. 5d).

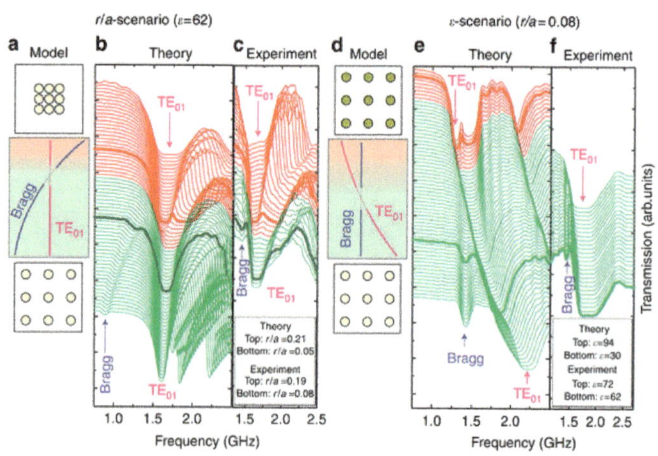


**Figure 5:** Transmission spectra for a square lattice of circular rods.

TE polarization. (**a–c**) r/a-scenario, ε=62. The calculated (**b**) and experimentally measured (**c**) transmission spectra of the metacrystal as a function of the filling ratio r/a. The calculated spectra are shown in the range $0.05 \le r/a \le 0.21$, the experimental spectra are shown in the range $0.08 \le r/a \le 0.19$. (**d–f**) ε-scenario, r/a=0.08. The calculated (**e**) and experimentally measured (**f**) transmission spectra of the metacrystal as a function of the dielectric permittivity. The calculated spectra are shown in the range $30 \le \varepsilon \le 94$, the experimental spectra are shown in the range $62 \le \varepsilon \le 94$. The spectra are shifted vertically by the constant value.

## DISCUSSION

Figure 6a shows our main result: the PhC–MM phase diagram for the 2D square lattice of circular rods. The axes represent the dielectric permittivity ε and filling ratio r/a. The phase diagram was obtained based on 50 gap maps (such as in Fig. 3a) calculated in the permittivity region $1 \le \varepsilon \le 100$ with the step r/a=0.01 for TE polarization. The diagram is divided into two areas, which represent the PhC and MM phases. The transition from PhC phase to MM phase can be identified by critical points, which are defined by a special combination of permittivity ε and filling ratio r/a. The curvature of the second dispersion curve around the Γ point decreases and finally becomes perfectly flat along the ΓX path (at ε=19.5, r/a=0.25, Fig. 2k). With future increase in ε or change in r/a, the $TE_{01}$ Mie gap completely splits from complicated coupled Mie–Bragg band becoming the lowest gap in the spectrum and the MM phase appears. Within this Mie gap, the magnetic permeability is negative and all waves in the medium are evanescent. Since all the Bragg gaps are located at higher frequencies, any diffraction losses in the vicinity of the Mie resonance are absent. The transition from the PhC phase to the MM phase is completed and the medium is classified as the MM with μ<0. A phase transition is usually characterized by an order parameter that basically arises from symmetry breaking but can also be defined for non-symmetry-breaking transitions[32]. Here for the unusual phase transition PhC→MM, we define the order parameter η as a measure of splitting in the energy scale between the $TE_{01}$ Mie and Bragg gaps. It is zero in the PhC phase and nonzero in the MM phase. Figure 6b presents the order parameter depending on both external variables ε and r/a.

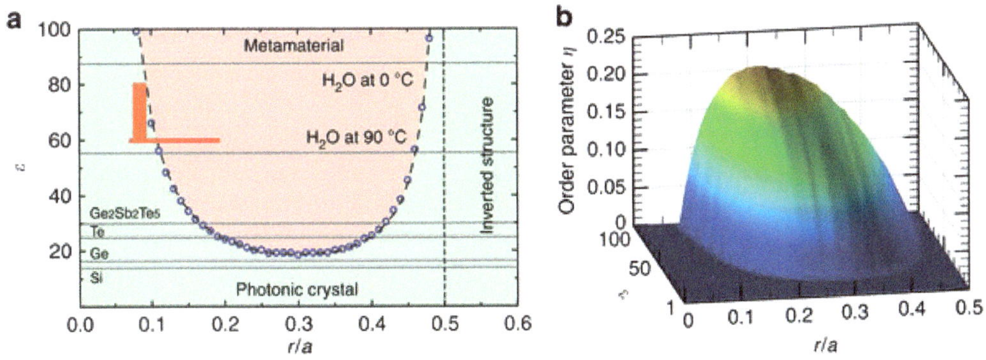


**Figure 6:** PhC–MM phase diagram.

(**a**) PhC–MM phase diagram for a square lattice of dielectric circular rods. TE polarization. Blue circles indicate the points where the $TE_{01}$ Mie gap splits from Bragg band becoming the lowest gap in the spectrum (Fig. 2). The experimentally investigated regions are marked by red. The dielectric permittivities εof different materials are marked by horizontal lines. (**b**) The dependence of the order parameter η on the dielectric permittivity ε and the filling factor r/a. The order parameter is defined as a difference in the energy between the $TE_{01}$ Mie and Bragg gaps in the MM phase.

Now, we review some features of the PhC–MM phase diagram. At low filling ratio r/a<0.05, the concentration of the completely isolated rods in the lattice is too small for achieving negative average permeability at reasonable values of ε. In contrast, when distance between rods become very small (r/a>0.45), the inter-rod interaction become strong and the hopping of the Mie photons from one rod to another dominates. This means that Mie resonances are not localized anymore and they finally exhibit the character of propagating Bloch waves. All requirements for the appearance of the MM phase are broken and the PhC phase extends its region of existence to very high ε. We notice that for the metacrystals with a small lattice constant a, the frequency of the Mie resonance is increasing. Therefore, there is no transition to the MM phase in this limit.

Identification of candidates for dielectric low-loss materials is critical for designing a practical MM for applications in different ranges of the electromagnetic spectrum. As an example, in Fig. 6a, we mark the regions of the MM phase for water in the microwave range (region $55 \le \varepsilon \le 87$ )[31], the chalcogenide glass $Ge_2Sb_2Te_5$ (ε≈30)[33] and tellurium (ε≈25)[34] in infrared. Thus, the well-studied 2D square lattice of germanium (ε≈16) and silicon (ε≈13)[34] rods has only PhC phase.

## CONCLUSION

We have studied a metacrystal composed of dielectric circular rods and introduced the concept of a phase diagram in the permittivity-filling ratio plane. The boundaries of the MM phase have been obtained theoretically and confirmed experimentally. On the basis of the proposed approach, one can obtain different PhC–MM phase diagrams by altering the dimension, symmetry, composition, size and geometry of the structural elements within a unit cell. The underlying mechanism of the PhC–MM transition provides unique criteria for goal-oriented search of novel low-loss MM with a wide range of applications at infrared and visible frequencies.

## REFERENCES

1. Joannopoulos, J. D., Johnson, S. G., Winn, J. N. & Meade, R. D. Photonic Crystals: Molding the Flow of Light 2nd edn (Princeton Univ. Press, 2008).
2. Limonov, M. F. & De La Rue, R. M. (eds) Optical Properties of Photonic Structures: Interplay of Order and Disorder (CRC Press, Taylor & Francis Group, 2012).
3. Enghata, N. & Ziolkowski, R. (eds) Electromagnetic Metamaterials: Physics and Engineering Exploration (Wiley-IEEE Press, 2006).
4. Smith, D. & Liu, R. (eds) Metamaterials: Theory, Design, and Applications (Springer, 2010).
5. Cai, W., Chettiar, U. K., Kildishev, A. V. & Shalaev, V. M. Optical cloaking with metamaterials. Nat. Photon. 1, 224–227 (2007).
6. Lapine, M., Shadrivov, I. V., Powell, D. A. & Kivshar, Y. S. Magnetoelastic metamaterials. Nat. Mater. 11, 30–33 (2012).
7. Silva, A. et al. Performing mathematical operations with metamaterials. Science 343, 160–163 (2014).
8. Smith, D. R., Pendry, J. B. & Wiltshire, M. C. K. Metamaterials and negative refractive index. Science 305, 788–792 (2004).
9. Soukoulis, C. M., Kafesaki, M. & Economou, E. N. Negative-index materials: New frontiers in optics. Adv. Mater. 18, 1941–1952 (2006).
10. Soukoulis, C. M. & Wegener, M. Past achievements and future challenges in the development of three-dimensional photonic metamaterials. Nat. Photon. 5, 523–530 (2011).
11. Zheludev, N. I. & Kivshar, Y. S. From metamaterials to metadevices. Nat. Mater. 11, 917–924 (2012).

12. Huang, K. C., Povinelli, M. & Joannopoulos, J. D. Negative effective permeability in polaritonic photonic crystals. Appl. Phys. Lett. 85, 543–545 (2004).
13. Kallos, E., Chremmos, I. & Yannopapas, V. Resonance properties of optical alldielectric metamaterials using two-dimensional multipole expansion. Phys. Rev. B 86, 245108 (2012).
14. Moitra, P. et al. Realization of an all-dielectric zero-index optical metamaterial. Nat. Photon. 7, 791–795 (2013).
15. Bohren, C. F. & Huffman, D. R. Absorption and Scattering of Light by Small Particles (Wiley-VCH, 1998).
16. Ginn, J. C. et al. Realizing optical magnetism from dielectric metamaterials. Phys. Rev. Lett. 108, 097402 (2012).
17. Zhao, Q., Zhou, J., Zhang, F. & Lippens, D. Mie resonance-based dielectric metamaterials. Mater. Today 12, 60–69 (2009).
18. Andryieuski, A. et al. Homogenization of resonant chiral metamaterials. Phys. Rev. B 82, 235107 (2010).
19. O'Brien, S. & Pendry, J. B. Photonic band-gap effects and magnetic activity in dielectric composites. J. Phys. Condens. Matter 14, 4035 (2002).
20. Yannopapas, V. & Moroz, A. Negative refractive index metamaterials from inherently non-magnetic materials for deep infrared to terahertz frequency ranges. J. Phys. Condens. Matter 17, 3717 (2005).
21. Wheeler, M. S., Aitchison, J. S. & Mojahedi, M. Three-dimensional array of dielectric spheres with an isotropic negative permeability at infrared frequencies. Phys. Rev. B 72, 193103 (2005).
22. Schuller, J. A., Zia, R., Taubner, T. & Brongersma, M. L. Dielectric metamaterials based on electric and magnetic resonances of silicon carbide particles. Phys. Rev. Lett. 99, 107401 (2007).
23. Vynck, K. et al. All-dielectric rod-type metamaterials at optical frequencies. Phys. Rev. Lett. 102, 133901 (2009).
24. Foteinopoulou, S. Photonic crystals as metamaterials. Physica B 407, 4056 (2012).
25. Notomi, M. Theory of light propagation in strongly modulated photonic crystals: Refractionlike behaviour in the vicinity of the photonic band gap. Phys. Rev. B 62, 10696 (2000).
26. Rybin, M. V., Filonov, D. S., Belov, P. A., Kivshar, Y. S. & Limonov, M. F. Switching from visibility to invisibility via Fano resonances: theory and experiment. Sci. Rep. 5, 8774 (2015).
27. Rybin, M. V. et al. Mie scattering as a cascade of Fano resonances. Opt. Express 21, 30107–30113 (2013).
28. Johnson, S. G. & Joannopoulos, J. D. Block-iterative frequency-domain methods for Maxwell's equations in a planewave basis. Opt. Express 8, 173 (2001).
29. Kaatze, U. Complex permittivity of water as a function of frequency and temperature. J. Chem. Eng. Data 34, 371–374 (1989).
30. Andryieuski, A., Kuznetsova, S. M., Zhukovsky, S. V., Kivshar, Y. S. & Lavrinenko, A. V. Water: Promising opportunities for tunable all-dielectric electromagnetic metamaterials. Sci. Rep. 5, 13535 (2015).
31. Ramakrishna, S. A. Physics of negative refractive index materials. Rep. Prog. Phys 68, 449 (2005).
32. Gooding, R., Ye, Y., Chan, C., Ho, K. & Harmon, B. Role of non-symmetrybreaking order parameters in determining the martensitic energy barrier: The bcc-to-9R transformation. Phys. Rev. B 43, 13626 (1991).
33. Ncˇmec, P. et al. Ge-Sb-Te thin films deposited by pulsed laser: An ellipsometry and Raman scattering spectroscopy study. J. Appl. Phys. 106, 103509 (2009).
34. Palik, E. D. Handbook of Optical Constants of Solids vol. 3 (Academic press, 1998).

# CHAPTER 10 DYNAMICALLY CONTROLLING THE EMISSION OF SINGLE EXCITONS IN PHOTONIC CRYSTAL CAVITIES

Francesco Pagliano[1] , YongJin Cho[1] , Tian Xia[1] , Frank van Otten[1] , Robert Johne[1,2] & Andrea Fiore[1]

[1]Applied Physics Department, COBRA Research Institute, Eindhoven University of Technology, 5600 MB Eindhoven, The Netherlands

[2] Max-Planck-Institute for the Physics of Complex Systems, No¨thnitzer Strae 38, 01187 Dresden, Germany

## ABSTRACT

**Single excitons in semiconductor microcavities represent a solid state and scalable platform for cavity quantum electrodynamics, potentially enabling an interface between flying (photon) and static (exciton) quantum bits in future quantum networks. While both single-photon emission and the strong coupling regime have been demonstrated, further progress has been hampered by the inability to control the coherent evolution of the cavity quantum electrodynamics system in real time, as needed to produce and harness charge–photon entanglement. Here using the ultrafast electrical tuning of the exciton energy in a photonic crystal diode, we demonstrate the dynamic control of the coupling of a single exciton to a photonic crystal cavity mode on a sub-nanosecond timescale, faster than the natural lifetime of the exciton. This opens the way to the control of single-photon waveforms, as needed for quantum interfaces, and to the real-time control of solid-state cavity quantum electrodynamics systems.**

## INTRODUCTION

An optical cavity modifies the electromagnetic environment of an emitter and thereby its radiative recombination dynamics. In the weak-coupling (WC) regime, that is, when the cavity loss rate exceeds the emitter–cavity coupling rate, the increased local optical density of states produces a change in the spontaneous emission rate[1]. This enables the efficient funneling of spontaneously emitted photons in a given mode, and can be used to realize efficient single-photon sources[2, 3, 4]. In the strong coupling (SC) regime, the emission becomes reversible and coherent Rabi oscillations of the emitter and cavity populations, resulting in a Rabi doublet, are observed[5, 6, 7]. A dynamic control of the emitter-field interaction is needed to take full advantage of cavity quantum electrodynamics (c-QED) processes. The real-time control of Rabi oscillations would enable the creation of entanglement between emitter and photon, a fundamental resource in the context of quantum communications[8]. In the WC regime, the ultrafast control allows for the generation of time-symmetric photon wavepackets, enabling the coherent transfer of quantum states between light and matter, a prerequisite for efficient quantum interfaces[9] and spin-memories in quantum dots (QDs)[10].

In atomic c-QED system at microwave frequencies, this control is achieved by changing the interaction time[11], while at optical frequencies it is realized by adiabatic passage techniques, which enable shaping a photon waveform on 100-ns timescales[1, 2, 9, 12]. The catch and release of single photons[13] as well as the photon waveform shaping[14] have

**Citation:** Francesco Pagliano,YongJin Cho,Tian Xia,Frank van Otten,Robert Johne& Andrea Fiore, Dynamically controlling the emission of single excitons in photonic crystal cavities, doi:10.1038/ncomms6786.

been recently demonstrated at microwave frequencies using superconducting circuit quantum electrodynamics. Semiconductor c-QED systems, for example, based on QDs in optical microcavities, present an evident potential in terms of integration and scalability. While both single-photon emission[15, 16]and the SC regime[5, 6] have been demonstrated, the progress in their application to quantum information has been strongly limited by the absence of fast control methods, which would enable the generation of symmetric photons[17, 18] as well as the implementation of quantum gates[19, 20]and of entangled photon–exciton states. While a number of approaches have been proposed for the control of the exciton–cavity interaction[21, 22, 23, 24, 25, 26], none of them has been applied to control c-QED on the ultrafast timescales typical of solid-state systems (tens of picoseconds in the SC regime and hundreds of picoseconds in the WC regime). A method for the all-optical ultrafast control of quality factor and mode volume has been very recently proposed by our group[27], but its application has so far remained limited to ensemble of QDs, eluding the most interesting single-emitter, single-photon regime.

Here we demonstrate dynamic control of the radiative interaction of a single solid-state emitter with an optical cavity, on a timescale faster than its natural decay (~ns). This is obtained by placing the QD in a photonic crystal cavity integrated with a diode designed to enable a >1 GHz modulation frequency. By the fast tuning of the exciton energy via the quantum confined Stark effect, we electrically change the QD–cavity coupling within~300 ps, which is sufficient to reshape the temporal waveform of the photon emitted in the WC regime. Further optimization of the structure will open the way to the control of SC c-QED systems.

## RESULTS

### Device Characterization

To be able to create a vertical electric field varying on a sub-nanoscale timescale, we integrate self-assembled QDs inside a GaAs micro-diode with vertical geometry where the top (bottom) part of the 400 nm-thick membrane region is p-(n-)doped. The vertical geometry, as compared to the lateral p–n junctions demonstrated in refs 28, 29, is essential to provide a large quantum confined Stark effect, since the permanent dipole moment in self-assembled QDs has a vertical orientation, which makes it more sensitive to vertical fields rather than later ones[30]. A small capacitanceC~0.3 pF and a total series resistance <200 Ω were obtained using an optimized design and fabrication (see Methods). A diode (Fig. 1) with a mesa diameter of 16 μm and a modified L3 photonic crystal cavity (PhCC) featuring a y-polarized mode centred ~978 meV at T=10 K with a quality factor Q~10,500 (Fig. 1d) is considered in the following. In Fig. 1b the 3 dB electrical bandwidth of this diode is shown to be 2 GHz (blue circles). For smaller diodes higher 3 dB frequencies of 4.2 GHz were observed (red circles) but with reduced Q. Micro photoluminescence (mPL) spectra (Fig. 2a) show a clear Stark tuning of the QD excitonic lines as a function of bias voltage V. From the field dependence of the excitonic $QD_1$ line in Fig. 2a, which shows Purcell enhancement when tuned inside the cavity mode at V=−1,200 mV, we estimate a permanent dipole moment p=(−0.04±0.02) e⁻nm and polarizability β=(−3.1±0.2) · $10^3$ e⁻nm kV$^{-1}$cm, in agreement with similar QD structures[31]. Time-resolved experiments (Fig. 2b) provide the decay time for $QD_1$ out of resonance (400 mV, blue triangles), $\tau^{QD1}_{off}$=(3.1±0.2) ns, and in resonance (−1,200 mV, red circles), $\tau^{QD1}_{on}$ =(0.27±0.03) ns. The slower component of the in-resonance bi-exponential decay, $\tau_{BG}$=(2.5±0.2) ns, can be attributed to the nonradiative recombination rate of dark excitons[32]. The emission rate into the mode can be derived correcting for the emission rate into leaky modes, providing a Purcell factor of $F_P=\tau^{BULK}\cdot(1/\tau^{QD1}_{on}-1/\tau^{QD1}_{off})$=(2.9±0.6), using the reference time $\tau^{BULK}$=(0.85±0.10) ns measured on the QD ensemble outside the PhC (green squares). The fraction of photons emitted into the cavity mode $\beta=1-(\tau^{QD1}_{on}/\tau^{QD1}_{off})$, is calculated as $\beta_{QD1}$=91% for $QD_1$, comparable with previous reports for L3 cavities[33]. Even higher beta factors may be achieved by positioning the QD at an anti-node of the cavity field by site-controlled growth[34], thus enhancing the Purcell effect and improving the funneling of the emitted photon into the desired mode.

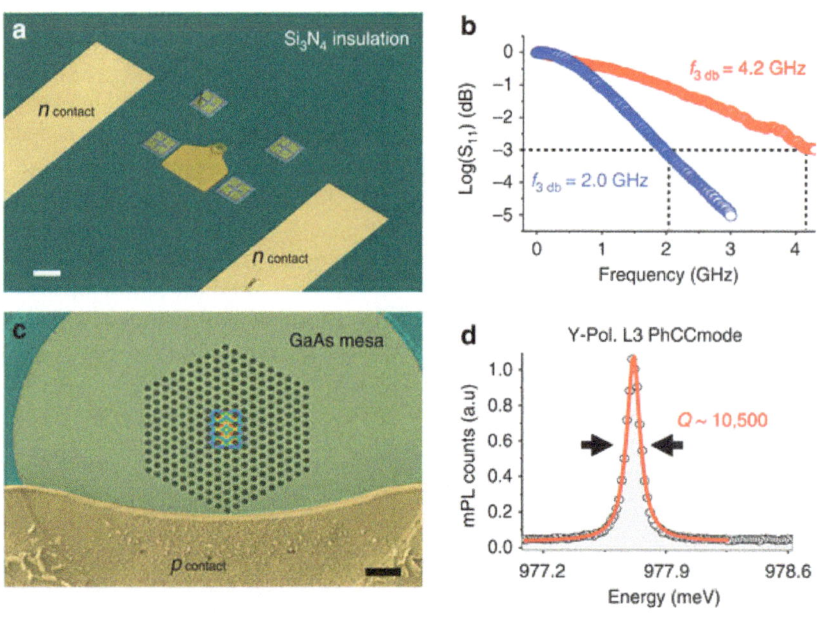


**Figure 1:** Ultrafast photonic crystal cavity diode.

Scanning electron microscope images of the photonic crystal cavity diode (scale bar, 10 μm) (**a**) and enlarged view (scale bar, 1 μm) of the L3 modified cavity with the electric field $|\boldsymbol{E}|^2$ distribution (over-imposed) of the $Y_1$ mode calculated from three-dimensional finite element (FE) simulation (**c**). (**b**) Network analyzer measurement of the $S_{11}$ one-port scattering parameter, showing the diode bandwidth of 2 GHz (blue circles) compared with the data of a smaller device with better electrical performance (red circles). (**d**) High resolution spectrum of the photonic crystal cavity (PhCC) L3 cavity $Y_1$ mode with Lorentzian fit (red curve) and estimated Q-factor of 10,500 (k/2π~22 GHz).

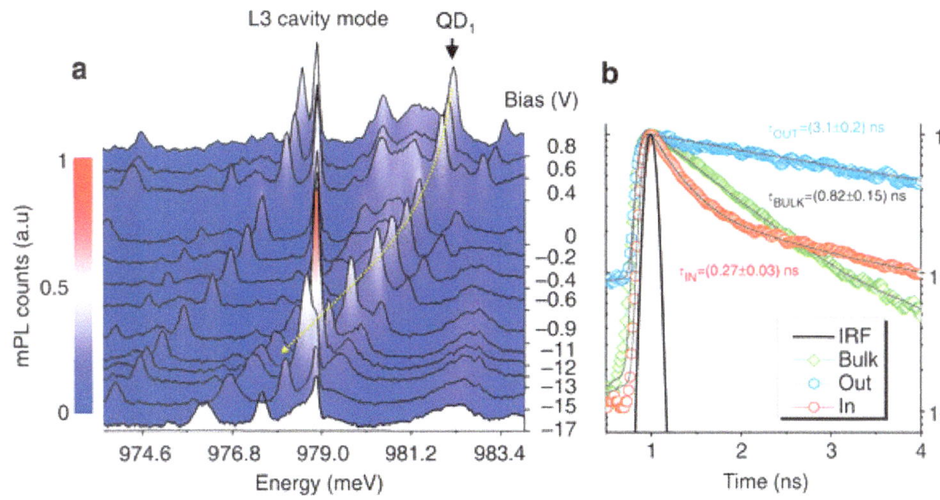


**Figure 2:** Static Stark tuning and spontaneous emission control.

(**a**) Low temperature micro photoluminescence (mPL) spectra for 800 nW excitation power for different direct-current (DC) voltages applied to the diode. A weakly coupled single-exciton line-labelled $QD_1$ is enhanced when tuned across the L3 cavity mode. The yellow line is a guide to the eye. (**b**) Time-resolved photoluminescence (TRPL) decay curves of the quantum dot $QD_1$ out of (light blue hexagons) and in (red circles) resonance with the cavity mode for T=10 K. The quantum dots (QDs) ensemble decay in the bulk is measured as reference (green curve) and the decay time constants are calculated from single and double exponential fits (grey lines) including the convolution with the IRF of the detector (black line).

## Ultrafast Modulation of the Exciton Energy

To infer information about the effective modulation of the electric field at the QD position, we study the time-integrated mPL under a square-wave modulation of the applied voltage, as shown in Fig. 3b for a modulation amplitude of 240 mV and frequency varying from 15 MHz to 3 GHz. As sketched in Fig. 3a, at low frequencies the voltage and correspondingly the exciton energy have sharp transitions between levels A (−1,210 mV) and B (−1,450 mV), resulting in two clear peaks (labelled as $A_1$, $B_1$ in Fig. 3b) for each QD line. The average bias is chosen so that the $QD_1$ line (yellow curves) is in resonance with the cavity in one half of the period ($A_1$). For modulation frequencies >200 MHz the field across the QDs is affected by the RC constant of the circuit and the two lines get closer to their mean central value. However, the fact that the two peaks are distinct and separated by more than the cavity linewidth at frequencies >2 GHz demonstrates the possibility to control the QD–cavity coupling on sub-nanoscale timescales.

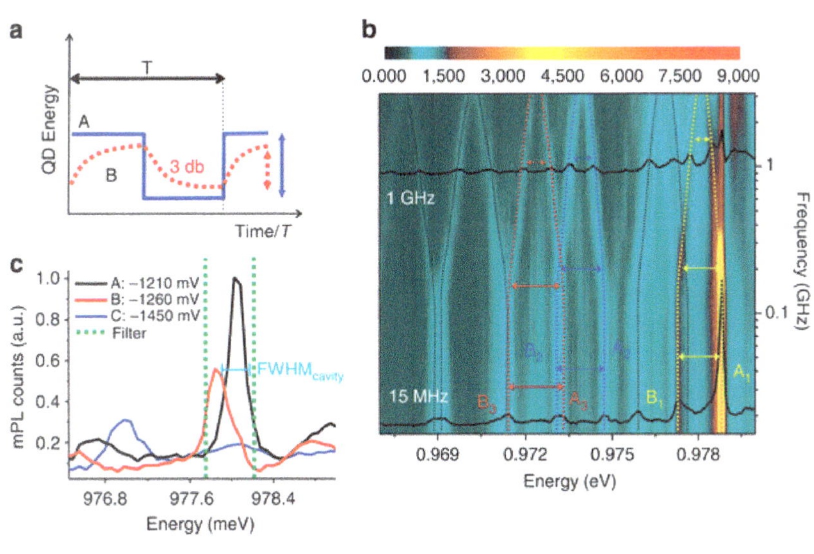


**Figure 3:** Square-wave bias modulation of exciton energy.

(**a**) Schematics for the low- (blue curve) and the high- (dotted red line) frequency exciton modulation between energies A and B due to the square-wave bias in the centre of the diode. (**b**) Micro photoluminescence (mPL) data collected from the quantum dots (QDs) inside the cavity and dynamically modulated by the square-wave bias between −1,210 and −1,450 mV at different frequencies up to 3 GHz. Two lines outside of the cavity mode spectral region are modulated between the energy states $A_2$–$B_2$(blue curves and arrows) and $A_3$–$B_3$ (red curves and arrows), while the coupled quantum dot $QD_1$ is tuned in($A_1$) and out($B_1$) of resonance with the cavity mode and enhanced. (**c**) Low-excitation spectra (~150 nW) for the quantum dot $QD_1$ at −1,210 mV (black), −1,260 mV (red) and −1,450 mV (blue). The green dotted lines show the filter window (with full-width at half maximum (FWHM)=0.65 nm) in which the fast modulated cavity signal is collected and sent to the superconducting single-photon detector (SSPD) for the measurements of Fig. 4.

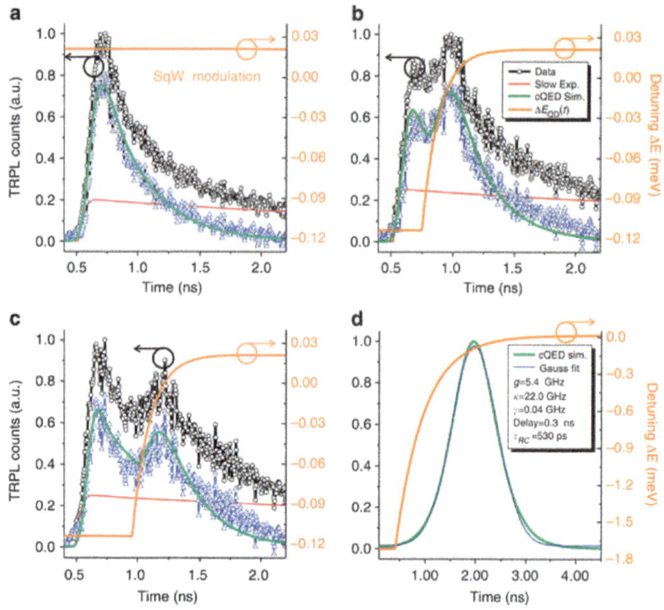


**Figure 4:** Ultrafast electrical control of $QD_1$–cavity coupling.

## Dynamic Control of the c-QED Interaction

We then investigated the dynamic modulation of the photon waveform by time-resolved PL experiments. For this purpose, we used a spectral filter to select the emission from the $QD_1$ line and we chose the square-wave bias to modulate $QD_1$ in and out of the cavity resonance, but inside the filter window. Figure 3c shows the low power (~150 nW) mPL spectra taken at the two modulation voltages of V=−1,210 mV (in resonance, black curve) and V=−1,260 mV (out of resonance, red curve). The background contribution to the mode feeding[35, 36] is negligible at this low pumping power, as shown by the very low PL intensity observed at the cavity resonance when the QD is detuned by ~1 meV (V=−1,450 mV, blue line in Fig. 3c). Depending on the delay between the laser and the electrical pulses, we observe different scenarios as shown in Fig. 4a–c. From each measured data (black circles), we subtracted the slow contribution due to the dark exciton (red line), as estimated from a bi-exponential fit, to best compare the relevant bright exciton dynamics with our c-QED simulation (green lines), based on a master equation formalism (see Methods). During the time interval of Fig. 4a $QD_1$ is fixed at its higher energy state, showing the natural on-resonance exponential decay. In Fig. 4b, the QD is initially detuned from the cavity and is brought in resonance at Δt~0.25 ns by the electric pulse, thus increasing the spontaneous emission rate. Changing the delay allows controlling the timing of the modulation and gives us the sub-nanoscale temporal control over the modulated cavity output, as shown in Fig. 4c for a larger Δt~0.47 ns delay. In this case, the modulated cavity signal is lower since the QD is brought in resonance with the cavity at a later time when its population has already partially decayed. The maximum modulation with respect to the unperturbed mono-exponential decay is a factor of 1.9 (at a time t=1.18 ns in Fig. 4c), and is limited by the maximum QD–cavity detuning of ~0.13 meV in this experiment (as seen in the limited ratio of 1.65 in the integrated PL peaks in resonance and out of resonance in Fig. 3c). The measurements show an excellent agreement with the c-QED simulations (green lines), taking into account the energy level modulation (orange lines) expected from the RC constant of the circuit and the QD and cavity parameters extracted from the measurements in Figs 1 and 2 (see Methods).

Time-resolved mPL experiments under quasi-resonant excitation. A pulsed laser at 970 nm with a pulse width of ~70 ps at a repetition rate of 60 MHz is used. The fast modulation of the normalized photon counts (black circles) collected from the cavity center during the dynamic tuning of the quantum dot $QD_1$ energy in (−1,210 mV) and out (−1,260 mV) of resonance with the cavity mode is shown for different delays between the laser excitation and the square-wave electric signal. The fast component of this filtered signal shows a good agreement with the theoretical simulations (green curves) not including the slow exponential term (red lines). The experimental delays can be more precisely estimated from the simulation and are found to be Δt>3 ns (**a**), Δt=(0.25±0.03) ns (**b**) and Δt=(0.47±0.01) ns (**c**). (**d**) Simulation of the photon symmetrization and Gaussian fit (blue line) with the same cavity quantum electrodynamics (c-QED) parameters, a larger tuning range and a square-wave bias filtered with a RC time constant of 530 ps.

## DISCUSSION

The present work demonstrates the dynamic control of the single-exciton spontaneous emission within about 300 ps (time between 10 and 90% of the total modulation amplitude in Fig. 4). An ultrafast modulation of the c-QED evolution is achieved when the emitter is dynamically tuned with respect to the cavity resonance. The dynamic control in our experiment is limited by the relatively high QD density, which forces us to use a narrow spectral filter and therefore a limited tuning range, so that the QD emission out of resonance is not negligible. The electrical bandwidth of our PhCC-diode would indeed be sufficient to control the rise time of the photon and achieve a complete symmetrization of its waveform, as estimated in the simulated dynamics of Fig. 4d, where the same c-QED parameters as our experiment and a larger tuning range have been assumed. We conclude that by combining our ultrafast diode structure with site-controlled QDs[34], a nearly perfect control of the waveform of emitted single photons can be achieved. Such symmetric photon waveform, in combination to phase modulation to compensate the chirp induced by the tuning[18], would enable efficient quantum state transfer in quantum networks with solid-state nodes[8, 10]. In addition, by increasing the cavity Q-factor up to 60,000, the maximum value allowed by the absorption in the p-doped layer, the control of the Rabi oscillation, would be possible for coupling constant g close to the present value, making the long-awaited control of solid-state c-QED processes possible and enabling the generation and control of charge–photon entanglement.

## METHODS

### Device Fabrication

Self-assembled InAs QDs were epitaxially grown on GaAs(001) in a Molecular Beam Epitaxy (MBE) reactor with Stranski–Krastanov method, with a density of few tens per $\mu m^2$, an ensemble peak wavelength at cryogenic temperature of about 1,270 nm and a mean single-exciton linewidth of about 100 μeV, using the low-growth rate procedure introduced in ref. 37. They are integrated in the middle of a GaAs intrinsic region that is embedded between two uniformly doped layers (80 nm-thick top GaAs doped $P=1.5 \times 10^{18}\,cm^{-3}$ and 80 nm-thick bottom GaAs doped $n=2 \times 10^{18}\,cm^{-3}$). The total thickness (400 nm) of the membrane is chosen as a compromise between low diode capacitance, low sheet resistance of the contact layers and high quality factor. The diode structure is grown on top of an undoped $Al_{0.8}Ga_{0.2}As$ sacrificial layer. Diode mesas with a diameter of 16 μm, have been defined using optical lithography and wet etched with a calibrated CitricAcid$H_2O$ solution. On the bottom n-doped layer a low-resistance multi-layer Ge/Ni/Au Ohmic contact is thermally evaporated and annealed at 415 °C to preserve the quality and the emission wavelength of the QDs. The $Si_3N_4$ insulation layer is deposited and reopened selectively with a Buffered HydroFluoric acid (BHF) solution in the regions of the mesas and of the n-contacts. A second $Si_3N_4$ hard mask of 380 nm is deposited for PhCC fabrication. The PhCC patterns are then defined by exposing a ZEP-520A resist with an electron beam lithography system using proximity effect correction. The PhCC holes are then transferred to the SiN hard mask by reactive ion etching and finally to the diode membrane using a $SiCl_4$:Ar recipe. The $Al_{0.8}Ga_{0.2}As$ sacrificial layer is undercut to a lateral extent of 2 μm around the PhCC pattern in 10% hydrofluoric acid. The SiN hard mask is finally removed by a low power $CF_4$:O2 plasma that preserves the underlying SiN insulation without damaging the GaAs surface. The top Zn/Au contact is then evaporated. This fabrication technique allows us to achieve a PhCC with quality factor $\sim 10^4$ and a GHz electrical bandwidth in the same device. The cavity resonance wavelength is lithographically tuned to match the ground state emission wavelength of the QD ensemble at T=10 K. Several modified L3 cavities, one for each device, are fabricated with lattice parameter a ranging from 290 to 320 nm and constant ratio r/a=0.32, with r being the non-modified hole radius. To enhance the quality factor of the cavity, a modified L3 design is used, shifting the first two holes adjacent to the cavity by about 0.15a and reducing their diameter by 20% ref. 38. The ideal $Q_{RAD}$ factor for this cavity, calculated from Finite Element Method (FEM) simulations, is ~27,000. From the scanning electron microscopy analysis, we have estimated a s.d. of $\sigma \sim 2.3$ nm in the hole-radii dimensions and positions, which may be partly responsible for the lower experimental Q value of 10,500. Moreover an absorption of $\alpha \sim 10^2\,cm^{-1}$ in the p-doped GaAs layer set a limit of~60,000 for the real$Q_{TOT}=Q_{RAD} \times Q_{ABS}/(Q_{RAD}+Q_{ABS})$.

### Electro-Optical Characterization

The sample is tested electrically and optically in a cryogenic probe-station setup where a RF micro-probe is used to bias the diodes one by one. A high numerical aperture objective is used to excite the devices and collect the emitted photons. The radiative signal is coupled to a single mode fibre for spectral characterization and time-resolved experiments. The diode IV curves have a turn-on voltage ~1.1 V at T=10 K, and a reverse bias current of order of few nanoamperes with breakdown voltages that range from −5 to −9 V for different devices. To estimate the impedances of the device and its 3 dB frequency, we use a network analyzer to measure the scattering parameter $S_{11}=(Z-Z_0)/(Z+Z_0)$, $Z_0$ being the 50-Ω reference resistance and $Z=R-j/(2\pi fC)$ the frequency dependent total impedance of the circuit. From the amplitude and the phase of $S_{11}$ both R and C can be estimated. A record 3-dB frequency of up to 4.2 GHz was obtained for optimized devices with a mesa diameter φ=12 μm (red circles of Fig. 1b). Although the larger diodes (φ=16 μm) show a lower electrical bandwidth, they represent a better compromise between speed and PhCC quality factor.

The time-resolved PL data in Fig. 4 is obtained in an experiment, where the photons emitted by $QD_1$ are collected by the confocal microscope, selected with a spectral filter (full-width half-maximum=0.65 nm) at the $QD_1$ energy and sent through a fibre to a superconducting single-photon detector, whose output signal is correlated with the trigger from the pulsed laser controller. To show the possibility to dynamically tune the QD at a specific time, we implemented a sub-nanoscale synchronization between the optical excitation pulse and the rising edge of the square-wave bias. The first channel of our fast bias generator provides the external trigger to the pulsed laser diode at 970 nm, while the second channel is used to generate the electrical function for the dynamic Stark tuning of $QD_1$. In this configuration, the two channels share the same trigger frequency of 60 MHz and a variable delay can be set between the optical and the electrical pulse, while the filtered signal is collected from the device and sent to the superconducting single-photon detector every 16.67 ns. While the optical and the electrical paths are kept fixed, the relative delay between the laser and the voltage pulses is changed by varying the internal delay between the first and the second channel of the bias generator with sub-nanoscale accuracy.

### c-QED Simulations

To theoretically describe the experimental situation, the QD is modelled as a three-level quantum emitter coupled to the field of an optical microcavity. In this scheme, the ground state of the QD (level 1) is incoherently pumped at time $t_0$ (operator $\sigma_{13}$) to the higher energy state (level 3), which then relaxes incoherently to the cavity coupled excitonic state of the QD (level 2) through the lowering operator $\sigma_{23}$. The dynamics of the system is simulated by the temporal evolution of the density matrix. The system Hamiltonian describing the coherent QD–cavity interaction in a frame rotating at $\omega_{FRAME}$ reads

$$H = \Delta_{CAV}\hat{a}^{\dagger}\hat{a} + \Delta_{QD}(t)\hat{\sigma}_{12}^{\dagger}\hat{\sigma}_{12} + i\frac{\hbar\Omega}{2}\left(\hat{a}^{\dagger}\hat{\sigma}_{12} - \hat{\sigma}_{12}^{\dagger}\hat{a}\right)$$

where $a(a^\dagger)$ is the bosonic annihilation (creation) operator for the cavity field, $\sigma_{12}$ is the lowering operator that couples the QD exciton state to the cavity mode transferring the quantum emitter to its ground state and $\Omega=2g$ is the Rabi frequency of the coupled system. The cavity frequency is fixed and the detunings are given by $\Delta_i=\omega_i-\omega_{FRAME}$ with $\Delta_{CAV}=0$. The time-dependent QD frequency $\omega_{QD}(t)$ follows the piecewise function (orange lines of Fig. 4) that best approximates the steady state response of the diode to a 50% duty cycle square-wave modulation, well below its cut-off frequency and is written as

$$\omega_{QD}(t)=\omega_A+(\omega_B-\omega_A)\times\begin{cases} C\,e^{\left(-\frac{T}{2\tau_{RC}}\right)}, & t<\delta \\ 1-Ce^{\left(\frac{\delta-t}{\tau_{RC}}\right)}, & \delta<t<\delta+\frac{T}{2} \\ Ce^{\left(\frac{\delta+\frac{T}{2}-t}{\tau_{RC}}\right)}, & t>\delta+\frac{T}{2} \end{cases}$$

where $C=1/(1+e^{-T/2\tau RC})$ and the temporal constant of the circuit $\tau_{RC}$ is set to 140 ps to reproduce the behaviour of the diode with an experimental electro-optical bandwidth ~1.2 GHz (estimated from Fig. 3b). The initial (final) QD frequency is denoted by $\omega_A$ ($\omega_B$), $\delta$ is the delay at which the tuning is occurring, and T=16.67 ns is the period of the electrical signal corresponding to the experimental rate of 60 MHz. The modulation is simulated between the two static energies$E_A$=978.02 meV and $E_B$=977.89 meV.

Including the incoherent pumping, the relaxation of the higher energy level to the bright exciton state and the system losses, the Master equation for the complete dynamics reads

$$\begin{aligned}\dot{\rho}=&-\frac{i}{\hbar}[H,\rho]+\kappa_{CAV}\left(\hat{a}\rho\hat{a}^\dagger-\frac{1}{2}\hat{a}^\dagger\hat{a}\rho-\frac{1}{2}\rho\hat{a}^\dagger\hat{a}\right)\\ &+\gamma_{QD}\left(\hat{\sigma}_{12}\rho\hat{\sigma}_{12}^\dagger-\frac{1}{2}\hat{\sigma}_{12}^\dagger\hat{\sigma}_{12}\rho-\frac{1}{2}\rho\hat{\sigma}_{12}^\dagger\hat{\sigma}_{12}\right)\\ &+\gamma_{relax}\left(\hat{\sigma}_{23}\rho\hat{\sigma}_{23}^\dagger-\frac{1}{2}\hat{\sigma}_{23}^\dagger\hat{\sigma}_{23}\rho-\frac{1}{2}\rho\hat{\sigma}_{23}^\dagger\hat{\sigma}_{23}\right)\\ &+\Phi_{PUMP}\left(\hat{\sigma}_{13}^\dagger\rho\hat{\sigma}_{13}-\frac{1}{2}\hat{\sigma}_{13}\hat{\sigma}_{13}^\dagger\rho-\frac{1}{2}\rho\hat{\sigma}_{13}\hat{\sigma}_{13}^\dagger\right)\end{aligned}$$

where $\kappa_{CAV}$ is the cavity loss rate, $\gamma_{QD}$ is the exciton radiative decay rate into the leaky modes of the photonic crystal, $\gamma_{relax}$ is the fast incoherent relaxation rate from the pumped high energy level to the exciton state and $\Phi_{PUMP}$ is the incoherent pump term. We solve the Master equation numerically. The cavity damping rate, the QD decay rate into leaky modes and the QD–cavity coupling rate have been extracted from the data in Figs 1 and 2 to be $\kappa_{CAV}/2\pi$=22 GHz,$\gamma_{QD}/2\pi$=0.036 GHz and $g/2\pi$=5.4 GHz, respectively. The precise electric field dependence of the QD energy and the electrical delay has been optimized to match the experimental data and is shown in Fig. 4. The emitted photon intensity from the QD–cavity system during the QD tuning is calculated as the mean value of the cavity loss rate using the relation $\kappa_{CAV}\hat{a}^\dagger\hat{a}=Tr\left[\kappa_{CAV}\hat{a}^\dagger\hat{a}\,\rho\right]$

## REFERENCES


1. Purcell, E. Spontaneous emission probabilities at radio frequencies. Phys. Rev. 69, 674 (1946).
2. Kuhn, A., Hennrich, M. & Rempe, G. Deterministic single-photon source for distributed quantum networking. Phys. Rev. Lett. 89, 067901 (2002).
3. McKeever, J. et al. Deterministic generation of single photons from one atom trapped in a cavity. Science 303, 1992 (2004).
4. Keller, M., Lange, B., Hayasaka, K., Lange, W. & Walther, H. Continuous generation of single photons with controlled waveform in an ion-trap cavity system. Nature 431, 1075–1078 (2004).
5. Reithmaier, J. P. et al. Strong coupling in a single quantum dot semiconductor microcavity system. Nature 432, 197–200 (2004).
6. Yoshie, Y. et al. Vacuum Rabi splitting with a single quantum dot in a photonic crystal nanocavity. Nature 432, 200–203 (2004).
7. Thompson, R. J., Rempe, G. & Kimble, H. J. Observation of normal-mode splitting for an atom in an optical cavity. Phys. Rev. Lett. 68, 1132 (1992).
8. Kimble, H. J. The quantum internet. Nature 453, 1023–1030 (2008).
9. Cirac, J. I., Zoller, P., Kimble, H. J. & Mabuchi, H. Quantum state transfer and entanglement distribution among distant nodes in a quantum network. Phys. Rev. Lett. 78, 3221 (1997).
10. Young, R. J. et al. Single electron-spin memory with a semiconductor quantum dot. N. J. Phys. 9, 365 (2007).
11. Raimond, J. M., Brune, M. & Haroche, S. Colloquium: Manipulating quantum entanglement with atoms and photons in a cavity. Rev. Mod. Phys. 73, 565–582 (2001).
12. Parkins, A. S., Marte, P. & Zoller, P. Synthesis of arbitrary quantum states via adiabatic transfer of Zeeman coherence. Phys. Rev. Lett. 71, 3095 (1993).
13. Yin, Y. et al. Catch and release of microwave photon states. Phys. Rev. Lett. 110, 107001 (2013).
14. Pechal, M. et al. Microwave-controlled generation of shaped single photons in circuit quantum electrodynamics. Preprint at http://arXiv.org/abs/1308.4094 (2013).
15. Michler, P. et al. A quantum dot single-photon turnstile device. Science 290, 2282 (2000).
16. Shields, A. J. Semiconductor quantum light sources. Nat. Photonics 1, 214 (2007).
17. Ferne´e, M., Rubinsztein-Dunlop, H. & Milburn, G. J. Improving single-photon sources with Stark tuning. Phys. Rev. A 75, 043815 (2007).
18. Johne, R. & Fiore, A. Single-photon absorption and dynamic control of the exciton energy in a coupled quantum-dot–cavity system. Phys. Rev. A 84, 053850 (2011).
19. Su, C. H., Greentree, A. D., Munro, W. J., Nemoto, K. & Hollenberg, L. C. L. High-speed quantum gates with cavity quantum electrodynamics. Phys. Rev. A 78, 062336 (2008).
20. Johne, R. & Fiore, A. Proposal for a two-qubit quantum phase gate for quantum photonic integrated circuits. Phys. Rev. A 86, 063815 (2012).
21. Faraon, A. et al. Local quantum dot tuning on photonic crystal chips,". Appl. Phys. Lett. 90, 213110 (2007).
22. Faraon, A., Majumdar, A., Kim, H., Petroff, P. & Vucˇkovic´, J. Fast electrical control of a quantum dot

strongly coupled to a photonic-crystal cavity. Phys. Rev. Lett. 104, 047402 (2010).

23. Midolo, L. et al. Spontaneous emission control of single quantum dots by electromechanical tuning of a photonic crystal cavity. Appl. Phys. Lett. 101, 091106 (2012).

24. Bose, R., Sridharan, D., Solomon, G. S. & Waks, E. All-optical tuning of a quantum dot in a coupled cavity system. Appl. Phys. Lett. 100, 231107 (2012).

25. Trotta, R. et al. Nanomembrane quantum-light-emitting diodes integrated onto piezoelectric actuators. Adv. Mater. 24, 2668–2672 (2012).

26. Bru¨ggemann, C. et al. Laser mode feeding by shaking quantum dots in a planar microcavity. Nat. Photonics 6, 30–34 (2012).

27. Jin, C. Y. et al. Ultrafast nonlocal control of spontaneous emission. Nat. Nanotech. 9, 886–890 (2014).

28. Tanabe, T., Nishiguchi, K., Kuramochi, E. & Notomi, M. Low power and fast electro-optic silicon modulator with lateral p-i-n embedded photonic crystal nanocavity. Opt. Express 17, 22505–22513 (2009).

29. Ellis, B. et al. Electrically pumped photonic crystal nanocavity light sources using a laterally doped pin junction. Appl. Phys. Lett. 96, 181103 (2010).

30. Reutera, D. et al. Influence of a lateral electric field on the optical properties of InAs quantum dots. Physica E 32, 73–76 (2006).

31. Fry, P. W. et al. «Inverted electron-hole alignment in inas-gaas self-assembled quantum dots». Phys. Rev. Lett. 84, 733 (2000).

32. Johansen, J., Julsgaard, B., Stobbe, S., Hvam, J. M. & Lodahl, P. Probing long-lived dark excitons in self-assembled quantum dots. Phys. Rev. B 81, 081304 (2010).

33. Balet, L. et al. Enhanced spontaneous emission rate from single InAs quantum dots in a photonic crystal nanocavity at telecom wavelengths. Appl. Phys. Lett. 91, 123115 (2007).

34. Atkinson, P. et al. Formation and ordering of epitaxial quantum dots. C.R. Phys. 9, 8 (2008).

35. Chauvin, N. et al. Controlling the charge environment of single quantum dots in a photonic-crystal cavity. Phys. Rev. B 80, 241306(R) (2009).

36. Winger, M. et al. Explanation of photon correlations in the far-off-resonance optical emission from a quantum-dot–cavity system. Phys. Rev. Lett. 103, 207403 (2009).

37. Alloing, B. et al. Growth and characterization of single quantum dots emitting at 1300 nm. Appl. Phys. Lett. 86, 101908 (2005).

38. Akahane, Y., Asano, T., Song, B. S. & Noda, S. Fine-tuned high-Q photoniccrystal nanocavity. Opt. Express 13, 1202–1214 (2005).

# CHAPTER 11 TOPOLOGICAL PHASES OF SOUND AND LIGHT

V. Peano,[1] C. Brendel,[1] M. Schmidt,[1] and F. Marquardt[1,2]

[1] Institute for Theoretical Physics, Friedrich-Alexander-Universität Erlangen-Nürnberg, Staudtstraße 7, 91058 Erlangen, Germany

[2] Max Planck Institute for the Science of Light, Günther-Scharowsky-Straße 1, 91058 Erlangen, Germany

## ABSTRACT

**Topological states of matter are particularly robust, since they exploit global features of a material's band structure. Topological states have already been observed for electrons, atoms, and photons. It is an outstanding challenge to create a Chern insulator of sound waves in the solid state. In this work, we propose an implementation based on cavity optomechanics in a photonic crystal. The topological properties of the sound waves can be wholly tuned *in situ* by adjusting the amplitude and frequency of a driving laser that controls the optomechanical interaction between light and sound. The resulting chiral, topologically protected phonon transport can be probed completely optically. Moreover, we identify a regime of strong mixing between photon and phonon excitations, which gives rise to a large set of different topological phases and offers an example of a Chern insulator produced from the interaction between two physically distinct particle species, photons and phonons.**

## SUMMARY

Recently, a new topology-based paradigm in the classification of the phases of matter has emerged. Topological states of matter have already been observed for electrons, atoms, and photons. It is an outstanding challenge to engineer a solid-state device on the nanoscale, supporting topologically protected sound waves. We show that such waves could emerge in a surprisingly simple setting: a suitably patterned slab of dielectric illuminated by a laser with an appropriately chosen phase pattern.

Our proposal takes advantage of the enhanced radiation pressure interaction in so-called optomechanical crystals. These crystals are the optomechanical analog of photonic crystals and support defects with co-localized optical and vibrational modes. In our analysis, we predict that creating an optomechanical array formed by a periodic arrangement of such defects will yield a Chern insulator when driven by a suitable laser field. The setup is easily tunable *in situ*by varying the laser drive amplitude and frequency. We show that the resulting chiral, topologically protected phonon transport along the edges can be probed completely optically. In addition to the phonon Chern insulator, we predict a second regime in which photons and phonons form hybrid topological bands, giving rise to a multitude of topological phases of sound and light. This regime represents a novel example of a Chern insulator produced from the interaction of two physically distinct particle species.

We expect that our results will motivate experimentalists and theoreticians to begin exploiting the possibilities offered

**Citation:** V. Peano, C. Brendel, M. Schmidt, and F. Marquardt, Topological Phases of Sound and Light, DOI:http://dx.doi.org/10.1103/PhysRevX.5.031011.

by engineering phonon-photon band structures using light. Experimental verification of our theory would be the first demonstration of topologically protected sound propagation on the nanoscale.

## INTRODUCTION

Recently, a new paradigm in the classification of the phases of matter has emerged that is based on topology [1]. The Hall conductance quantization in a 2D electron gas placed inside a magnetic field is so precise that it serves as a standard to define the Planck constant. The precision is due to the current being carried by chiral edge states which are robust against scattering by disorder. It was realized that at the heart of this effect there is the nontrivial topology of the bulk electron band structure encoded in topological invariants, the Chern numbers [2]. The modern exploration of new topological phases started with the prediction of the anomalous quantum Hall effect [3]. This is a so-called Chern-insulator state that is realized in a staggered magnetic field that has a vanishing average. The subsequent discovery of the quantum spin Hall effect [4,5] then proved that even time-reversal symmetry breaking is not necessary. In this case, the nontrivial topology is induced by the spin-orbit coupling. A third pathway to a nontrivial topology is the time-dependent modulation of the band structure in Floquet topological insulators [6–9].

Inspired by these new developments in our understanding of electronic systems, researchers have begun to extend the concept to other settings. Proposals and first experiments on topological phases exist for cold atoms and ions (see, e.g., Refs. [10–14]). More closely related to our setting is the theoretical [15–21] as well as experimental [9,22–25] investigation of topologically nontrivial phases of light (see Ref. [26] for a recent review). Unlike electrons, photons are electrically neutral. Nevertheless, they mimic the dynamics of charged particles while hopping on a lattice, e.g., when the time-reversal symmetry is broken by synthetic gauge fields [15,16,18–20] or when an effective spin-orbit coupling is engineered [17,21,23].

At present, it remains an outstanding challenge to engineer topological phases for sound waves (phonons) in the solid state, with the resulting robust chiral edge state transport that is useful for applications in phononics. So far, topological properties have been conjectured to be present in the vibrations of individual microtubule macromolecules in biophysics [27], although the precise mechanism requires further investigation. Moreover, recently it was pointed out that masses connected by springs or rigid links in special networks (related to isostatic lattices) show topological features of vibrations. These include zero modes localized at some sample edges of an appropriate geometry [28], propagation of topologically protected nonlinear solitary waves [29] in 1D chains, and topologically robust defect modes bound to dislocations inside a 2D lattice [30]. In contrast to those works, here we propose a 2D phonon metamaterial of the Chern-insulator class that shows chirally propagating edge states robust against disorder. Very recently, there have been steps in this direction for macroscopic systems, employing circulating fluid currents [31] to break time-reversal invariance, or wiring up pendula [32] in the appropriate way to create a topological insulator. Our goal is to propose a fully tunable nanoscale system.

It is not trivial to engineer the required nonreciprocal phases for the transport of phonons in a tunable solid-state platform. Although it would be conceivable to employ local time-dependent modulation of the stress, e.g., using electrodes and piezoelectric materials (essentially emulating the route towards photonic magnetic fields proposed in Ref. [19]), this is not very practical, since the number of wires would scale with the system size.

The tool we employ instead is cavity optomechanics [33], a rapidly evolving field that studies the interaction between radiation and nanomechanical motion, with possible applications in sensing, classical and quantum communication, and tests of foundational questions in quantum physics. In particular, we consider the flexible and scalable platform of optomechanical (OM) crystals [34–38]. These systems are based on freestanding photonic crystals, where engineered defects support colocalized optical and vibrational modes interacting via radiation pressure. Recently, it has been proposed that an array of such point defects would form an optomechanical array "metamaterial" where the resulting optical and mechanical band structures could be tuned *in situ* by a driving laser [39]. Here, we show how to implement a nontrivial topology for sound waves in a solid-state device, based on such optomechanical arrays. This can be achieved when a suitable lattice geometry is chosen and the driving laser imprints an appropriate phase pattern on the optomechanical interaction. The light field then induces an effective Hamiltonian for the sound waves that leads to a Chern insulator with robust edge modes. We emphasize that a single laser field (with a suitable phase pattern) is enough; no time-dependent modulation of any kind is required in our approach. Our proposal not only presents a practicable route towards phonon Chern insulators in the solid state, but its realization would also represent the first example of a topological state of matter produced using optomechanics.

In addition, we find that upon sweeping the laser frequency one can also enter a regime where it is no longer possible to view phonons and photons as separate. Instead, a whole series of topological phase transitions arises where both sound and light are involved. This would be an example of a topologically nontrivial hybrid band structure made of two physically distinct particle species, with corresponding edge states for the emerging hybrid excitations. In contrast to the recently proposed photon-exciton topological polaritons [40–42], the interaction in our case is tunable *in situ* over a wide range via the laser amplitude.

## RESULTS

### A. Opt Mechanical Arrays

In the field of cavity optomechanics [33] the basic interaction between light and mechanical motion comes about because any deformation of an optical cavity's boundaries will lead to a shift of the cavity's optical mode frequencies. Focusing on a single cavity mode, its energy may therefore be expressed as $\hbar\omega_{\text{cav}}(\hat{x})\hat{a}^{\dagger}\hat{a}$, where $\hat{x}$ represents the mechanical displacement and $\hat{a}^{\dagger}\hat{a}$ is the photon number. Expanding to leading order in x^, which is usually an excellent approximation [33], this yields the basic interaction $\hbar\omega'_{\text{cav}}\hat{x}\hat{a}^{\dagger}\hat{a}$. The mechanical motion is very often dominated by a single harmonic vibration mode, such that $\hat{x} = x_{ZPF}(\hat{b} + \hat{b}^{\dagger})$, with $x_{ZPF}$ the mechanical zero-point fluctuations and $\hat{b}$ the phonon annihilation operator. Thus, one arrives at the fundamental optomechanical interaction

$$-\hbar g_0 \hat{a}^{\dagger}\hat{a}(\hat{b} + \hat{b}^{\dagger}), \tag{1}$$

where $g_0 = -\omega'_{cav} x_{ZPF}$ is the bare coupling constant. The optomechanical coupling rate $g_0$ represents the optical shift due to a mechanical zero-point displacement, and it is typically much smaller than the photon decay rate κ. However, by illuminating the sample with laser light, one can effectively enhance the optomechanical interaction. When the system is driven by a laser, one can write $\hat{a} = \alpha + \delta\hat{a}$, where α is the complex amplitude set by the laser drive and $\delta\hat{a}$ represents the quantum fluctuations on top of that. Keeping the leading nontrivial terms, one obtains a quadratic Hamiltonian (the so-called linearized optomechanical interaction),

$$-\hbar g_0(\alpha^* \delta\hat{a} + \alpha\delta\hat{a}^{\dagger})(\hat{b} + \hat{b}^{\dagger}). \tag{2}$$

This is the well-tested basis for the description of almost all quantum-optomechanical experiments to date [33]. The new, effective coupling constant $g=g_0\alpha$ is laser tunable and may be complex, containing a phase factor set by the laser phase, which becomes crucial in our scheme. Equation (2) describes the interconversion between phonons and photon excitations at the cavity mode frequency (terms $\delta\hat{a}^{\dagger}\hat{b}$ and $\delta\hat{a}\hat{b}^{\dagger}$). . Physically, these conversion processes can be understood as anti-Stokes Raman transitions, where the driving photons impinging on the cavity are inelastically scattered into higher-frequency photons by absorbing a phonon (enabling, e.g., laser cooling of mechanical motion). Depending on the laser frequency, there can also be Stokes processes, where driving photons are scattered to lower frequencies while creating a phonon $(\delta\hat{a}^{\dagger}\hat{b}^{\dagger})$, although these are not important for our scheme. For notational simplicity (and following convention), from now on we replace $\delta\hat{a}$ by $\hat{a}$.

In the solid state, the largest values of $g_0$ have been reached in OM crystals [34–38]. These are freestanding photonic crystals, i.e., dielectric slabs with an appropriate pattern of holes that creates complete optical and mechanical band gaps. A local modification of the pattern of holes generates a point defect where optical and mechanical modes can become localized. The OM interaction between such a localized optical and mechanical mode is described by Eq. (1), with$g_0$ on the order of around 1 MHz.

Future optomechanical arrays [43–46] can be produced by fabricating a periodic array of such point defects, in 1D or 2D. The localized modes on adjacent lattice sites will have an evanescent overlap, leading to tunneling of photons and phonons between sites i and j with rates $J_{ij}$ and$K_{ij}$, respectively [43,46,47]. For photons, such tunneling-induced transport between localized modes has been demonstrated experimentally in photonic crystal coupled resonator waveguide arrays [48].

Combining both the optomechanical interaction at each site j as well as the tunneling between sites, the generic optomechanical array Hamiltonian [46,49] reads

$$\hat{H}/\hbar = \sum_j \Omega\hat{b}_j^{\dagger}\hat{b}_j - \Delta\hat{a}_j^{\dagger}\hat{a}_j - (g_j\hat{a}_j^{\dagger}\hat{b}_j + \text{H.c.}) + \hat{H}_{\text{hop}}. \tag{3}$$

The annihilation operators of photons and phonons are denoted by $\hat{a}_j$ and $\hat{b}_{j}$, where the site index $j = (j_1, j_2, s)$ will include a sublattice label s for a non-Bravais lattice. The $\hat{a}_j$ are already displaced by the classical steady-state light amplitude $\alpha_j$, set by the laser amplitude (as explained above), and the $\hat{b}_j$ are displaced by the static mechanical displacement $\beta_j$, set by the constant radiation force; see Appendix A. The term $\hat{H}_{\text{hop}}/\hbar = -\sum_{i,j} J_{ij}\hat{a}_i^{\dagger}\hat{a}_j - \sum_{i,j} K_{ij}\hat{b}_i^{\dagger}\hat{b}_j$ incorporates the hopping of photons and phonons between different sites, and Δ is the laser detuning from the optical resonance, $\Delta \equiv \omega_L - \omega_{cav}$, as we switched to a frame rotating at the laser frequency. To be clear, we note that the detuning Δ defined here already includes a static effective shift of the optical resonance due to the mechanical displacement $\beta_j$, which depends on the laser intensity and is found from the self-consistent classical solution. This is already known well for the standard optomechanical system [33].

The optomechanical interaction displayed in Eq. (3) converts phonons into photons propagating inside the array (and vice versa). The strength of these processes is described by a laser-tunable coupling constant $g_j=g_0\alpha_j$ that is parametrically enhanced by the light amplitude, where $|\alpha_j|^2$ would be the steady-state photon number in mode j. The amplitude $\alpha_j$depends on the site j for the case of an inhomogeneous driving field, which we consider below. For the sake of simplicity, we omit the Stokes transitions of the type $\hat{a}_j^{\dagger}\hat{b}_j^{\dagger}$, where photon-phonon pairs are emitted (or annihilated). Stokes processes are strongly suppressed in the parameter regime that will turn out to be suitable for the topologically nontrivial phase (where Δwill be negative, corresponding to a "red-detuned" laser drive), so we neglect them at first. Since the anti-Stokes processes considered here conserve the total excitation number, Hamiltonian Eq. (3) is equivalent to a single-particle Hamiltonian, and we are able to use the standard classification of topological phases [1].

## CHERN INSULATOR IMPLEMENTED IN AN OPTOMECHANICAL ARRAY

In multimode OM systems, the optical backaction can be used to engineer the effective mechanical interaction, which has been suggested to pave the way to phononic quantum information processing (see, e.g., Refs. [50,51]). In this context, the phonon hopping amplitudes are modified by the new pathways that are opened by the OM interaction. A phononic excitation can be virtually converted into a photon on site i, hop to site j, and be converted back into a phonon on site j. From standard perturbation theory, the probability amplitude associated with this pathway is $J_{ij}g_i g_j^*/(\Omega + \Delta)^2$. Hence, a pattern of phases in the optomechanical coupling $g_j$can lead to a synthetic gauge field for phonons in the form of effective hopping rates $K_{ij}^{(\text{eff})} = K_{ij} + K_{ij}^{(\text{opt})}$ that contain an optically induced component with complex phases [51].

Inspired by previous studies that have indicated that a staggered magnetic field for particles on a kagome lattice yields topologically nontrivial phases [16,52–54], our investigations focus on a kagome optomechanical array. We choose this geometry since it can be naturally implemented in 2D optomechanical crystals based on the snowflake design [47,55], which have been demonstrated in an experiment recently [38]. The general approach described here is of course applicable to other lattice geometries as well. The kagome optomechanical array is sketched in Fig. 1. The idea is to have hexagonal patches of periodically arranged snowflake-shaped holes, with linear dislocation defects forming at the edges between those patches. As has been shown in the experiment [38], a suitable modification of the hole pattern inside the linear defect then creates a pointlike defect with localized modes. The nearest-neighbor coupling between those modes will generate the connectivity of a kagome lattice. Its unit cell contains three sites (s=A,B,C) forming an equilateral triangle (we set the side to 1). Thus, the optomechanical band structure will comprise altogether six bands, three of them photonlike and three phononlike.

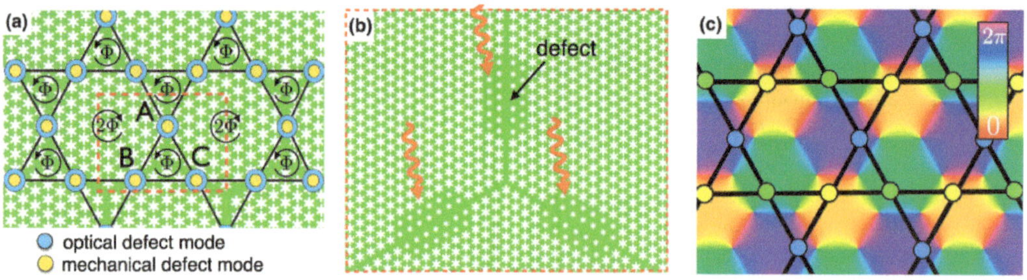


**Figure. 1:** A kagome optomechanical array. (a) Sketch of the overall arrangement of optical and vibrational modes, with nearest-neighbor hopping on a kagome lattice. The effective magnetic fluxes (indicated) add up to zero, realizing a Chern insulator. (b) Schematic representation of the elementary building block in a possible realization based on a 2D snowflake optomechanical crystal [for clarity, the snowflake size in (a) had been exaggerated in comparison]. The picture shows three linear defects that form at the interfaces between hexagonal domains of a periodic snowflake hole pattern. In the center of each, there is an engineered localized pointlike defect mode (as in the experiment of Ref. [38]). (c) Suitable optical phase pattern, generated by the superposition of three beams meeting at 120° angles in the plane of the sample (illustrated here for a scalar field).

Some general properties of the band structure can be deduced purely from the symmetry of the kagome lattice geometry, without assuming anything about the range of the hopping or other details. The hopping term H^hop $\hat{H}_{\text{hop}}$ maintains the time reversal T, the inversion symmetry I with respect to any corner of the triangle, and the symmetry C3 (rotations by n2π/3 around the triangle center, n∈Z). Then, in the absence of the laser drive there is no optical or mechanical band gap: For both the photons and the phonons, the central band touches one of the remaining bands (top or bottom) at the center of the Brillouin zone, $\vec{\Gamma} = (0,0)$, and the other one at the symmetry points, $\vec{K} = (2\pi/3, 0)$ and $\vec{K}' = (\pi/3, \pi/\sqrt{3})$, where Dirac cones form.

We now assume the lattice to be driven by a laser, with an optical phase that depends on the site within the unit cell, leading to $g_j = ge^{i\varphi_s}$. To retain the C3 symmetry, we choose a phase mismatch of 2π/3 between the sublattices, $\varphi_B - \varphi_A = \varphi_C - \varphi_B = \varphi_A - \varphi_C = 2\pi/3$. Physically, the driving laser phase pattern has to form a vortex around each triangle center (Fig. 1). This can be achieved via wave front engineering of the impinging laser field. As opposed to our recently proposed optomechanical generation of arbitrary synthetic magnetic fields for photons[39], we emphasize that here (i.e., for a Chern insulator of sound waves) only a single laser frequency is needed and that the imprinted optical phase field is periodic in space, greatly simplifying its generation. In fact, one does not even need the versatility of a spatial light modulator for this task. Superimposing three plane waves impinging on the sample automatically creates the required pattern of optical phases, if they are at 120° angles with respect to each other within the plane of the sample. This is illustrated in Fig. 1(c) for the slightly simplified case of three interfering plane waves of a scalar field, and we confirm that it also works when taking into account the vector nature of the electromagnetic field.

## BAND STRUCTURE AND TOPOLOGICAL CLASSIFICATION

The resulting band structure in the presence of such a drive is shown in Fig. 2, and we explain it in more detail below. The site-dependent optomechanical interaction breaks the time-reversal symmetry, thereby opening large gaps between the mechanical bands. As we show, these gaps are topologically nontrivial and lead to topologically protected sound waves propagating at the edge of a finite system.

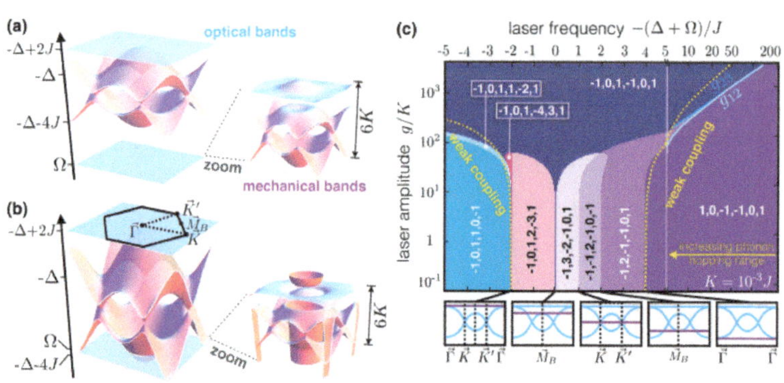


**Figure. 2:** (a) Band structure of a kagome optomechanical array, shown here in the case of well-separated optical and mechanical bands ("weak-coupling limit"). The three mechanical bands appear flat on the scale of the optical bands. A zoom-in shows the resulting phonon insulator.

(b) The "strong-coupling" limit where photon and phonon excitations mix. (c) Topological phase diagram. The different topological phases are marked by the color code and by the set of six Chern numbers, corresponding to the bands ordered by frequency, uniquely identifying each phase. The schematic band structures below the phase diagram indicate the symmetry points where a pair of bands touch at the corresponding phase transition. The OM coupling g (set by the driving laser amplitude) is displayed on a logarithmic scale. The scale of the laser frequency (expressed via the detuning Δ) is linear but switches to logarithmic for large negative detunings, for clarity. The onset of weak coupling is indicated by the line g=0.1Δ34, where Δ34 is the gap between optical and mechanical bands. The approximate analytical expressions (see Appendix D) for the boundaries g12 and g23 of the intermediate topological phase introduced by the long-range hopping are also shown.

As discussed before, Hamiltonian Eq. (3) can be translated from its second-quantized form into a single-particle version, where the nature of the excitation (photon versus phonon) is treated as an internal state. Translational invariance permits us to rewrite it in momentum space, using a plane wave ansatz:

$$\hat{H}(\vec{k})/\hbar = \bar{\omega} - \delta\omega\hat{\sigma}_z/2 - (\bar{t} + \delta t\hat{\sigma}_z/2)\hat{\tau}(\vec{k}) - g(\hat{\mu}\hat{\sigma}_x + \hat{\nu}\hat{\sigma}_y). \quad (4)$$

Here, we assume that only nearest-neighbor sites are coupled, although that (reasonable) approximation could be lifted without destroying any of the essential physical properties discussed in the following; see Appendix B. The binary degree of freedom expressed by $\sigma_z = \pm 1$ denotes photon ( +1) versus phonon ( −1) excitations, and $\hat{\sigma}_{x,y,z}$ are the Pauli matrices in this subspace. Furthermore, we introduce the parameters $\bar{\omega} = (\Omega - \Delta)/2$, $\delta\omega = \Omega + \Delta$, $\bar{t} = (J + K)/2$, and $\delta t = (J - K)$.

The 3×3 matrices $\hat{\mu}$, $\hat{\nu}$, and $\hat{\tau}(\vec{k})$ in Eq. (4) act on the sublattice degree of freedom, referring to the three sites s=A,B,C of the unit cell. The Hermitian hopping matrix $\hat{\tau}(\vec{k})$ encodes motion on the kagome lattice, with $\tau_{AB} = 1 + e^{-i\vec{k}\vec{a}_1}$, $\tau_{AC} = 1 + e^{i\vec{k}\vec{a}_3}$, and $\tau_{BC} = 1 + e^{-i\vec{k}\vec{a}_2}$, where $\vec{a}_1 = (-1, -\sqrt{3})$, $\vec{a}_2 = (2, 0)$, $\vec{a}_3 = (-1, \sqrt{3})$ are the lattice basis vectors. At the symmetry points in the Brillouin zone, the eigenbasis of the $C_3$ rotations diagonalizes $\hat{\tau}(\vec{k})$: the eigenvectors are the $|\circlearrowleft\rangle \equiv |1, e^{i2\pi/3}, e^{-i2\pi/3}\rangle/\sqrt{3}$, the antivortex $|\circlearrowright\rangle \equiv |1, e^{-i2\pi/3}, e^{i2\pi/3}\rangle/\sqrt{3}$,, and the vortex-free state $|\oslash\rangle \equiv |1, 1, 1\rangle/\sqrt{3}$. The matrices $\hat{\mu}, \hat{\nu}$ describe the conversion between photons and phonons. When the OM interaction converts a phonon into an array photon, a driving photon is absorbed and its angular momentum is transferred to the array photon. For example, a vortex-free phonon $|M, \oslash\rangle$ is converted into a photon with a vortex, $|O, \circlearrowleft\rangle$ (M, mechanical; O, optical). The remaining allowed transitions are $|M, \circlearrowleft\rangle \leftrightarrow |O, \circlearrowright\rangle$ and $|M, \circlearrowright\rangle \leftrightarrow |O, \oslash\rangle$. All allowed transitions have matrix element −g, and this fully specifies μ^ and ν^ in the $C_3$ eigenbasis (see also Appendix B).

The eigenfrequencies of Hamiltonian Eq. (4) form six photon-phonon polariton bands. The admixture between photon and phonon bands is weak for all quasimomenta if the highest mechanical and lowest optical bands are separated by a gap larger than the OM coupling, $\Delta_{34} \equiv -\Delta - 4J - \Omega + 2K \gg g$, which we call the "weak-coupling limit." Then, the photons can be adiabatically eliminated, arriving at an effective description for the phonons that incorporates the optical backaction.

In the limit of both weak coupling ( $\Delta_{34} \gg g$) and large detuning, $\Delta_{34} \gg J$, the optically induced effective phonon hopping $K_{ij}^{(\mathrm{eff})}$ will be restricted to nearest neighbors, whence we arrive at the model investigated in Ref. [16,52,53]. A phonon hopping three times anticlockwise around a triangle, at each step with probability amplitude $K_{ij}^{(\mathrm{eff})} \approx K + e^{-i2\pi/3}Jg^2/(\Delta + \Omega)^2 \equiv K^{(\mathrm{eff})}e^{i\Phi/3}$, picks up the phase (see Appendix C)

$$\Phi = -\frac{3\pi}{2} + 3\arctan\frac{2K(\Delta + \Omega)^2 - Jg^2}{\sqrt{3}Jg^2}. \quad (5)$$

Keeping in mind that a vector potential $\vec{A}(\vec{r})$ imprints the phase $q\hbar^{-1}\int_{\vec{r}_i}^{\vec{r}_f} \vec{A}(\vec{r}) \cdot d\vec{r}$ on a particle with charge q hopping on a lattice from $\vec{r}_i$ to $\vec{r}_f$, f, we interpret Φ as the (dimensionless) flux of a synthetic gauge field piercing a triangle. Notice that there is no net average magnetic field as the flux piercing a hexagon is −2Φ; see Fig. 1. The flux Φ decreases monotonically from 0 to−2π with the laser amplitude g. We emphasize that in realistic implementations the photon hopping rate J is much larger than the phonon hopping rate K. It is precisely in this limit that the construction adopted here works well (with interference between direct phonon transport of amplitude K and virtual transport via the photonic route). Indeed, values for the phase all the way down to −2π can be reached, staying well within the weak-coupling limit where Eq. (5) has been derived.

In the opposite, large-bandwidth limit, $J \gg \Delta_{34}$, only a small quasimomentum region close to the $\vec{\Gamma}$-point contributes to the optically induced mechanical hopping. Away from $\vec{\Gamma}$, the OM interaction is suppressed, as the energetic distance between the lowest optical band and the mechanical bands rapidly increases. Thus, the effective mechanical hopping $K_{ij}^{(\mathrm{eff})}$ is long range in this limit, and this will change the topological properties that we discuss below. In general, the range of $K_{ij}^{(\mathrm{eff})}$ is governed by the ratio J/Δ34 and can be tuned by changing the gap Δ34 via the laser frequency.

Finally, going away from weak coupling, one can enter a regime where photon and phonon bands cross and hybridize strongly. In the following, we discuss the topological properties of the optomechanical band structure for all of these regimes.

For systems in the quantum Hall state class A, which is realized here, the topological state is uniquely identified by the bands' Chern invariants (or TKNN invariant [2] after Thouless, Kohmoto, Nightingale and den Nijs). They are defined as the integral over the Brillouin zone of the Berry curvature of each band [2]:

$$C_l = \frac{1}{2\pi}\int_{\mathrm{BZ}} d^2k[\nabla_{\vec{k}} \times \vec{A}_l(\vec{k})] \cdot \vec{e}_z, \qquad l = 1, \ldots, 6. \quad (6)$$

The Berry connection $\vec{A}_l = i\langle \vec{k}_l|\nabla_{\vec{k}}|\vec{k}_l\rangle$ depends on the eigenstates $|\vec{k}_l\rangle$ of Hamiltonian Eq. (4), describing hybrid excitations of photons and phonons. The full topological phase diagram calculated numerically as a function of the laser parameters is shown in Fig. 2. Whenever two (or more) bands touch, their Chern numbers may change, signaling a topological phase transition.

We start our analysis from the regime of weak coupling, when optical and vibrational bands are separated sufficiently (this regime is delimited by the dotted yellow lines in Fig. 2). For concreteness, we focus on negative detunings Δ, to the right of the diagram. At the far right, we are in the limit of nearest-neighbor hopping. There, the phonons are made to realize the kagome Chern-insulator model, with the flux Φ given by Eq. (5). In this model, all three mechanical bands are separated by complete band gaps. Both band gaps close simultaneously for special values of the flux where time-reversal symmetry is unbroken [16]. In our case, this happens when the laser is switched off, where Φ=0, and when it reaches a critical amplitude $g = g_{\text{tp}} \equiv (\Delta + \Omega)\sqrt{K/J}$, where $\Phi = -\pi$. The Chern numbers are $C_{1/3} = \pm\text{sgn}[\sin(\Phi)]$, $C_2 = 0$, where the bands are ordered by increasing energy [52]. Hence, a topologically nontrivial phase arises as soon as the driving is switched on, and the system changes to a different topological phase above the threshold gtp. The photons also experience a synthetic gauge field, whose flux can be obtained from Eq. (5) by exchanging K and J and changing the sign. This flux is therefore small and has opposite direction. The photon band Chern numbers thus turn out to be $C_{4/6} = \mp\text{sgn}[\Phi]$, $C_5 = 0$, without any transition at gtp.

When the photon and phonon bands come closer by changing the laser detuning, an effective long-range hopping of phonons is induced optically, as discussed above. Then, a new topological phase appears for intermediate laser amplitudes, not predicted in the simple nearest-neighbor model. The reason is that the mechanical band gaps no longer close simultaneously but instead at two different critical couplings $g_{12}$ and $g_{23}$ (Fig. 2). In the previously discussed limit of short-range hopping, $J/\Delta_{34} \to 0$, these would again coalesce to become gtp. The Chern numbers for long-range hopping can be computed analytically (see Appendix D).

We now turn to the regime where the photon and phonon bands overlap and interact strongly [see the center of the phase diagram Fig. 2(c)]. There, the topological phases cannot be understood any more as induced by an effective staggered synthetic gauge field for the phonons. They give rise to a phase diagram that is unique for optomechanical arrays. In this regime, a number of different phases appear. By inspecting the limit of small coupling ( g→0), one notices that the topological phase transitions occur whenever bands touch at the symmetry points $\Gamma$, $\vec{K}$, and $\vec{K}'$ or at the special points $\vec{M}_A$, $\vec{M}_B$, and $\vec{M}_C$ [see the sketch of the band structures at the phase transitions, bottom of Fig. 2(c)]. This remains true for arbitrary coupling, and can be explained as follows. Topological phase transitions can happen whenever bands touch each other (instead of repelling), which is possible if there are selection rules preventing them from interacting. At the symmetry points $\vec{k} = \Gamma, \vec{K}, \vec{K}'$, this is guaranteed by angular momentum conservation, whereas at $\vec{M}_A$ the optical and mechanical kagome sublattice sites Aare decoupled from the remaining sublattices $B$ at $\vec{M}_B$ and $C$ at $\vec{M}_C$). The bands actually touch simultaneously at $\vec{K}, \vec{K}'$, due to inversion symmetry, while rotational symmetry makes them touch simultaneously at $\vec{M}_A$, $\vec{M}_B$, and $\vec{M}_C$. From these considerations, we can predict the transitions to occur at the laser detunings $\Delta + \Omega \approx -4J, -2J, -J, 0, 2J$, for small coupling g and small mechanical hopping K. The resulting set of Chern numbers for all six bands is displayed in Fig. 2, for each of the various topological phases.

## CHIRAL EDGE STATE TRANSPORT

A fundamental consequence of the topological nature of the optomechanical band structure is the appearance of chiral edge states at the boundaries of a finite-size system. These excitations are topologically protected against scattering if the bands are separated by a complete band gap. They are thus very distinct from the type of edge states that are produced in graphene-type systems with Dirac dispersion, which are not robust against disorder and whose existence even depends on the details of the boundary. The net number of such edge states (right movers minus left movers) within a given band gap is directly determined by the sum of the Chern numbers of all lower-lying bands. While in the effective short-range kagome model each pair of subsequent bands is separated by such a gap, this is not generally true in the full optomechanical model. Large gaps are desirable because they are more robust against dissipation, disorder, and Stokes scattering, described by additional terms $\hat{H}_{\text{st}} = -\hbar(g_j \hat{a}_j^\dagger \hat{b}_j^\dagger + \text{H.c.})$ in the Hamiltonian. However, topological band gaps cannot be larger than the mechanical bandwidth ~K, since they arise by the interplay of intrinsic and optically induced hopping; see Appendix E. For example, in the realistic regime where the optical bandwidth is larger than the mechanical frequency, the largest topological gap ωgap is given $\omega_{\text{gap}} \approx g\sqrt{2J/K}$ (we consider a laser drive at the mechanical red sideband of the lowest frequency optical mode); see Appendix E. In order to show the feasibility of a topological optomechanical array, we include all of these aspects in the remainder of the paper, with a photon decay rate κ and a mechanical damping rate γ; see Appendixes A and F. In Fig. 3, we show that for realistic parameters the topological gaps are surprisingly resilient to dissipation. The bulk band structure in Fig. 3(d) has a topological gap between the second band (a hybrid photon-phonon band) and the third band. The band gaps in the bulk photon and phonon LDOS, shown in Figs. 3(e) and 3(f) are weakly smeared by dissipation, although the band gap is much smaller than the photon decay rate κ. Such a robustness, which is related to the optomechanically induced transparency phenomenon [56–58], has been noticed in a different context already for a 1D optomechanical array [49]. It occurs because the excitations of the hybrid band have strong phononic components at the band edge, making them less sensitive to photon decay; see the color code in Fig. 3(d). In a strip of finite width (30 unit cells), one can observe that the phononic wave functions form well-localized chiral edge states [Figs. 3(b) and3(c)]. The residual bulk DOS inside the band gap [in Figs. 3(f) and 3(g)] is produced when the mechanical dissipation smears the nearby large peak in the DOS (the height of this peak is larger by a factor of ≈1600 than the residual bulk DOS inside the band gap). That peak is formed by the localized excitations in the flat mechanical band of the kagome lattice. We have checked that the transport is still chiral even in the presence of such residual bulk DOS; see Appendix G. Our analysis shows that the coupling to the localized excitations causes injection losses but not backscattering.

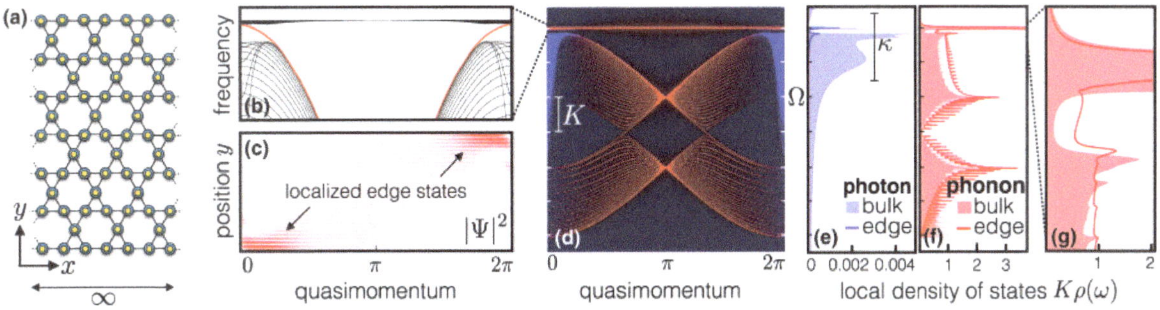


**Figure. 3:** Edge states in a kagome optomechanical array. (d) Band structure in the center of a finite-width strip, whose geometry is shown in (a), as a function of the wave number along the longitudinal direction of the strip. Blue (red) indicates large photonic (phononic) components. Optical and mechanical dissipation, as well as the Stokes interaction, have all been included (see Appendix F). The indicated band gap is of topological nature. (b) Zoom-in of the strip's band structure (here for clarity without dissipation). The dispersion of the edge states is highlighted. The corresponding phonon probability density as a function of position across the strip is shown in (c), demonstrating localization at the edges for quasimomenta where the frequency lies in the bulk band gap. The photonic component (not shown) is small. (e),(f) Local density of states for photons and phonons, respectively, both in the bulk and at the edge of such a strip (of 30 unit cell width). The band gap is much smaller than the photon decay rate κ. (g) Zoom-in of the phonon local density of states. The parameters are $\Omega = 0.1J$, $\Delta = -4.02J$, $K = 0.005J$, $g = 0.007J$, $\kappa = 0.01J$, and $\gamma = 8 \times 10^{-5}J$. .

Finally, we study transport in a finite-size array, for an experimentally realistic setting that reveals the robustness against backscattering by disorder. Additional robustness against mechanical dissipation in the form of clamping losses can be provided by engineering "phonon shields," as demonstrated in Ref. [37]. Since the gapless excitations at the sample edge are phononic in nature, they could be excited by applying local oscillating stress. On the other hand, in the current setting, it is experimentally most straightforward to shine light on the sample edge. Even though the *photon* states are not localized at the edge (unlike the *phonon*edge modes), this simple approach works surprisingly well. Effectively, the beat note between the weak, local probe laser and the strong, global driving laser creates an oscillating radiation pressure force, launching phonons. In Fig. 4, we show a simulation of topologically protected chiral sound waves excited locally by a laser, traveling along the sample edge around an obstruction. In addition, we have checked that moderate random on-site disorder also does not affect the transport. Moreover, it turns out that spatially resolved imaging of the light field emanating from the sample [Fig. 4(a)] can be used to map out the phonon edge state. This is because the local vibrations will imprint sidebands on the strong drive laser, and one of these sidebands appears at the probe laser frequency, which then can be spectrally filtered and imaged.

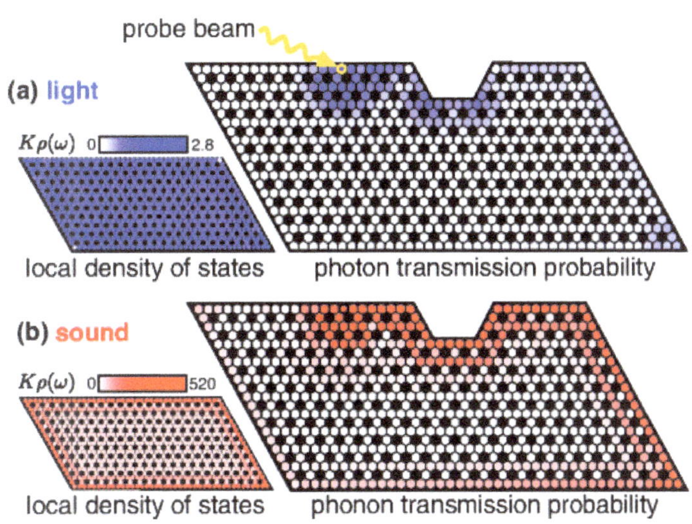


**Figure. 4:** Simulation of transport in a finite system. A probe laser is injected locally at a site on the sample edge, at a fixed frequency, launching phonons that are transported along the chiral edge state. The larger figures depict the probability map of finding a photon (a) or a phonon (b), demonstrating transmission around an obstacle. Because of the optomechanical interaction, the light intensity in (a), which could be imaged locally, represents a faithful probe of the chiral phonon transport in (b). The probe frequency lies in a bulk band gap (ωL−ωprobe=1.06J) that just permits mechanical edge states. The parameters correspond to large detuning between optical and mechanical bands $(\Delta/J = -5.7,\ \Omega/J = 1,\ g/J = 0.2,\ K/J = 0.05,\ \kappa/J = 0.1,\text{ and }\gamma/J = 0.002)$.

In the strong-coupling regime discussed above, where photons and phonons mix completely, one obtains chiral transport of photon-phonon polariton excitations, which can also be excited and read out in the manner discussed here.

The phonons will eventually decay, since the topological protection prevents disorder-induced backscattering but not dissipation (the same is true as well for all topological photon systems, for example). The number of sites over which the phonons propagate along the chiral edge state is given by their speed (the slope of the edge state dispersion) divided by the overall mechanical decay rate (including both intrinsic and optically induced dissipation). In the simplest case, the typical scale of the propagation length is given by the ratio of the mechanical hopping K over the mechanical decay rate $\gamma$. For parameters compatible with state-of-the-art devices, the phonons can propagate for about 100 sites before decaying; see Appendix G for a more detailed analysis. This is completely sufficient for connecting phonon reservoirs and other applications in phononics.

We now address briefly the most relevant issues for the experimental realization of optomechanical Chern insulators. The most important constraint in our model is the need to avoid a mechanical lasing instability [33] that may appear for larger bandwidths ($J>\Omega/3$) due to the Stokes terms. For laser frequencies below the blue sideband of the lowest frequency optical mode, the instability threshold is reached when the cooperativity equals (see Appendix H)

$$C \equiv 4\frac{g^2}{\kappa\gamma} = 1 + \left(\frac{\Delta + 4J - \Omega - 2K}{\kappa/2}\right)^2. \tag{7}$$

The laser intensity (proportional to g2) therefore has to remain below this threshold, which is accounted for when selecting parameters in our simulation results displayed here.

Regarding experimental parameters, in general, in our figures we remain compatible with those of the recent 2D snowflake crystal single-defect experiment [38,55], where they report$\Omega\approx 2\pi\times 9$ GHz and a single-photon coupling strength of $g_0\approx 2\pi\times 250$ kHz. To obtain the$g = g_0\sqrt{n_{\text{phot}}}$ employed in Fig. 3 (where we assume $J=10\ \Omega$ and $g=0.007J$) would require on the order of $n_{\text{phot}} \sim 10^6$ circulating photons. Although challenging, this should be doable, especially since any possible increase of the phonon number due to unwanted heating (and finite temperatures in general) does not affect measurements of the band structure and transmission amplitudes, since the fluctuations do not contribute to the average signal amplitudes.

In all future experiments on transport in optomechanical arrays, it will be important to minimize disorder due to fabrication fluctuations, and efforts to characterize and optimize this are only now starting. In particular, postfabrication processing techniques, such as local oxidation [59], can be employed in the future in order to drastically reduce the disorder by orders of magnitude. In numerical simulations, we have seen that the topological effects persist robustly up to disorder strengths of 2% of $\Omega$ in the mechanical on-site frequencies and up to about J in the optical on-site frequencies (at $J=10\ \Omega$). More generally, we observe in our simulations that there is a wide latitude in parameter combinations to obtain the effects discussed here. For example, it may be more convenient experimentally to use larger photon hopping rates J. Then, the instability is reached for smaller g, the band gaps are smaller, and the edge states' penetration length is larger. We checked that for $J=100\ \Omega$ and $g=5\times 10^{-2}\ \Omega$ (and the other parameters in the same range as Fig. 3), one can still find a topological band gap. The corresponding edge states are well localized on a strip of width 60 unit cells.

## CONCLUSION

Apart from its fundamental interest, chiral phonon transport, robust against disorder, could be useful for many settings. Among them are the transport of phonons between localized long-lived vibrational modes (forming robust "phononic networks") and the study of quantized heat transport [60] in an unconventional setting (with a "one-way" connection between heat reservoirs). The realization of a phonon Chern insulator would thus also enable the observation of new physical phenomena relevant to phononics. In addition, the mechanism we employ is conceptually distinct from anything that has been considered for photons, to the best of our knowledge. In fact, the optomechanical route towards Chern insulators has major advantages over other proposals that have been put forward for photons and which one might try to translate to phonons: The optomechanical concept is more flexible than geometry-based approaches [17,23], since the properties can be tuned quickly *in situ*, and in contrast to settings based on local electrical modulation [19], it does not require local wiring of any kind (which is hard to scale up).

The flexibility of the approach proposed here, where the pattern of the laser field determines the band structure, could be exploited to generate more general layouts *in situ*, where arbitrarily shaped regions of different topological phases are produced, studying the transport through the edge states that form at their interfaces, possibly arranged in interesting interferometer configurations. Moreover, the time-dependent local control of the band structure could be the basis for quench experiments on topological phases of light and sound. Finally, if future improvements in the coupling g0 between single photons and phonons were to permit entering the strong single-photon coupling regime (with $g0\sim\kappa, g0\sim\Omega$), optomechanical fractional Chern insulators could be realized, being governed by strong quantum correlations.

## ACKNOWLEDGEMENTS

We acknowledge support by the ERC Starting Grant OPTOMECH, by the DARPA project ORCHID, and by the European Marie-Curie ITN network cQOM. We thank Aashish Clerk, Alexander Altland, and Oskar Painter for discussions.

## REFERENCES

1. M.Z. Hasan and C.L. Kane, Colloquium: Topological Insulators, Rev. Mod. Phys. **82**, 3045 (2010).
2. D.J. Thouless, M. Kohmoto, M.P. Nightingale, and M. den Nijs, Quantized Hall Conductance in a Two-Dimensional Periodic Potential, Phys. Rev. Lett. **49**, 405 (1982).
3. F.D.M. Haldane, Model for a Quantum Hall Effect without Landau Levels: Condensed-Matter Realization of the "Parity Anomaly," Phys. Rev. Lett. **61**, 2015 (1988).

4. C.L. Kane and E.J. Mele, Z2 Topological Order and the Quantum Spin Hall Effect, Phys. Rev. Lett. **95**, 146802 (2005).

5. B. Andrei Bernevig and S.-C. Zhang, Quantum Spin Hall Effect, Phys. Rev. Lett. **96**, 106802 (2006).

6. T. Oka and H. Aoki, Photovoltaic Hall Effect in Graphene, Phys. Rev. B **79**, 081406 (2009).

7. Z. Gu, H.A. Fertig, D.P. Arovas, and A. Auerbach, Floquet Spectrum and Transport through an Irradiated Graphene Ribbon, Phys. Rev. Lett. **107**, 216601 (2011).

8. N.H. Lindner, G. Refael, and V. Galitski, Floquet Topological Insulator in Semiconductor Quantum Wells, Nat. Phys. **7**, 490 (2011).

9. T. Kitagawa, M.A. Broome, A. Fedrizzi, M.S. Rudner, E. Berg, I. Kassal, A. Aspuru-Guzik, E. Demler, and A.G. White, Observation of Topologically Protected Bound States in Photonic Quantum Walks, Nat. Commun. **3**, 882 (2012).

10. A. Bermudez, T. Schaetz, and D. Porras, Photon-Assisted-Tunneling Toolbox for Quantum Simulations in Ion Traps, New J. Phys. **14**, 053049 (2012).

11. N. Goldman, G. Juzeliūnas, P. Öhberg, and I.B. Spielman, Light-Induced Gauge Fields for Ultracold Atoms, Rep. Prog. Phys. **77**, 126401 (2014).

12. M. Atala, M. Aidelsburger, J.T. Barreiro, D. Abanin, T. Kitagawa, E. Demler, and I. Bloch, Direct Measurement of the Zak Phase in Topological Bloch Bands, Nat. Phys. **9**, 795 (2013).

13. L. Duca, T. Li, M. Reitter, I. Bloch, M. Schleier-Smith, and U. Schneider, An Aharonov-Bohm Interferometer for Determining Bloch Band Topology, Science **347**, 288 (2015).

14. G. Jotzu, M. Messer, R. Desbuquois, M. Lebrat, T. Uehlinger, D. Greif, and T. Esslinger, Experimental Realization of the Topological Haldane Model with Ultracold Fermions, Nature (London) **515**, 237 (2014).

15. F.D.M. Haldane and S. Raghu, Possible Realization of Directional Optical Waveguides in Photonic Crystals with Broken Time-Reversal Symmetry, Phys. Rev. Lett. **100**, 013904 (2008).

16. J. Koch, A.A. Houck, K. Le Hur, and S.M. Girvin, Time-Reversal-Symmetry Breaking in Circuit-QED-Based Photon Lattices, Phys. Rev. A **82**, 043811 (2010).

17. M. Hafezi, E.A. Demler, M.D. Lukin, and J.M. Taylor, Robust Optical Delay Lines with Topological Protection, Nat. Phys. **7**, 907 (2011).

18. R.O. Umucalılar and I. Carusotto, Artificial Gauge Field for Photons in Coupled Cavity Arrays, Phys. Rev. A **84**, 043804 (2011).

19. K. Fang, Z. Yu, and S. Fan, Realizing Effective Magnetic Field for Photons by Controlling the Phase of Dynamic Modulation, Nat. Photonics **6**, 782 (2012).

20. M. Hafezi and P. Rabl, Optomechanically Induced Non-Reciprocity in Microring Resonators, Opt. Express **20**, 7672 (2012).

21. A.B. Khanikaev, S. Hossein Mousavi, W.-K. Tse, M. Kargarian, A.H. MacDonald, and G. Shvets, Photonic Topological Insulators, Nat. Mater. **12**, 233 (2012).

22. Z. Wang, Y. Chong, J.D. Joannopoulos, and M. Soljacic, Observation of Unidirectional Backscattering-Immune Topological Electromagnetic States, Nature (London) **461**, 772 (2009).

23. M. Hafezi, S. Mittal, J. Fan, A. Migdall, and J.M. Taylor, Imaging Topological Edge States in Silicon Photonics, Nat. Photonics **7**, 1001 (2013).

24. M.C. Rechtsman, J.M. Zeuner, Y. Plotnik, Y. Lumer, D. Podolsky, F. Dreisow, S. Nolte, M. Segev, and A. Szameit, Photonic Floquet Topological Insulators, Nature (London) **496**, 196 (2013).

25. L.D. Tzuang, K. Fang, P. Nussenzveig, S. Fan, and M. Lipson, Non-Reciprocal Phase Shift Induced by an Effective Magnetic Flux for Light, Nat. Photonics **8**, 701 (2014).

26. L. Lu, J.D. Joannopoulos, and M. Soljacic, Topological Photonics, Nat. Photonics **8**, 821 (2014).

27. E. Prodan and C. Prodan, Topological Phonon Modes and Their Role in Dynamic Instability of Microtubules, Phys. Rev. Lett. **103**, 248101 (2009).

28. C.L. Kane and T.C. Lubensky, Topological Boundary Modes in Isostatic Lattices, Nat. Phys. **10**, 39 (2014).

29. B. Gin-ge Chen, N. Upadhyaya, and V. Vitelli, Nonlinear Conduction via Solitons in a Topological Mechanical Insulator, Proc. Natl. Acad. Sci. U.S.A. **111**, 13004 (2014).

30. J. Paulose, B. Gin-ge Chen, and V. Vitelli, Topological Modes Bound to Dislocations in Mechanical Metamaterials, Nat. Phys. **11**, 153 (2015).

31. Z. Yang, F. Gao, X. Shi, X. Lin, Z. Gao, Y. Chong, and B. Zhang, Topological Acoustics,Phys. Rev. Lett. **114**, 114301 (2015).

32. R. Süsstrunk and S.D. Huber, Observation of Phononic Helical Edge States in a Mechanical Topological Insulator, Science **349**, 47 (2015).

33. M. Aspelmeyer, T.J. Kippenberg, and F. Marquardt, Cavity Optomechanics, Rev. Mod. Phys. **86**, 1391 (2014).

34. M. Eichenfield, J. Chan, R.M. Camacho, K.J. Vahala, and O. Painter, Optomechanical Crystals, Nature (London) **462**, 78 (2009).

35. A.H. Safavi-Naeini, T.P. Mayer Alegre, M. Winger, and O. Painter, Optomechanics in an Ultrahigh-Q Two-Dimensional Photonic Crystal Cavity, Appl. Phys. Lett. **97**, 181106 (2010).

36. E. Gavartin, R. Braive, I. Sagnes, O. Arcizet, A. Beveratos, T.J. Kippenberg, and I. Robert-Philip, Optomechanical Coupling in a Two-Dimensional Photonic Crystal Defect Cavity, Phys. Rev. Lett. **106**, 203902 (2011).

37. J. Chan, T.P. Mayer Alegre, A.H. Safavi-Naeini, J.T. Hill, A. Krause, S. Groblacher, M. Aspelmeyer, and O. Painter, Laser Cooling of a Nanomechanical Oscillator into Its Quantum Ground State, Nature (London) **478**, 89 (2011).

38. A.H. Safavi-Naeini, J.T. Hill, S. Meenehan, J. Chan, S. Gröblacher, and O. Painter, Two-Dimensional

Phononic-Photonic Band Gap Optomechanical Crystal Cavity, Phys. Rev. Lett.**112**, 153603 (2014).

39. M. Schmidt, V. Peano, and F. Marquardt, Optomechanical Dirac physics, New J. Phys. **17**, 023025 (2015). M. Schmidt, S. Kessler, V. Peano, and F. Marquardt, Optomechanical creation of magnetic fields for photons on a lattice, Optica **2**, 635 (2015).
40. T. Karzig, C.-E. Bardyn, N.H. Lindner, and G. Refael, Topological Polaritons from Quantum Wells in Photonic Waveguides or Microcavities, Phys. Rev. X **5**, 031001 (2015).
41. A.V. Nalitov, D.D. Solnyshkov, and G. Malpuech, Polariton Z Topological Insulator, Phys. Rev. Lett. **114**, 116401 (2015).
42. C.-E. Bardyn, T. Karzig, G. Refael, and T.C.H. Liew, Topological Polaritons and Excitons in Garden-Variety Systems, Phys. Rev. B **91**, 161413 (2015).
43. G. Heinrich, M. Ludwig, J. Qian, B. Kubala, and F. Marquardt, Collective Dynamics in Optomechanical Arrays, Phys. Rev. Lett. **107**, 043603 (2011).
44. D.E. Chang, A.H. Safavi-Naeini, M. Hafezi, and O. Painter, Slowing and Stopping Light Using an Optomechanical Crystal Array, New J. Phys. **13**, 023003 (2011).
45. A. Xuereb, C. Genes, and A. Dantan, Strong Coupling and Long-Range Collective Interactions in Optomechanical Arrays, Phys. Rev. Lett. **109**, 223601 (2012).
46. M. Ludwig and F. Marquardt, Quantum Many-Body Dynamics in Optomechanical Arrays,Phys. Rev. Lett. **111**, 073603 (2013).
47. A.H. Safavi-Naeini and O. Painter, Proposal for an Optomechanical Traveling Wave Phonon-Photon Translator, New J. Phys. **13**, 013017 (2011).
48. M. Notomi, E. Kuramochi, and T. Tanabe, Large-Scale Arrays of Ultrahigh-Q Coupled Nanocavities, Nat. Photonics **2**, 741 (2008).
49. W. Chen and A.A. Clerk, Photon Propagation in a One-Dimensional Optomechanical Lattice, Phys. Rev. A **89**, 033854 (2014).
50. M. Schmidt, M. Ludwig, and F. Marquardt, Optomechanical Circuits for Nanomechanical Continuous Variable Quantum State Processing, New J. Phys. **14**, 125005 (2012).
51. S.J.M. Habraken, K. Stannigel, M.D. Lukin, P. Zoller, and P. Rabl, Continuous Mode Cooling and Phonon Routers for Phononic Quantum Networks, New J. Phys. **14**, 115004 (2012).
52. K. Ohgushi, S. Murakami, and N. Nagaosa, Spin Anisotropy and Quantum Hall Effect in the Kagome Lattice: Chiral Spin State Based on a Ferromagnet, Phys. Rev. B **62**, R6065 (2000).
53. D. Green, L. Santos, and C. Chamon, Isolated Flat Bands and Spin-1 Conical Bands in Two-Dimensional Lattices, Phys. Rev. B **82**, 075104 (2010).
54. H. Katsura, N. Nagaosa, and P.A. Lee, Theory of the Thermal Hall Effect in Quantum Magnets, Phys. Rev. Lett. **104**, 066403 (2010).
55. A.H. Safavi-Naeini and O. Painter, Design of Optomechanical Cavities and Waveguides on a Simultaneous Bandgap Phononic-Photonic Crystal Slab, Opt. Express **18**, 14926 (2010).
56. G.S. Agarwal and S. Huang, Electromagnetically Induced Transparency in Mechanical Effects of Light, Phys. Rev. A **81**, 041803 (2010).
57. S. Weis, R. Rivière, S. Deléglise, E. Gavartin, O. Arcizet, A. Schliesser, and T.J. Kippenberg, Optomechanically Induced Transparency, Science **330**, 1520 (2010).
58. A.H. Safavi-Naeini, T.P. Mayer Alegre, J. Chan, M. Eichenfield, M. Winger, Q. Lin, J.T. Hill, D. Chang, and O. Painter, Electromagnetically Induced Transparency and Slow Light with Optomechanics, Nature (London) **472**, 69 (2011).
59. C.J. Chen, J. Zheng, T. Gu, J.F. McMillan, M. Yu, G.-Q. Lo, D.-L. Kwong, and C.W. Wong, Selective Tuning of High-Q Silicon Photonic Crystal Nanocavities via Laser-Assisted Local Oxidation, Opt. Express **19**, 12480 (2011).
60. K. Schwab, E.A. Henriksen, J.M. Worlock, and M.L. Roukes, Measurement of the quantum of thermal conductance, Nature (London) **404**, 974 (2000).
61. M. Kohomoto, Topological Invariant and the Quantization of the Hall Conductance, Ann. Phys. (N.Y.) **160**, 343 (1985).

CHAPTER 12

# TRAPPED ATOMS IN ONE-DIMENSIONAL PHOTONIC CRYSTALS

C-L Hung[1,2], S M Meenehan[2,3], D E Chang[4], O Painter[2,3] and H J Kimble[1,2]

[1] Norman Bridge Laboratory of Physics 12-33, California Institute of Technology, Pasadena, CA 91125, USA

[2] Institute for Quantum Information and Matter, California Institute of Technology, Pasadena, CA 91125, USA

[3] Thomas J Watson, Sr, Laboratory of Applied Physics, California Institute of Technology, Pasadena, CA 91125, USA

[4] ICFO—Institut de Ciencies Fotoniques, Mediterranean Technology Park, E-08860 Castelldefels, Barcelona, Spain

## ABSTRACT

**We describe one-dimensional (1D) photonic crystals that support a guided mode suitable for atom trapping within a unit cell, as well as a second probe mode with strong atom–photon interactions. A new hybrid trap is analyzed that combines optical and Casimir–Polder forces to form stable traps for neutral atoms in dielectric nanostructures. By suitable design of the band structure, the atomic spontaneous emission rate into the probe mode can exceed the rate into all other modes by more than tenfold. The unprecedented single-atom reflectivity $r_0 \gtrsim 0.9$ for the guided probe field should enable diverse investigations of photon-mediated interactions for 1D atom chains and cavity quantum electrodynamics.**

## INTRODUCTION

New opportunities in atomic, molecular and optical (AMO) physics and quantum information science emerge from the capability to achieve strong radiative interactions between single atoms and the fields of nanoscopic optical waveguides and resonators [1]. For example, strong atom–photon interactions in lithographic structures [2–6] could be used to create quantum optical circuits with long-range atom–atom interactions mediated by single photons [7, 8]. Moreover, linear arrays of atoms radiatively coupled to nanophotonic waveguides exhibit a wide range of remarkable phenomena, including coherent transport of atomic emission [9–11], guided superradiance and polaritons [12–14], as well as highly reflecting atomic mirrors [15, 16]. The interplay of atomic emission into the waveguide and photon-mediated forces can lead to self-organization of atoms into exotic spatial configurations along the waveguide [17, 18].

A long-standing obstacle to this scientific frontier is the challenge of trapping atoms in vacuum near dielectric surfaces (~ 100 nm) while at the same time achieving strong interactions between one atom and photon. A far-off resonance dipole-force trap (FORT) [19] can provide atomic localization by using modes of the dielectric for optical trapping [20–22] and has been used to trap cold atoms within hollow-core optical fibers [23–26] and external to fiber-taper waveguides [27–29].

Motivated by these advances, in this paper we present principles for the design of optical traps and strong atom–photon interactions in one-dimensional (1D) photonic crystal waveguides. Our choice of 1D waveguides is based on their simple photonic band structures and their compatibility with current laser cooling and trapping technology [27–29].

We analyze the trap potential $U_{tot}(\mathbf{r})$ due to light-shifts from a FORT [30–32] together with Casimir–Polder (CP)

**Citation:** C-L Hung, S M Meenehan, D E Chang, O Painter and H J Kimble, Trapped atoms in one-dimensional photonic crystals, doi:10.1088/1367-2630/15/8/083026.

interactions with the dielectric [33–36] (figures 1(a) and (e)). Despite the proximity of the surfaces, stable potentials $U_{tot}(\mathbf{r})$ are achieved for modest optical intensities (~ 5 mW $\mu m^{-2}$) for blue-detuned FORTs operated at a 'magic' wavelength for theD2 line of atomic cesium [19]. A new possibility for trapping is also identified for which vacuum forces from CP interactions are exploited to close the trap perpendicular to the plane of structure, which would otherwise be unstable with either the FORT or the CP potential alone [37].

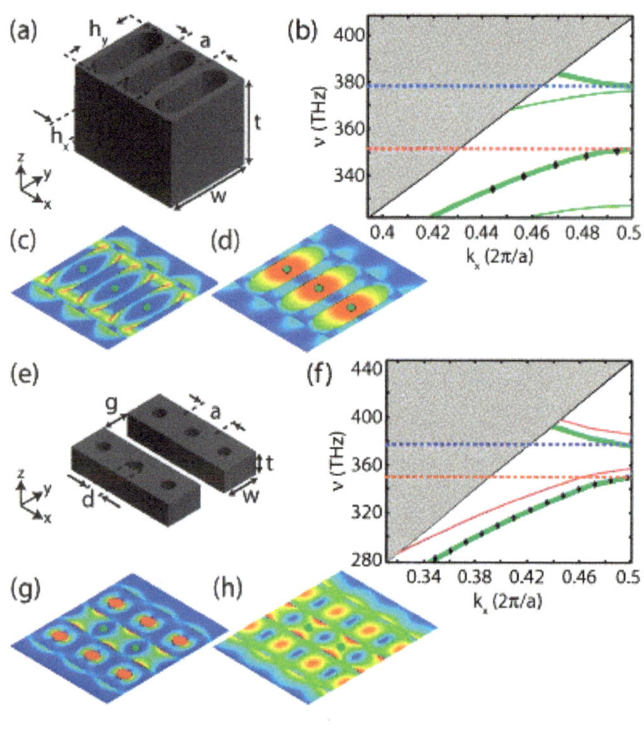


**Figure 1:** (a) Schematic for the single nanobeam structure with dimensions (a,w,t,hx,hy) = (367,845,825,246,745) nm. (b) Band diagram for the single nanobeam in (a) showing only bands with even vector symmetry about the y and z symmetry planes. The trapping and probing bands are shown as thicker lines, with the trap$\omega_T/2\pi$ (probe $\omega_A/2\pi$) frequency as a blue (red) dashed line. (c) Field intensity of the blue trapping mode and (d) field amplitude of the probe mode in the center plane z = 0 for the single nanobeam in (a). Green spheres mark the locations of minima of the trapping potential. (e) Schematic for the double nanobeam structure with (a,w,t,d,g) = (335,335,200,116,250) nm. (f) Band diagram for the double nanobeams in (e) displaying only modes of even vector symmetry in z. The proximity of the two nanobeams results in a band structure composed of even (green) and odd (magenta) superpositions of single nanobeam modes. We focus on the even parity supermodes due to their large field amplitude in the gap. (g) Field intensity of the blue trapping mode and (h) field amplitude of the probe mode in the center plane z = 0 of the double nanobeams in (e). The black diamonds in (b) (f) mark resonances for finite structures of 81 unit cells from figures 4 and 5).

In addition to the waveguide trapping properties, strong near-resonance atom–photon interactions of trapped atoms are found to arise for waveguides with properly tailored band structure [38–42]. For practically realizable structures, we find $\gamma_{1D}/\gamma\rangle \gtrsim 10$, where $\gamma_{1D}$ is the atomic decay rate into the (guided) probe mode and $\gamma\rangle$ the rate into all other modes. One atom trapped within the structure could thereby attenuate a resonant probe with transmission $|1 - r_0|^2 \lesssim 10^{-2}$ [9, 16].

## DESIGN PRINCIPLES FOR THE ONE-DIMENSIONAL PHOTONIC CRYSTAL WAVEGUIDES

As illustrated in figure 1, we focus on two of the simplest quasi-1D photonic crystal geometries. The first waveguide consists of a single silicon-nitride nanobeam (refractive index n = 2) with a 1D array of filleted rectangular holes along the propagation direction; atoms are trapped in the centers of the holes (figure 1(a)). The second waveguide consists of two parallel silicon nitride nanobeams, each with a periodic array of circular holes, with atoms trapped in the gap between the beams (figure 1(e)).

The design of a 1D-photonic crystal waveguide with distinct modes for optical trapping and strong atom–photon interactions is constrained by the region of the optical band structure containing a continuum of unguided optical modes (i.e. the light cone indicated in gray in figures 1(b) and (f)). Modes within the light cone can still have large amplitude in the structure but radiate energy into the surrounding vacuum leading to unacceptable loss. The top of the vacuum light line is at the Brillouin zone boundary (X-point, where kxa = π), so the lattice constant a is constrained by a < λ/2, where λ is the smaller of the (vacuum) wavelengths for trapping and probe fields. Here, kx is the Bloch wavevector along the waveguide axis x.

Once a is fixed, additional guided modes can be ‹pulled› below the light line by increasing the width and thickness of the structure. With appropriate modes below the light line for probing and trapping, the spacing of these modes at the X-point can be tuned by altering the size of the holes, which enables the probe mode to be resonant with the frequency $\omega_A$ of the atomic transition while simultaneously matching the optical frequency $\omega_T$ of the trap mode to a 'magic' frequency for the atom [19].

## OPTICAL TRAPPING AND SURFACE CASIMIR–POLDER POTENTIAL INSIDE THE 1D-WAVEGUIDES

Within this general context, here we consider only blue-detuned FORTs for which the trapping mode has an intensity minimum at the trapping site7. Our analysis is for the D2 line of atomic Cs with probe wavelength near the atomic resonance $\lambda_A$ = 852 nm and with a blue-detuned FORT at the magic wavelength $\lambda_T$ = 793 nm [19]. Note that our results are readily transcribed to other atomic transitions by way of the scale invariance of Maxwell's equations [43].

The photonic crystals are assumed to be suspended in vacuum and composed of SiN. Band structures are calculated using the MIT photonic-bands (MPB) software package [44]. Field profiles for guided modes are calculated using the finite-element-method simulations8. Results for the single and double nanobeam structures are presented in figure 1.

With suitable guided modes for trapping in hand, we have developed numerical tools for evaluating the FORT and CP potentials inside the waveguide, and hence the total potential$U_{tot} = U_{FORT} + U_{CP}$. The adiabatic potential $U_{FORT}(\mathbf{r})$ is readily calculated using the electric field distribution of the trap mode, $\mathbf{E}(\mathbf{r}) = \mathbf{u}kx(\mathbf{r})e^{ikxx}$ [30–32]; $\mathbf{u}k\,x(\mathbf{r})$ is the periodic Bloch wave function at kx.

The surface potential $U_{CP}$(**ra**) is determined from the formalism in [35] for the imaginary component of the scattering Green's tensor $\mathbf{G}_{sc}(\mathbf{r}_a,\mathbf{r}_a,\omega)$ from Maxwell›s equations for a point dipole at the atomic location $\mathbf{r}_a$ with the vacuum contribution (i.e. no dielectric structure) subtracted. We evaluate $\mathbf{G}_{sc}(\mathbf{r}_a,\mathbf{r}_a,\omega)$ numerically by adapting the procedures from [36] (see appendix A).

Figures 2 and 3 display numerical results for $U_{CP}(\mathbf{r})$, $U_{FORT}(\mathbf{r})$ and $U_{tot}(\mathbf{r})$ for the single and double nanobeams for kx below the X-point. The calculations are for the $6S_{1/2}$,F = 4 hyperfine ground state of Cs for the FORT modes indicated in figures 1(c) and (g)9. For these initial calculations, we make the reasonable assumption for SiN that the dielectric constant $\epsilon$ is frequency independent, $\epsilon(\mathbf{r},\omega) \rightarrow \epsilon(\mathbf{r})$; see appendix A.1 for the justification by numerical evaluation of $U_{CP}(\mathbf{r})$ using the full frequency dependence of $\epsilon(\omega)$ for SiN at several locations.

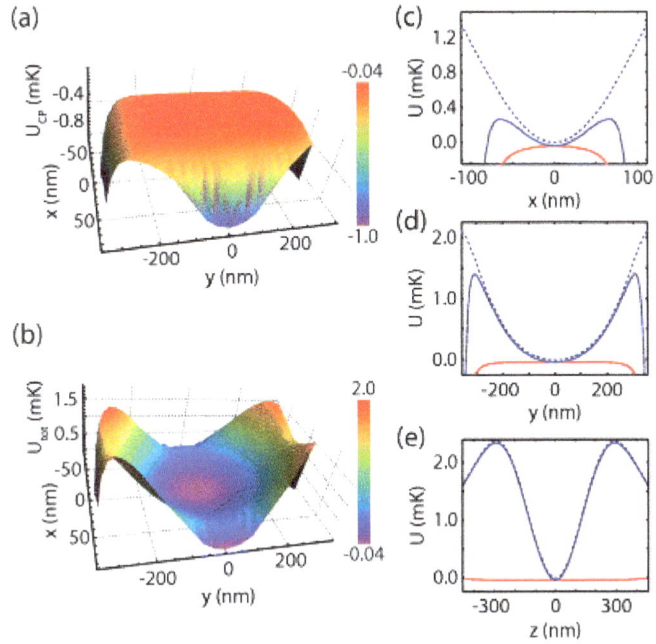


**Figure 2:** Trapping potentials for the single nanobeam structure in figure 1(a) for Cs $6S_{1/2}$, F = 4 level and $\lambda_T$ = 793 nm. (a) CP potential $U_{CP}(\mathbf{r})$ and (b) total potential$U_{tot}(\mathbf{r}) = U_{CP}(\mathbf{r}) + U_{FORT}(\mathbf{r})$ in the central z = 0 plane. Panels (c)–(e) show line cuts of$U_{CP}$ (red solid), $U_{FORT}$ (blue dashed) and $U_{tot}$ (blue solid) along the (c) x-, (d) y- and (e) z-axis. Average trap intensity for a unit cell is 4.9 mW $\mu m^{-2}$.

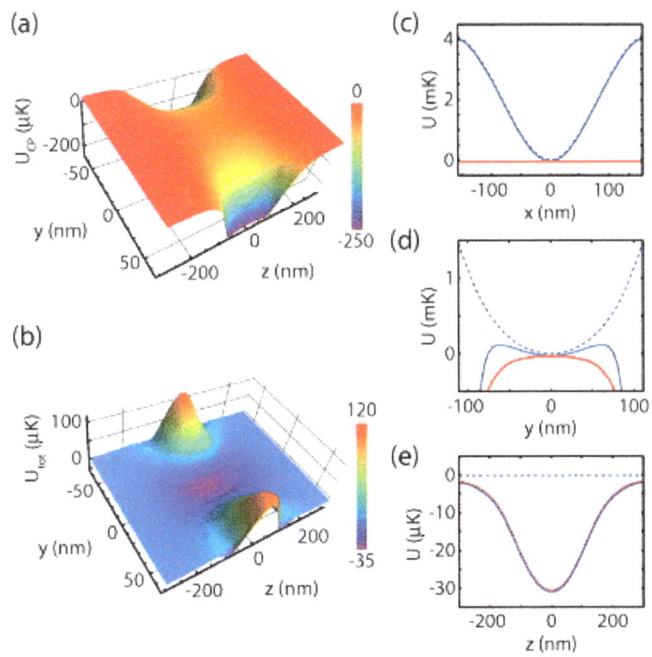


**Figure 3:** Trapping potentials for the double nanobeam structure in figure 1(e) for Cs $6S_{1/2}$,F = 4 level and $\lambda_T$ = 793 nm. (a) CP potential $U_{CP}(\mathbf{r})$ and (b) total potential$U_{tot}(\mathbf{r}) = U_{CP}(\mathbf{r}) + U_{FORT}(\mathbf{r})$ in the transverse x = 0 plane. Panels (c)–(e) show line cuts of $U_{CP}$ (red solid), $U_{FORT}$ (blue dashed) and $U_{tot}$ (blue solid) along the (c) x-, (d) y- and (e) z-axis. Average trap intensity for a unit cell is 3.5 mW $\mu m^{-2}$.

## Stable Optical Trapping in a Single Nanobeam

$U_{tot}(\mathbf{r})$ for the single nanobeam in figure 2 reveals that modest optical intensity is sufficient to overcome the attractive CP interactions and create a stable potential in the center of the vacuum space at $\mathbf{r}_{min} = 0$ within a unit cell. An atom would be localized at distances (dx,dy) = (123,373) nm from the walls of the dielectric. The trap oscillation frequencies for a Cs atom would be (fx,fy,fz) $\simeq$ (612,180,484) kHz.

## A New Hybrid trap in Double Nanobeams: Utilizing Vacuum Casimir–Polder force to realize a Stable Trap Condition

For the double nanobeams [45], the FORT alone is insufficient to trap the atom, as the mode has a (weak) local intensity maximum along the z direction that repels an atom. However, the CP potential $U_{CP}(\mathbf{r})$ along z provides the force necessary to overcome the repulsive optical force and to form a stable trap. The result is a hybrid optical–vacuum trap that circumvents the 'no-go' theorem for vacuum trapping alone [37].

Potentials for our hybrid trap are illustrated in figure 3. At the trap minimum $\mathbf{r}_{min} = 0$, an atom would be localized at distance dy = 125 nm from adjacent surfaces of the dielectric beams. Oscillation frequencies for a Cs atom would be (fx,fy,fz) $\simeq$ (1013,390,57) kHz.

## Achieving strong atom–photon interactions via guided modes near a photonic band edge

As concerns strong radiative interactions, our structures trap an atom in a region of large amplitude for the probe field, leading to small mode volume per unit cell (appendix E). Atom–photon interactions can be further enhanced near a band edge [38–42], where the density of states diverges due to a van Hove singularity. To quantify the radiative coupling, we determine the decay rate $\gamma_{tot}$ for a point dipole located at $\mathbf{r}_a = 0$ for a structure with Nunit cells10. Finite-difference-time-domain (FDTD) calculations are performed to evaluate the classical Green's tensor $\mathbf{G}(\mathbf{r}_a,\mathbf{r}_a,\omega)$ and thence $\gamma_{tot}$ following [34, 35, 46] (appendix B).

## Green's Tensor and Atomic Spontaneous Emission Inside the 1D-Waveguide

Figures 4(a) and 5(a) display the diagonal components Im[Gii($\nu_d$)] of the Green's tensor as functions of dipole frequency $\nu_d = \omega_d/2\pi$ and relate to the emission rate of resonant point dipoles polarized along the i = x-, y- or z-axis for the single and double nanobeams. Firstly, in figure 4(a), Im[Gii] is enhanced along the x (periodic) direction across a broad frequency range, and is suppressed in the y and z directions, as can be explained by the orientation of the induced array of image dipoles along the single nanobeam.

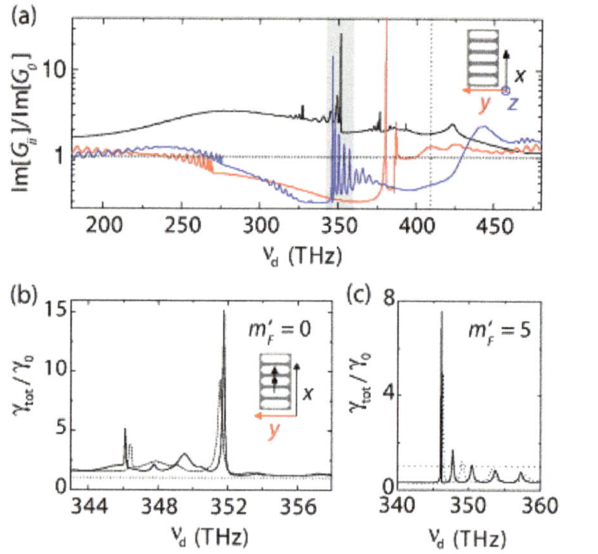


**Figure 4:** Green's tensor and total atomic decay rate $\gamma_{tot}$ versus source dipole frequency$\nu_d$ for the single nanobeam at $\mathbf{r}$ = 0. Panel (a) shows the diagonal components of the Green's tensor, Im[Gxx] (solid black), Im[Gyy] (red) and Im[Gzz] (blue), normalized to the free space value Im[$G_0$] (dashed line). The number of unit cells is N = 81. CesiumD2-line frequency $\nu_A$ = 352 THz is centered in the shaded area. The vertical dotted line marks the light line, beyond which all decay channels are lossy. Panels (b) and (c) show $\gamma_{tot}$ (black curves), normalized to the free space value $\gamma_0$ (dotted line), in the frequency range marked by the shaded area in (a). The solid (dashed) curve is evaluated using 81(61) unit cells. The atomic spin is aligned to the x-axis, with the spin projection quantum number (b) $m'_F$ = 0 and (c) $m'_F$ = 5.

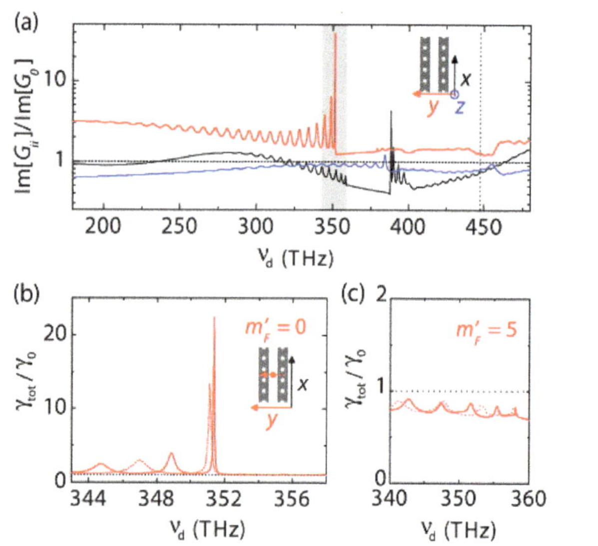


**Figure 5:** As in figure 4 now for the double nanobeams at $\mathbf{r}$ = 0. Panel (a) shows the diagonal components of the Green's tensor, Im[Gxx] (solid black), Im[Gyy] (red), and Im[Gzz] (blue), normalized to Im[$G_0$] for free space (dashed line). N = 81 unit cells. Cesium D2-line frequency $\nu_A$ = 352 THz is centered in the shaded area. The vertical dotted line marks the light line, beyond which all decay channels are lossy. Panels (b) and (c) show $\gamma_{tot}$ (red curves), normalized to the free space value $\gamma_0$ (dotted line), in the frequency range marked by the shaded area in (a). The solid (dashed) curve is evaluated using 81(61) unit cells. The atomic spin is aligned to the y-axis, with the spin projection quantum number (b) and (c) m'F = 5.

Secondly, figures 4(a) and 5(a) display a series of resonant peaks due to strong emission into various guided modes. In the region near the Cs D2 line (i.e. $\nu_d \simeq \nu_A = \omega_A/2\pi = 352$ THz), we find peaks in Im[Gxx] for the single nanobeam and in Im[Gyy] for the double nanobeams. These peaks are due to emission into our designated probe modes for the respective structures, where for the single (double) nanobeam(s), the probe mode is principally polarized along the x-(y-) axis. Each peak is from a discrete set of propagation constants$k^{(n)}x \simeq \pi n/aN$ imposed by the boundary conditions for the finite structures. Here, N is the total number of cells in the single (double) beam, and $n \leq N$ is an even (odd) integer. We find excellent agreement between the frequencies of these resonances and the band diagram of the probe mode ('diamonds' in figures 1(b) and (f)) for various (n,N).

The peaks become larger and narrower as $k^{(n)}x$ approaches the X-point, owing to the diminishing group velocity [40, 42]. Beyond the X-point, the probe resonances disappear, leaving a broad background corresponding to coupling into lossy (radiation) modes.

On an expanded frequency scale around $\nu_d \simeq \nu_A$, figures 4(b) and 5(b) show calculated atomic decay rates $\gamma_{tot}$ for the $6P_{3/2}$, F' = 5 → $6S_{1/2}$, F = 4 transition in atomic Cs (appendix B). When the atomic dipole is aligned along the principal polarization of the designated probe mode ($\hat{x}$—single beam; $\hat{y}$—double beam), the emission rate $\gamma_{1D}$ into the probe mode is strongly enhanced at frequencies corresponding to $k^{(n)}x$ near the X-point. Specifically, for $\nu_d = \nu_A$ large enhancements in $\gamma_{1D}$ occur for the initial excited state $6P_{3/2}$, F' = 5, $m'_F = 0$, while $\gamma_{1D}$ is suppressed for the initial state $m'_F = 5$. This is because the probe mode predominantly supports π-polarization and hence $\Delta m_F = 0$. Coupling between states with $\Delta m_F \neq 0$ is small. Of course, additional guided modes can contribute to $\gamma_{tot}$, as is evidenced for the $m'_F = 5$ state due to field polarizations perpendicular to the atomic spin, such as $\hat{z}$ ($\hat{x}$) for the single (double) beam(s) in figure 4(a) (figure 5(a)).

## Enhanced Probe-Mode Coupling rate $\gamma_{1D}$ and the Resulting Strong atom–Photon Interaction

From $\gamma_{tot}(\nu_d)$ and an analytic model of coupling to the guided-mode near the X-point, we estimate the contributions of $\gamma_{1D}$ and γ› to $\gamma_{tot} = \gamma_{1D} +$ γ› near the largest resonances in figures 4(b) and 5(b) (appendix D.1). For $m'_F = 0$ and N = 81, we find that $\gamma_{1D}/\gamma_0 \simeq 15$ and$\gamma›/\gamma_0 \simeq 1.2$ for the single nanobeam, while $\gamma_{1D}/\gamma_0 \simeq 21$ and $\gamma›/\gamma_0 \simeq 1.0$ for the double nanobeams11. Here, $\gamma_0/2\pi = 5.2$ MHz, the free-space Cs decay rate.

The ratios $\gamma_{1D}/\gamma_{tot}$ and $\gamma_{1D}/\gamma›$ serve as metrics for the strength of atom–photon interactions for our 1D photonic crystals. For example, the resonant reflectivity $r_0$ of a trapped atom for the probe field should scale as $r_0 = \gamma_{1D}/\gamma_{tot}$ [9, 16], which for the double nanobeams leads to$r_0 \simeq 0.95$. For a cavity quantum electrodynamics (QED) system with one 'impurity' atom surrounded by $N_A$ 'mirror' atoms along a 1D-lattice [16], the ratio of the coherent coupling rate $g_1 = \sqrt{N_A}\gamma_{1D}/2$ to the effective dissipative rate γ› would exceed unity even for $N_A = 1$ atom. For conventional cavity QED, we estimate a 1-photon Rabi frequency ~ 2π × 2 GHz for figure 5(b) with N = 81.

## CONCLUSION

Certainly there are challenges in the implementation of our designs, including atom loading into the small trap volumes and light scattering from device imperfections. The latter can reduce the expected coupling rate $\gamma_{1D}$ and introduce trap imperfections. The former challenge may be resolved either by loading atoms continuously from an atomic reservoir surrounding the nanostructure [27–29] to improve the loading efficiency, or by using a moving optical lattice to transfer atoms precisely into the trapping sites. We are working to address these issues by numerical simulation, device fabrication and cold-atom experiments with nanoscopic structures. Our efforts are motivated by the prediction $\gamma_{1D}/\gamma_{tot} \gtrsim 0.9$ in figures 4 and 5, which is unprecedented in AMO physics and which could create new scientific opportunities (e.g. quantum many-body physics for 1D atom chains with photon-mediated interactions, and high-precision studies of vacuum forces). Moreover, our double nanobeam structure provides proof-of-principle for a promising new concept that combines optical and vacuum forces to form stable traps for neutral atoms in dielectric nanostructures.

Our designs of 1D periodic structures represent a basis for extensions to more complicated geometries, including two-dimensional (2D) photonic crystals with trapped atoms. A step in this direction is with structures that are still 1D in nature (e.g. a line of defect patterns embedded in a 2D photonic band gap crystal). Such structures should offer larger values for$\gamma_{1D}$ and further suppressions of emission rate γ› to other modes (including the vacuum).

## ACKNOWLEDGMENT

We gratefully acknowledge D J Alton, K S Choi, D Ding and A Goban. Funding is provided by the IQIM, an NSF Physics Frontier Center with support of the Moore Foundation, by the AFOSR QuMPASS MURI, by the DoD NSSEFF program (HJK) and by NSF grant no. PHY0652914 (HJK). DEC acknowledges funding from Fundacio Privada Cellex Barcelona.

Note added in proof.

After submission of our paper, related work in [47] was published.

Appendix A.: Calculation of Casimir–Polder potentials

The CP potential $U_{CP}(\mathbf{r}_a)$ is calculated from the following integral [35]:

$$U_{CP}(\mathbf{r}_a) = -\frac{\hbar\mu_0}{2\pi}\mathrm{Im}\left\{\int_0^\infty d\omega\omega^2\,\mathrm{Tr}[\boldsymbol{\alpha}^0(\omega)\cdot\mathbf{G}_{sc}(\mathbf{r}_a,\mathbf{r}_a,\omega)]\right\}, \tag{a1}$$

where Tr[.] denotes the trace, $\boldsymbol{\alpha}^0$ is the dynamic polarizability tensor of ground-state cesium atom, and $\mathbf{G}_{sc}(\mathbf{r}_a,\mathbf{r}_a,\omega) = \mathbf{G}(\mathbf{r}_a,\mathbf{r}_a,\omega) - \mathbf{G}_0(\mathbf{r}_a,\mathbf{r}_a,\omega)$ is the scattering Green›s tensor, that is, the Green›s tensor **G** subtracted by the vacuum contribution $\mathbf{G}_0$ evaluated at atomic location $\mathbf{r}_a$; $2\pi\hbar$ is Planck's constant, and $\mu_0$ is the vacuum permeability. The Green's tensor is the solution to the Maxwell equation $[\nabla\times\nabla\times-\frac{\omega^2}{c^2}\epsilon(\mathbf{r},\omega)]\mathbf{G}(\mathbf{r},\mathbf{r}',\omega) = \mathbf{I}\delta^{(3)}(\mathbf{r}-\mathbf{r}')$, corresponding to the electric field response to a point dipole current source. $\epsilon(\mathbf{r},\omega)$ is the dielectric function and **I** is the unity tensor.

We employ FDTD calculations [48] to solve numerically for the Green's tensors of our structures. The integral of equation (A.1) is evaluated by adapting a procedure established in [36] and by using a deformed contour $[\nabla \times \nabla \times -\frac{\omega^2}{c^2}\epsilon(\mathbf{r},\omega)]\mathbf{G}(\mathbf{r},\mathbf{r}',\omega) = \mathbf{I}\delta^{(3)}(\mathbf{r}-\mathbf{r}')$, in the upper half of the complex frequency plane, parameterized by a real number $\xi \geq 0$ and a constant $\sigma > 0$. As explained in [36], this is equivalent to solving the Green's tensor at real frequencies $\xi$ with a fictitious global conductivity applied to the dielectric function $\epsilon'(\mathbf{r},\xi) = (1 + i\sigma/\xi)\epsilon(\mathbf{r},\xi)$. The integration can then be performed in the time domain (via the convolution theorem) and converges quickly due to fast decay from $\sigma$.

Specifically, equation (A.1) is numerically evaluated using

$$U_{\mathrm{CP}}(\mathbf{r}_{\mathrm{a}}) = \frac{\hbar}{2\pi}\int_0^{\infty} dt\, \mathrm{Im}\left[g_{\mu\nu}(-t)\right]\hat{x}_{\mu}\cdot\mathbf{E}_{\mathrm{sc},\nu}(\mathbf{r}_{\mathrm{a}},t), \tag{a2}$$

where $\mathbf{E}_{\mathrm{sc},}\nu(\mathbf{r}_{\mathrm{a}},t)$ is the (real) electric field generated by a point dipole current source $\mathcal{J} = \delta(t)\hat{x}_{\nu}\ (\hat{x}_{\nu} = \hat{x},\hat{y},\hat{z})$ located at the position $\mathbf{r}_{\mathrm{a}}$ and scattered by a structure with a dielectric function $\epsilon'(\mathbf{r},\xi)$.12 Here, the indices $\mu$ and $\nu$ are repeated for summation convention, $g\mu\nu(t)$ is the Fourier transform of $g_{\mu\nu}(\xi) = -i\xi\sqrt{1+\frac{i\sigma}{\xi}}(1+\frac{i\sigma}{2\xi})\Theta(\xi)\alpha^{0}_{\mu\nu}(\omega(\xi))$ and $\Theta(\xi)$ and $\Theta(\xi)$ is the Heaviside step function.

## Material Dispersion

For the initial calculations, we take the dielectric constant $\epsilon$ to be frequency independent, $\epsilon(\mathbf{r},\omega) \rightarrow \epsilon(\mathbf{r})$. In principle it is important to take into account the frequency dependence of the refractive index when calculating the CP potential. However, for the trapping of atoms at distances of 100 nm from dielectric structures, the dominant contribution to the force comes from the material response near the atomic resonance frequencies. This is dictated by the smooth frequency cutoff coming from the atomic polarizability; see equation (A.1), as well as [49] for more detailed discussions. At distances on the order of the resonant wavelengths or more (not relevant to our calculations), the low-frequency material response would become increasingly important [49]. Since the refractive index of silicon nitride is approximately constant across the optical domain, we adopt the approximation based on a frequency independent dielectric constant $\epsilon(\mathbf{r})$.

In order to fully justify our assumption, we have performed calculations using full material dispersion, and have found the results well approximated by calculations using constantn = 2. For the simulated material in FDTD calculations, we have taken the measured dispersion in dc, THz [50] and optical [51] regimes and use a Lorentz oscillator model to fit the complex refractive index $\tilde{n} = n + ik$; see figures (A.1a)–(b). The resulting CP potentials are shown in figure A.1(c) for the single nanobeam structure and in figure A.1(d) for the double nanobeam structure. Each of them differs from the values calculated with n = 2 by no more than 8% both at the trap center and closer to the walls.

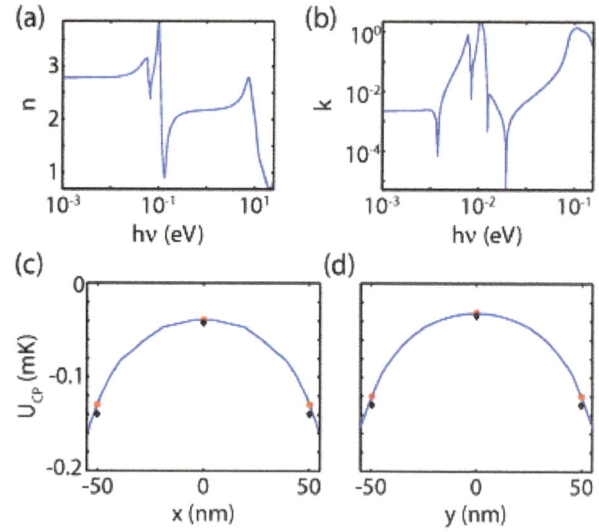


**Figure A.1:** Panels (a) and (b) show the real (n) and imaginary (k) refractive indices, respectively, of silicon nitride obtained from a model of 9 Lorentz oscillators. Panel (c) shows the CP potential calculated for the single nanobeam structure along the x-axis. The blue line shows the calculation using a frequency independent n = 2. At the discrete points x = −50, 0 and 50 nm, the frequency independent results (red circles) were compared with results using the full frequency dependent material response (black diamonds). Panel (d) shows a similar plot along the y-axis of the double nanobeam structure.

The time required for computation increases linearly with the number of Lorentz oscillators used to model the material, and, for the nine oscillator model used here, the computation time increases by an order of magnitude. Due to limited computational resources, we have used the frequency independent results as a reasonable approximation when calculating the CP potential throughout both structures.

Appendix B.: Calculation of $\gamma_{\mathrm{tot}}$

To determine the total spontaneous decay rate $\gamma_{\mathrm{tot}}$ for an atom in our structures, we also solve for the classical Green's tensors and evaluate $\gamma_{\mathrm{tot}}$ via [34, 35, 46]

$$\gamma_{\mathrm{tot}} = \frac{2\mu_0\omega_j^2}{\hbar}\mathrm{Im}\left\{\sum_{\{0\}}\mathrm{Tr}[\mathbf{D}_j\cdot\mathbf{G}(\mathbf{r}_{\mathrm{a}},\mathbf{r}_{\mathrm{a}},\omega_j)]\right\}, \tag{b2}$$

where $\mathbf{D}_j = \langle\{0\}|\mathbf{d}^{\dagger}|j\rangle\langle j|\mathbf{d}|\{0\}\rangle$ is the dipole matrix element between the ground state manifold and the excited state j, and ωj is the transition frequency. The total decay rate$\gamma_{\mathrm{tot}} = \gamma_{\mathrm{1D}} + \gamma'$ includes the decay rate $\gamma_{\mathrm{1D}}$ to a guided mode of interest as well as the rate $\gamma'$ to all other modes of the structure, including lossy modes. As discussed below, the contributions of $\gamma_{\mathrm{1D}},\gamma'$ to $\gamma_{\mathrm{tot}}$ can be estimated from the global frequency dependence$\gamma_{\mathrm{tot}}(\omega)$.

To obtain $\gamma_{\mathrm{tot}}(\omega)$, we evaluate the Green's tensor for the real dielectric function $\epsilon(\mathbf{r})$ using the FDTD method, followed by a discrete Fourier analysis.

## Appendix C.: Validation

To validate our numerical procedures, we have performed calculations of $U_{CP}$ for several geometries where analytical solutions are available, including an atom near an infinite dielectric or metal half-space [52] and an atom located above an infinite dielectric grating [53], and found excellent agreement between our simulations and the exact results.

We have validated our calculations of $\gamma_{tot}$ for the cases of an (atomic) dipole near an infinite dielectric, metallic parallel plates, a nanofiber [54] and 2D-photonic band-gap microcavities [55].

Appendix D.: Guided mode resonances

For our structure with an infinite number of unit cells, a guided mode (denoted by λ) contribution to the imaginary part of the Green›s tensor can be calculated as [46, 56], $\mathrm{Im}\left[\mathbf{G}^{\lambda}_{1D}(\mathbf{r}_a, \mathbf{r}_a, \omega)\right] = ac^2\mathbf{u}_{\lambda}(\mathbf{r}_a; k_x) \otimes \mathbf{u}^*_{\lambda}(\mathbf{r}_a; k_x)/2\omega v_{\lambda}$, when the frequency ω intersects the frequency band ωλ at a propagation constant kx below the light line. Here, **u**λ(**r**;kx) is the orthonormal mode function, and vλ is the group velocity, both available via numerical calculations [44] (see also footnote 8). As we scan the frequency ω, $\mathrm{Im}\left[\mathbf{G}^{\lambda}_{1D}\right]$ increases monotonically and diverges as kx approaches the X-point, where vλ → 0. The guided mode Green's tensor vanishes when ω lies beyond the frequency of the band edge.

Based on this analysis, we can evaluate the decay rate $\gamma^{(\infty)}_{1D}(\omega)$ into the designated probe mode for an infinite structure, and compare it with the heights of resonant features in $\gamma_{tot}(\omega)$ for finite structures with different numbers of unit cells, as shown in figures 4(b) and 5(b). Indeed, the actual $\gamma_{1D}$ of a finite-size structure must deviate from $\gamma^{(\infty)}_{1D}(\omega)$ due to boundary conditions that transform a continuous spectrum into a discrete set of resonant peaks [40], as shown in figures 4(b) and 5(b). When the number of unit cells is increased in our calculation of $\gamma_{1D}$ over the range N = 11 to 81, we find that the frequencies $\omega^{(n)}$ of the resonant peaks shift in position and the peaks change height. As documented by the black diamonds in figures 1(b), (f), the $\omega^{(n)}$ arise from the discrete set of propagation constants $k_x^{(n)} \simeq \pi n/aN$ imposed by the boundary conditions for the finite structures with n either even or odd.

The peaks in $\gamma_{tot}$ at the set of frequencies $\omega^{(n)}$ build up on top of a fairly constant background within the frequency range displayed in figures 4(b) and 5(b). We assume that this background represents the contribution of γ› to $\gamma_{tot}$, and subtract the background to estimate $\gamma_{1D}$. The resulting form for $\gamma_{1D}\cdot\omega^{(n)}$ consists of a set of resonant peaks whose heights at discrete $\omega^{(n)}$ qualitatively map out $\gamma^{(\infty)}_{1D}(\omega)$ calculated for the infinite structure, with the maximum peak height for $\gamma_{1D}$ occurring for the peak closest to the band edge. Moving further away from the band edge, we find that our numerical estimate of $\gamma_{tot}(\omega) - \gamma›(\omega)$ asymptotes to the calculated value of $\gamma^{(\infty)}_{1D}(\omega)$ reasonably well.

## Estimation of $\gamma_{1D}$ and γ›

From the previous discussions, we identify that the decay rate into other modes γ› can be read off from the broad background in $\gamma_{tot}$. Specifically, we estimate $\gamma› = \gamma_{tot}(\omega›)$ at a frequency ω› just across the band edge and away from any resonant peak for a guided mode. The decay rate into the probe mode $\gamma_{1D}$ can then be estimated using $\gamma_{1D} = \gamma_{tot} - \gamma›$.

For the single-beam structure and the atomic spin orientation shown in figure 4(b), we find a peak total decay rate $\gamma_{tot}/\gamma_0 \approx 15$ and a background level $\gamma›/\gamma_0 \approx 1.2$ near the cesium D2-line frequency $\nu_A$ = 352 THz. We estimate the coupling to the resonant probe mode$\gamma_{1D} = \gamma_{tot} - \gamma› \approx 14\gamma_0$. For the double-beam structure and spin orientation shown in figure 5(b), we find $\gamma_{tot}/\gamma_0 \approx 22$, $\gamma›/\gamma_0 \approx 1$, and, therefore, $\gamma_{1D}/\gamma_0 \approx 21$. Here,$\gamma_0/2\pi$ = 5.2 MHz is the free-space (vacuum) decay rate for the D2 line.

## Appendix E.: Effective area and mode Volume for Probe

Both the single and double nanobeam structures lead to atom localization in a region of large amplitude for the probe field. One measure of the strength of the atom-field coupling is the effective mode volume $V_m$ per unit cell, where

$$V_{\mathrm{m}} = \int \epsilon(\mathbf{r})|\mathbf{E}(\mathbf{r})|^2 d^3r/\epsilon(\mathbf{r}_{\min})|\mathbf{E}(\mathbf{r}_{\min})|^2. \tag{e1}$$

Here the integration is carried out over the volume of a unit cell. That is, the integration domain along propagation direction x extends over the distance a (i.e. the lattice constant), while in the transverse y,z directions, the integration domain is from −∞ to +∞.

For the single nanobeam, the probe mode has a global maximum at $\mathbf{r}_{\min} = 0$ and an effective mode volume $V_m \sim 0.13\ \mu m^3$. For the double nanobeams, the probe mode has a saddle-like intensity distribution around $\mathbf{r}_{\min} = 0$, resulting in $V_m \sim 0.11\ \mu m^3$ for a unit cell.

## REFERENCES


1. Kimble H J 2008 Nature 453 1023
2. For a review see Vahala K J 2003 Nature 424 839
3. Lev B, Srinivasan K, Barclay P, Painter O and Mabuchi H 2004 Nanotechnology 15 S556
4. Dayan B, Parkins A S, Aoki T, Ostby E P, Vahala K J and Kimble H J 2008 Science 319 1062
5. Lepert G, Trupke M, Hartmann M J, Plenio M B and Hinds E A 2011 New J. Phys. 13 113002
6. Volz J, Gehr R, Dubois G, Estve J and Reichel J 2011 Nature 475 210
7. Cirac J I, Zoller P, Kimble H J and Mabuchi H 1997 Phys. Rev. Lett. 78 3221
8. Duan L-M and Kimble H J 2004 Phys. Rev. Lett. 92 127902
9. Shen J T and Fan S 2005 Opt. Lett. 30 2001
10. Kien F L, Gupta S D, Nayak K P and Hakuta K 2005 Phys. Rev. A 72 063815
11. Dzsotjan D, Sørensen A S and Fleischhauer M 2010 Phys. Rev. B 82 075427
12. Kien F L and Hakuta K 2008 Phys. Rev. A 77 013801
13. Chang D E, Sørensen A S, Demler E A and Lukin M D 2007 Nature Phys. 3 807

14. Zoubi H and Ritsch H 2010 New J. Phys. 12 103014
15. Chang Y, Gong X R and Sun C P 2011 Phys. Rev. A 83 013825
16. Chang D E, Jiang L, Gorshkov A V and Kimble H J 2012 New J. Phys. 14 063003
17. Asboth J K, Ritsch H and Domokos P 2008 Phys. Rev. A 77 063424
18. Chang D E, Cirac J I and Kimble H J 2013 Phys. Rev. Lett. 110 113606
19. Ye J, Kimble H J and Katori H 2008 Science 320 1734
20. Ovchinnikov Y B, Shulga S V and Balykin V I 1991 J. Phys. B: At. Mol. Opt. Phys. 24 3173
21. Vernooy D W and Kimble H J 1997 Phys. Rev. A 55 1239
22. Burke J, Chu S T, Bryant G W, Williams C J and Julienne P S 2002 Phys. Rev. A 65 043411
23. Renn M J, Montgomery D, Vdovin O, Anderson D Z, Wieman C E and Cornell E A 1995 Phys. Rev. Lett. 75 3253
24. Ito H, Nakata T, Sakaki K, Ohtsu M, Lee K I and Jhe W 1996 Phys. Rev. Lett. 76 4500
25. Christensen C A, Will S, Saba M, Jo G B, Shin Y I, Ketterle W and Pritchard D 2008 Phys. Rev. A 78 033429
26. Bajcsy M, Hofferberth S, Balic V, Peyronel T, Hafezi M, Zibrov A S, Vuletic V and Lukin M D 2009 Phys. Rev. Lett. 102 203902
27. Vetsch E, Reitz D, Sague G, Schmidt R, Dawkins S T and Rauschenbeutel A 2010 ´ Phys. Rev. Lett. 104 203603
28. Dawkins S T, Mitsch R, Reitz D, Vetsch E and Rauschenbeutel A 2011 Phys. Rev. Lett. 107 243601
29. Goban A, Choi K S, Alton D J, Ding D, Lacrozte C, Pototschnig M, Thiele T, Stern N P and Kimble H J 2012 ˇ Phys. Rev. Lett. 109 033603
30. Rosenbusch P, Ghezali S, Dzuba V A, Flambaum V V, Beloy K and Derevianko A 2009 Phys. Rev. A 79 013404
31. Kien F L, Schneeweiss P and Rauschenbeutel A 2013 Eur. Phys. J. D 67 92
32. Ding D, Goban A, Choi K S and Kimble H J 2012 arXiv:1212.4941
33. Hinds E A, Lai K S and Schnell M 1997 Phil. Trans. R. Soc. A 355 2353
34. Agarwal G S 1975 Phys. Rev. A 12 1475
35. Buhmann S Y, Knoll L, Welsch D G and Dung H T 2004 ¨ Phys. Rev. A 70 052117
36. Rodriguez A W, McCauley A P, Joannopoulos J D and Johnson S G 2009 Phys. Rev. A 80 012115
37. Rahi S J, Emig T and Kardar M 2010 Phys. Rev. Lett. 105 070404
38. Soljacic M, Johnson S G, Fan S, Ibanescu M, Ippen E and Joannopoulos J D 2002 J. Opt. Soc. Am. B 19 2052
39. Koenderink A F, Kafesaki M, Soukoulis C M and Sandoghdar V 2006 J. Opt. Soc. Am. B 23 1196
40. Manga Rao V S C and Hughes S 2007 Phys. Rev. Lett. 99 193901
41. For a review see Baba T 2008 Nature Photon. 2 465
42. Hoang T B, Beetz J, Midolo L, Skacel M, Lermer M, Kamp M, Hsfling S, Balet L, Chauvin N and Fiore A ˇ 2012 Appl. Phys. Lett. 100 061122
43. Joannopolous J D, Johnson S G, Winn J N and Meade R D 2008 Photonic Crystals: Molding the Flow of Light (Princeton, NJ: Princeton University Press)
44. Johnson S G and Joannopolous J D 2001 Opt. Express 8 173
45. Eichenfeld M, Camacho R, Chan J, Vahala K J and Painter O 2009 Nature 459 550–6
46. Søndergaard T and Tromborg B 2001 Phys. Rev. A 64 033812
47. Thompson J D, Tiecke T G, de Leon N P, Feist J, Akimov A V, Gullans M, Zibrov A S, Vuletic V and ´ Lukin M D 2013 Science 340 1202–5
48. Oskooi A F, Roundy D, Ibanescu M, Bermel P, Joannopoulos J D and Johnson S G 2010 Comput. Phys. Commun. 181 687
49. Buhmann S Y and Welsch D G 2007 Prog. Quantum Electron. 31 51
50. Cataldo G, Beall J A, Cho H M, McAndrew B, Niemack M D and Wollack E J 2012 Opt. Lett. 20 4200
51. Djurisic A B and Li E H 1998 Appl. Opt. 37 5291
52. Dzayloshinkii I E, Lifshitz E M and Pitaevskii L P 1961 Sov. Phys.—Usp. 4 153
53. Contreras-Reyes A M, Guerout R, Neto P A M, Dalvit D A R, Lambrecht A and Reynaud S 2010 ´ Phys. Rev. A 82 052517
54. Kien F L, Gupta S D, Balykin V I and Hakuta K 2005 Phys. Rev. A 72 032509
55. Hwang J K, Ryu H Y and Lee Y H 1999 Phys. Rev. B 60 4688–95
56. Yao P, Manga Rao V S C and Hughes S 2010 Laser Photon. Rev. 4 499–516

# CHAPTER 13 PHOTONIC NANOSTRUCTURES DESIGN AND OPTIMIZATION FOR SOLAR CELL APPLICATION

Qian Liu [1], Eric Sandgren [1], Miles Barnhart [2], Rui Zhu [2] and Guoliang Huang[2]

[1]Department of System Engineering, University of Arkansas at Little Rock, 2801 S University Ave, Little Rock, AR 72204, USA

[2]Department of Mechanical & Aerospace Engineering, University of Missouri, E3422, Lafferre Hall, Columbia, MO 65211, USA

## ABSTRACT

**In this paper, a semiconducting photonic nanostructure capable of wide range absorption and tunable optical resonance has been designed with a proposed theoretical optimization model. The design consists of ZnO/CdS core-shell nanowire arrays as well as multilayer thin films that act to absorb incident electromagnetic (EM) waves over a broad frequency range. Theoretical, as well as numerical, studies of the nanostructure inside a solar cell plate have been conducted in order to validate the proposed microstructural design. Excellent energy absorption rates of EM waves have been achieved in the high frequency range by using the optical resonance of the nanowire array. By combining multilayer thin film with the core-shell nanowire in the unit cell of a photonic solar cell, a broadband high absorption has been achieved. Moreover, the geometry of the proposed photonic nanostructure is obtained through the implementation of a genetic algorithm. This avoids local minima and an optimized absorption rate of ~90% over the frequency range of 300 to 750 THz has been obtained in the solar cell.**

## INTRODUCTION

Light trapping via the chemical and geometric design of solar cells promises to dramatically increase solar cell efficiency by increasing light absorption without increasing the exciton or free carrier collection length. Various approaches have been proposed to enhance the overall solar cell efficiency in the thin film or nano-based solar cells. These include: putting a photonic crystal or distributed Bragg reflector in the back of thin film solar cells [1,2,3], using a disperse carbon interlayer and textured photonic crystal backside reflector to prompt back reflected effects [4,5], and combining holey double-layer photonic crystals with ultrathin c-Si solar cells [6]. Recently, solar cells utilizing nanowire arrays, overlapping nanospheres, nano-colloids, and self-assembled silver nanoparticles have shown promise by exhibiting enhanced light trapping effect and greater energy absorption rates than conventional planar semiconductors [7,8,9,10]. Grandidier *et al.* [8] experimentally and theoretically demonstrated efficiency enhancements in solar cells from dielectric nanosphere arrays. Mendes *et al.* [9] developed novel plasmonic back-reflector structures using spherical gold colloids for pronounced far-field scattering with a remarkable broadband enhancement. Morawiec *et al.* [10] demonstrated the broadband absorption enhancement of thin solar cells due to the light scattering effects of plasmonic back reflectors composed of self-assembled silver nanoparticles. Specifically, nanowire arrays have drawn a great deal of

**Citation:** Qian Liu, Eric Sandgren, Miles Barnhart, Rui Zhu and Guoliang Huang, Photonic Nanostructures Design and Optimization for Solar Cell Application, doi:10.3390/photonics2030893.

attentions due to their optical absorption properties which can enhance the light trapping effect. Hu and Chen [11] demonstrated that Si nanowire is capable of significantly improving light absorption when compared to a typical Si thin film within a high frequency region. ZnO has also been studied extensively as a semiconductor in solar cell applications due to its wide band gap and high electron mobility [7,12,13]. Ordered arrays of nanowires can benefit from diffraction effects, collective resonances, and changes in the optical density of states which are absent in single nanowire or random arrays of nanowires. Garnett and Yang [14] measured the photocurrent of solar cells made with ordered silicon nanowire arrays with different silicon absorber thicknesses to determine the light trapping path length enhancement factor. The current study of nanowire arrays has been focused mainly on Si and ZnO type compositions. Myung *et al.* [15] synthesized a high density ZnO/CdS core-shell nanowire array by depositing the CdS outer layers on pre-grown, vertically-aligned ZnO nanowire, through chemical vapor deposition. Seol *et al.* [16] created a CdSe/CdS/ZnO nanowire array through a three-step solution-based method which was used as a photoanode of a quantum dot sensitized solar cell capable of generating a maximum power conversion efficiency of 4.15%. Although experimental observations of the enhanced absorptions have been reported in those solar cells with nanowire arrays, few theoretical and simulation works have been conducted which can be very useful to provide guidelines for designing high efficiency solar cell plates with outstanding light trapping effect.

Moreover, photonic nanowire arrays assembled with thin film structures have exhibited an increase in light trapping in a large frequency range. Li *et al.* [17] found that optical absorption can be greatly enhanced with the addition of a Si nano-pillar to a Si thin film owing to suppressed reflection and transmission. Lin *et al.* [18] demonstrated an improvement in the photovoltaic performance of a ZnO/$P_3$HT hybrid nanostructure through interface molecular modification on the ZnO nanowire surface. Wu *et al.* [19] reported a 74% increase in efficiency of ZnO nanowire dye-sensitized (DSSC) thin film solar cells. Mokkapati *et al.* [20] provided criteria for optimizing light trapping of metal nanoparticles in periodic arrays in a long wavelength range, allowing for a higher order diffraction mode while maintaining the highest possible fill factor. Therefore, designing a photonic crystal unit cell that combines multilayer thin film structures and multi-shell nanowires with optimized geometric properties can lead to solar cells with a broadband, high-energy absorption rate.

The objective for this paper is to investigate a solar cell consisting of multilayer thin film structures, as well as periodic core-shell nanowire photonic crystals, as an optical absorber. First, theoretical and numerical analyses are carried out on a simple five-layered thin film solar cell. Next, a microstructural design consisting of ZnO/CdS nanowire arrays and multilayer thin films is proposed. Light absorbing performance of the proposed design is evaluated numerically. In order to yield the highest absorption rate possible in both low and high frequency ranges, non-linear optimization is carried out to generate structural geometries that exhibit excellent absorption performance in a broad frequency range.

## THEORETICAL ANALYSIS OF ONE-DIMENSIONAL SOLAR CELL WITH MULTILAYER NANO FILMS

To achieve broadband optical absorption in real structures such as thin solar cell plates, photonic nanostructures designed with complex geometries are considered. Theoretical investigation of multilayer nanoscale thin film inside the solar cell plate can be a good starting point to understand the energy absorption inside the solar cell nanostructures. In addition, the periodicity of the solar cell nanostructures can be designed along with the reflection and transmission rates. Furthermore, numerical simulations should be performed to yield the optimal structural parameters and, therefore, increase the photon absorption rate.

First, theoretical investigations of the one-dimensional (1D) multilayer solar cell plate are conducted, with the goal of achieving a guideline for the microstructural design that is focused on absorption in a high frequency region. This model will then be extended to the three-dimensional (3D) case in the next section to investigate and design a solar cell with both nanowire arrays and composite thin films. In order to understand the underlying physical mechanism for photon absorption in optical resonance, analytical modeling as well as numerical simulations are carried out. An analytical model of the multilayer planar solar cell is initially constructed by using the matrix transfer formulation to calculate the reflection and transmission in a 1D multilayered planar structure, as shown in Figure 1. The structure consists of $(N-2)$ layers sandwiched together by two semi-infinite layers (layer 1 and layer $N$).

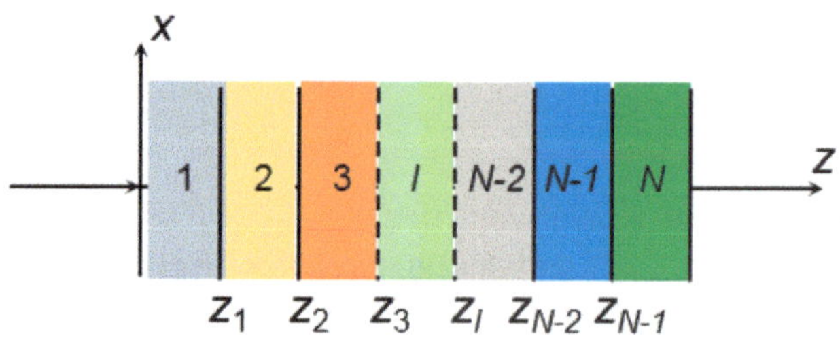


**Figure 1:** 1D multilayer planar structure.

The relative permittivity, permeability, and refractive index of the $l$-th layer is denoted by $_l$, $\mu_l$ and $n_l$ ($l = 1, 2,\ldots, N$), respectively. Each layer of the multilayer structure is a medium with complex permittivity and magnetic permeability given by [21]:

$$\varepsilon_l^* = \varepsilon_l + i\varepsilon_l'$$
$$\mu_l^* = \mu_l + i\mu_l^i \tag{1}$$

where the imaginary portion represents optical absorption behavior in the medium and the dielectric loss tangent can be written as:

$$tan\delta_l = \frac{\varepsilon_l'}{\varepsilon_l} \tag{2}$$

In this paper, we use relative permittivity when carrying out analysis and simulations, which is defined as the ratio of dielectric permittivity over the permittivity of vacuum. For the transverse electric (TE) wave, the electric and magnetic fields in the region $l$ are given by [22]:

$$\boldsymbol{E}_l = \hat{\boldsymbol{y}}A_l e^{ik_x x + ik_{lz}(z-z_{l-1}) - i\omega t} + \hat{\boldsymbol{y}}B_l e^{ik_x x + ik_{lz}(z-z_{l-1}) - i\omega t}$$

$$\boldsymbol{H}_l = \left[-\frac{k_{lz}}{\omega\mu_0\mu_l}\hat{\boldsymbol{x}} + \frac{k_x}{\omega\mu_0\mu_l}\hat{\boldsymbol{z}}\right] A_l e^{ik_x x + ik_{lz}(z-z_{l-1}) - i\omega t} + \left[\frac{k_{lz}}{\omega\mu_0\mu_l}\hat{\boldsymbol{x}} + \frac{k_x}{\omega\mu_0\mu_l}\hat{\boldsymbol{z}}\right] B_l e^{ik_x x - ik_{lz}(z-z_{l-1}) - i\omega t} \quad (3)$$

where $z_0 = z_l = 0$, $z_l = z_{l-1} + d_l$ and $d_l$ is the thickness of the $l$-th layer ($l = 2, \ldots, N-1$). The constants $A_l$ and $B_l$ represent wave components with the propagating velocity components along the positive and negative $z$ directions, respectively. The variables $_0$and $\mu_l$ are the permittivity and permeability of vacuum, respectively. The total fields for transverse magnetic (TM) waves can then be obtained by interchanging $\mathbf{E}_l \rightarrow \mathbf{H}_l, \mathbf{H}_l \rightarrow -\mathbf{E}_l$, and $\mu_l \leftrightarrow \varepsilon_l$ in Equation (3). The constants $A_l$ and $B_l$ can be determined by using interfacial conditions at z=zl ($l = 1, 2, \ldots, N-1$), which lead to:

$$\begin{bmatrix} A_l \\ B_l \end{bmatrix} = \boldsymbol{P}_l \boldsymbol{D}_l^{-1} \boldsymbol{D}_{l+1} \begin{bmatrix} A_{l+1} \\ B_{l+1} \end{bmatrix} \quad (4)$$

where D−1l is the inverse matrix of Dl. The matrices Pl and Dl are given by

$$\boldsymbol{P}_l = \begin{bmatrix} e^{\phi_l} & 0 \\ 0 & e^{-\phi_l} \end{bmatrix} \quad (5)$$

$$\boldsymbol{D}_l = \begin{bmatrix} 1 & 1 \\ \eta & -\eta \end{bmatrix} \quad (6)$$

respectively. In Equations (5) and (6),

$$\phi_1 = 0 \qquad \phi_l = \begin{cases} (\omega/c) d_l \alpha_l, & k_x^2 > k_l^2 \\ -i(\omega/c) d_l \alpha_l, & k_l^2 > k_x^2 \end{cases} \quad (l = 2, 3, \ldots, N-1) \quad (7)$$

and

$$\eta = \begin{cases} i\gamma_l, k_x^2 > k_l^2 \\ \gamma_l, k_l^2 > k_x^2 \end{cases} \quad (8)$$

with γl=αl/μl for a TE wave and γl=αl/εl for a TM wave. Note that

$$\alpha_l = \begin{cases} \sqrt{(n_1 sin\theta_1)^2 - n_l^2}, k_x^2 > k_l^2 \\ \sqrt{n_l^2 - (n_1 sin\theta_1)^2}, k_l^2 > k_x^2 \end{cases} \quad (9)$$

where $n_l^2 =$ $_l$ ($l = 1, 2, \ldots, N-1$) is the refraction index. From Equation (4), we can derive the following relation

$$\begin{bmatrix} A_1 \\ B_1 \end{bmatrix} = \boldsymbol{M} \begin{bmatrix} A_N \\ B_N \end{bmatrix} \quad (10)$$

where the 2×2 matrix $\boldsymbol{M} = \prod_{l=1}^{N-1} \boldsymbol{P}_l \boldsymbol{D}_l^{-1} \boldsymbol{D}_{l+1}$.

One can set $B_N$=0, since there is no reflection in the region $N$. Therefore, the transmission and reflection coefficients can be calculated as $\tilde{t} = 1/M_{11}$ and $\tilde{r} = M_{21}/M_{11}$, respectively. $M_{ij}$ ($i, j = 1, 2$) are the components of the 2 × 2 matrix **M**. Finally, the power transmittance $T$ and reflectance $R$ can be calculated as:

$$T = \frac{n_N cos\theta_N}{n_1 cos\theta_1} \widetilde{tt^*} \quad (11)$$

$$R = \widetilde{rr^*} \quad (12)$$

respectively, where θN=sin−1(n1sinθ1/nN).

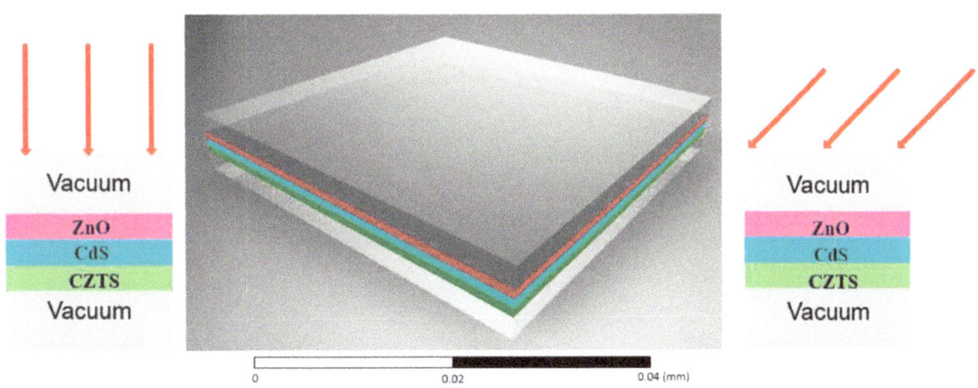


**Figure 2:** Five-layer thin film solar cell under different angles of incidence.

In Figure 2, a five-layer thin film solar cell with vacuums as the top and bottom outer layers are studied with different angles of incidence. Both a theoretical model and finite element (FE) model are used to calculate the reflection, transmission, and absorption rates of the five-layer thin film. The relative permittivity and dielectric loss tangents of ZnO, CdS, and CZTS layers are all frequency-dependent and determined through experimental measurement. Figure 3a,b demonstrate the dielectric loss tangents and the real parts of the relative permittivity of the three layers, respectively.

The thicknesses of ZnO, CdS, and CZTS layers are 400, 200 and 200 nm, respectively. Three different angles of incidence (0°, 30°, and 60°) are studied. In order to validate the analytical model, a numerical simulation is also conducted on the five-layer thin film with the same dispersive material parameters by using Ansoft HFSS14. In Figure 4, the reflection, transmission, and absorption results obtained from the theoretical model and the FE model are plotted as solid curves and small triangles, respectively. Excellent agreement between the two results can be observed, which confirms that the developed theoretical model can describe the exact wave behavior in the multilayer thin film solar cell.

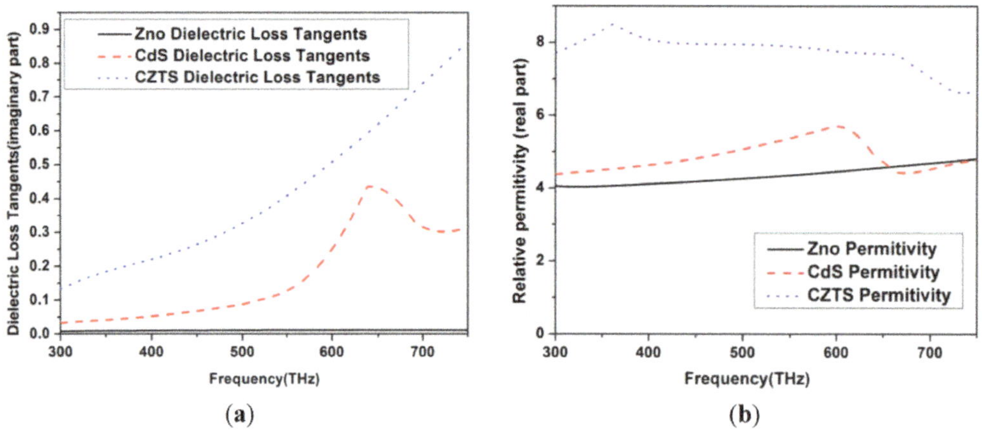


**Figure 3:** (**a**) Imaginary parts of the dielectric loss tangent. (**b**) Real parts of the relative permittivity of ZnO, CdS, and CZTS layers.

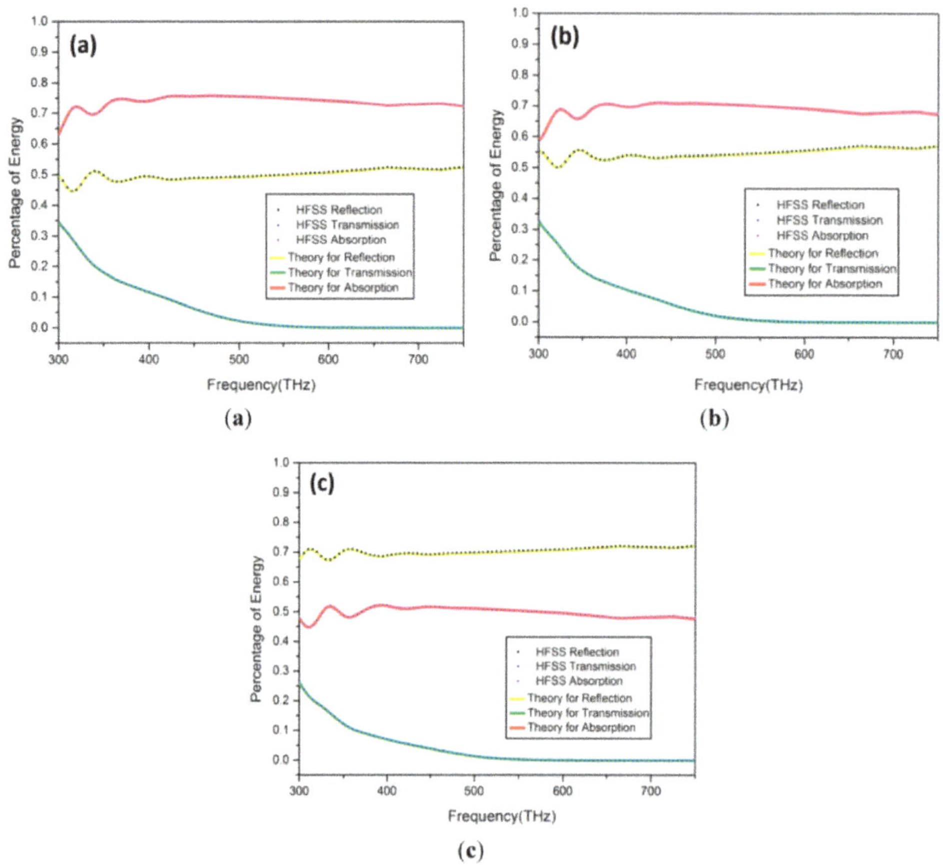


**Figure 4:** Theoretical and numerical results of the reflection, transmission and absorption of the five-layer thin film solar cell under (**a**) 0°. (**b**) 30°. (**c**) 60° angle of incident.

Also, by comparing the results of the three different angles of incidence, it can be found that the reflection increases with larger angle of incident while the few changes in transmission can be observed especially in the high frequency range. The absorptions for 0° and 30° incidences are almost the same, while decreased absorption can be found when the angle of incidence increases to 60°.

## NUMERICAL SIMULATION AND OPTIMIZATION ON SOLAR CELL CONSISTING OF CORE-SHELL NANOWIRE ARRAYS AND MULTI-LAYER THIN FILMS

Nanowire arrays have demonstrated unusual optical absorption properties in solar cell applications. Current studies of nanowires are mostly concerned with Si and ZnO compositions. Hu and Chen's study [11] mentioned that nanowire arrays with moderate filling ratio could exhibit greater absorption rates and lower reflectance compared with thin films, especially in a high frequency regime. The nanostructure design is of great concern in the research and development of novel solar cell structures. For nanostructures with complex geometry, the aids of numerical methods such as the finite element method (FEM) are required. In our study, a core-shell nanowire is used in order to obtain enhanced absorption. The inner core is ZnO and the outer shell is made of CdS. The core-shell nanowire is then situated within a vacuum. Simulations are conducted by using Ansoft High Frequency Structure Simulator (HFSS) with a Floquet port feature, which is capable of modeling the oblique incidence and polarization of the wave at the boundary of the periodic-planar structure. A photonic structure is created by using MATLAB in combination with HFSS's Visual Basic scripting. Results of the magnitude and phase of the reflected and transmitted waves, as well as absorption rate through the periodic nanostructures, are then obtained. Figure 5 shows the simulation result of the energy flux density of a unit cell with core-shell nanowire structure and Figure 6 shows the absorption rate of the core-shell nanowire structure; the coefficient describes intensity or total power percentage absorbed relative to an incident wave. In Figure 6, we also draw the reflection and transmission coefficient curves; these two amplitude coefficients are represented by lower case r and t in Equations (11) and (12).

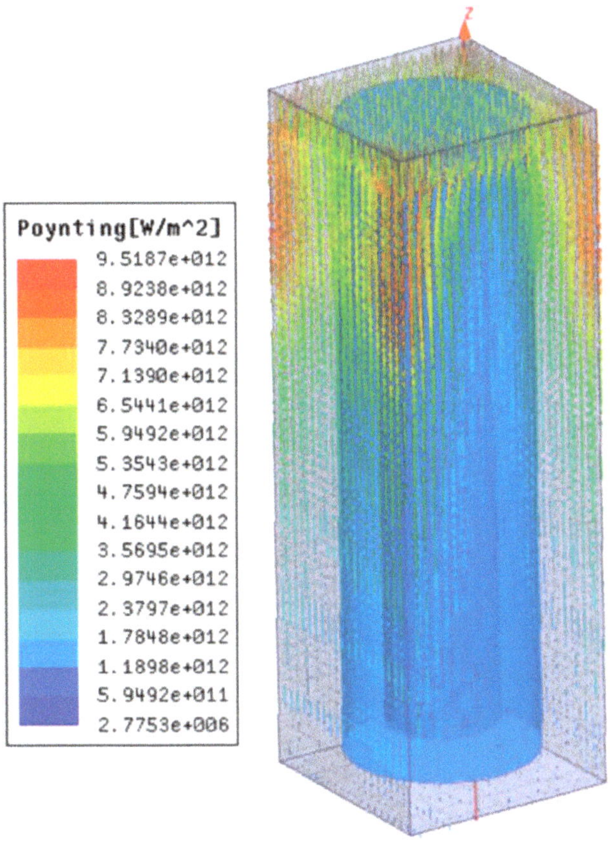


**Figure 5:** The energy field of the unit cell with core-shell nanowire structure.

It can be observed from Figure 6 that obviously higher absorption and relatively lower reflectance can be achieved at a high frequency range compared with those at a low frequency range. The influence of the optical resonance on the increased absorption can be found from 600 to 750 THz, as shown in Figure 6. However, it will be ideal if the absorption rate is consistently high across a broad frequency range. To accomplish this, the original core-shell model needs to be modified through the optimization of structural dimensions, as well as the material compositions inside each unit cell of the nanowire array.

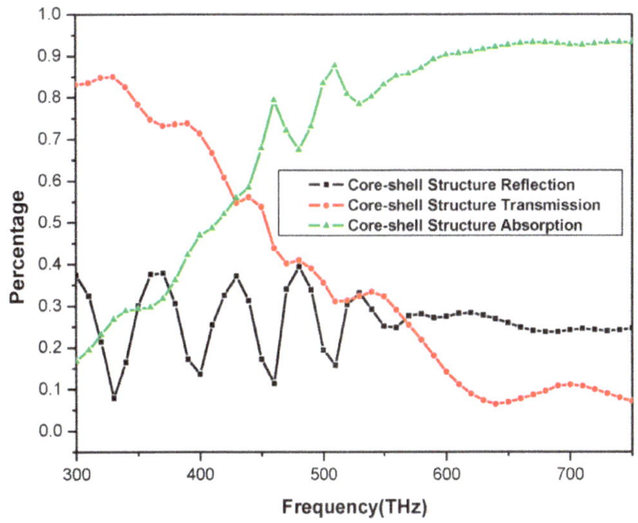


**Figure 6:** Absorption rate of the core-shell nanowire structure.

Furthermore, a nanowire array combined with a nanofilm can significantly enhance the optical absorption of a solar cell in a broad frequency range and has become a promising direction for future research of high-efficiency solar cells. By adding the Si nanowires on Si thin film, Li *et al.* [12] found that optical absorption of the combined structure could be greatly enhanced. To begin the microstructural design of the optimized 3D solar cell, we first consider a periodic array of double shelled (ZnO/CdS/CZTS) nanowire structures that are embedded in a multilayer thin film. Details of the initial unit cell design can be found in Figure 7. A ZnO nanowire is first grown on the top of a ZnO thin film, which is then encapsulated by a layer of CdS via vapor deposition. Finally, a top layer of CZTS is added via vapor deposition to envelope the middle layer of CdS in the unit cell. Analysis on the initial design will be conducted in order to get baseline absorption rates prior to optimization.

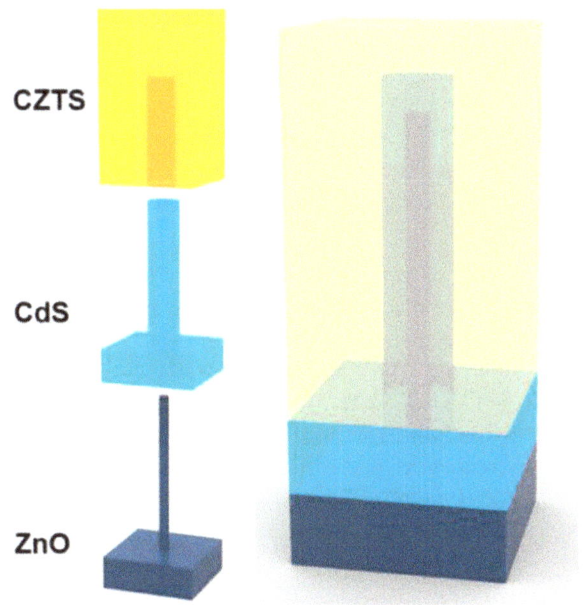


**Figure 7:** Model of the proposed nanowire-nanofilm unit cell configuration.

The 3D solar cell is created and the optical wave simulation in the nanowire and layered medium is conducted by using Ansoft HFSS14. Periodicity of the structure is utilized in the simulation. As a result, analysis is only carried out on a single unit cell where periodic boundary conditions are applied. The Floquet port is used exclusively with planar-periodic structures in HFSS. Chief examples are planar phased arrays and frequency-selective surfaces which may be idealized as infinitely large. The analysis of the infinite structure is then accomplished by analyzing a unit cell. Linked boundaries most often form the side walls of a unit cell but, in addition, a boundary condition is required to account for the infinite space above. The Floquet port is designed for this purpose. The Floquet port is closely related to a wave port in that a set of modes ("Floquet modes") represent the fields on the port boundary. Fundamentally, Floquet modes are plane waves with the propagation direction set by the frequency, phasing, and geometry of the periodic structure. Just like wave modes, Floquet modes also have propagation constants and experience cut-off at low frequency. Ansoft HFSS is used, of which the Floquet port can also deal with the oblique incident wave with certain polarization illuminating the periodic material structure problem. This was done in order to get the magnitude and phase of the reflection and transmission waves. The HFSS solution includes a model decomposition of which all information is obtained in the form of an S-matrix interrelating the Floquet modes. Figure 8a shows the dimensions of a single unit cell in the simulation. The three layers in the middle of the FE model are the dispersive materials CZTS, CdS, and ZnO from top to bottom, respectively. The value of permeability are all equal to unity ($\mu$ = 1) since the materials are nonmagnetic over the entire optical frequency range of interest. The permittivity of vacuum is 1 and the others' permittivity and dielectric loss tangents are all dispersive, experimentally-determined optical constants of those material components and are used in the FE simulation. The thicknesses of the three thin films are set to 400, 200, and 200 nm for CZTS, CdS, and ZnO, respectively. The incident light is set normal to the film substrate, *i.e.*, it is parallel to the nanowires. Periodic conditions are employed at a pair of surfaces parallel to the *x-z* and *y-z* planes by using the master/slave boundary. The final structure is realized through infinitely extending the structural units having the height of 1800 nm and a width dimension equal to the nanowire array periodicity, as indicated by the $d_1$ in Figure 8a,b showing the meshed unit cell which is sandwiched between two bodies of vacuum with finite length.

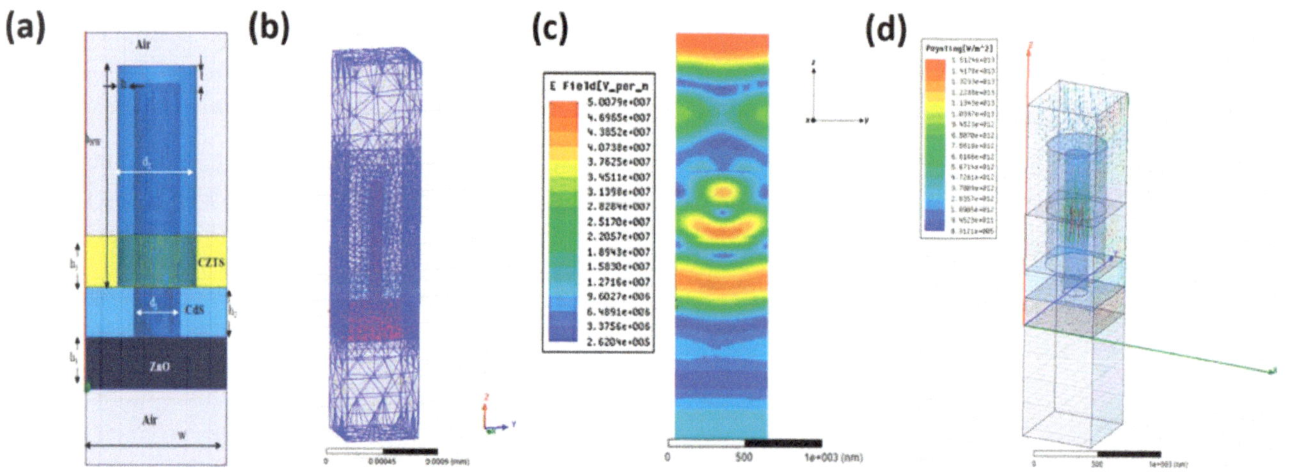


**Figure 8:** (**a**) Dimensions of a single unit cell. (**b**)The meshed FE model. (**c**) The simulated electric field flux. (**d**) The simulated Poynting energy field in the unit cell.

The optical information is obtained from the spatial distribution of the electric field flux which results from the interaction between the structure and the incident light, as shown in Figure 8c. The frequency range of the incident light varies between 300 to 750 THz corresponding to the wavelength from 400 to 1000 nm. These values are selected because they occupy the frequency spectrum of solar radiation which is of particular interest. The results of Poynting flux distribution is shown in Figure 8d. It can be found from Figure 8c,d that the electric field in the core-shell nanowire regime is enhanced and energy is gathered by the nanostructure due to light trapping effect. Therefore, very little energy is transmitted out, but still a large of part is reflected back by the nanostructure.

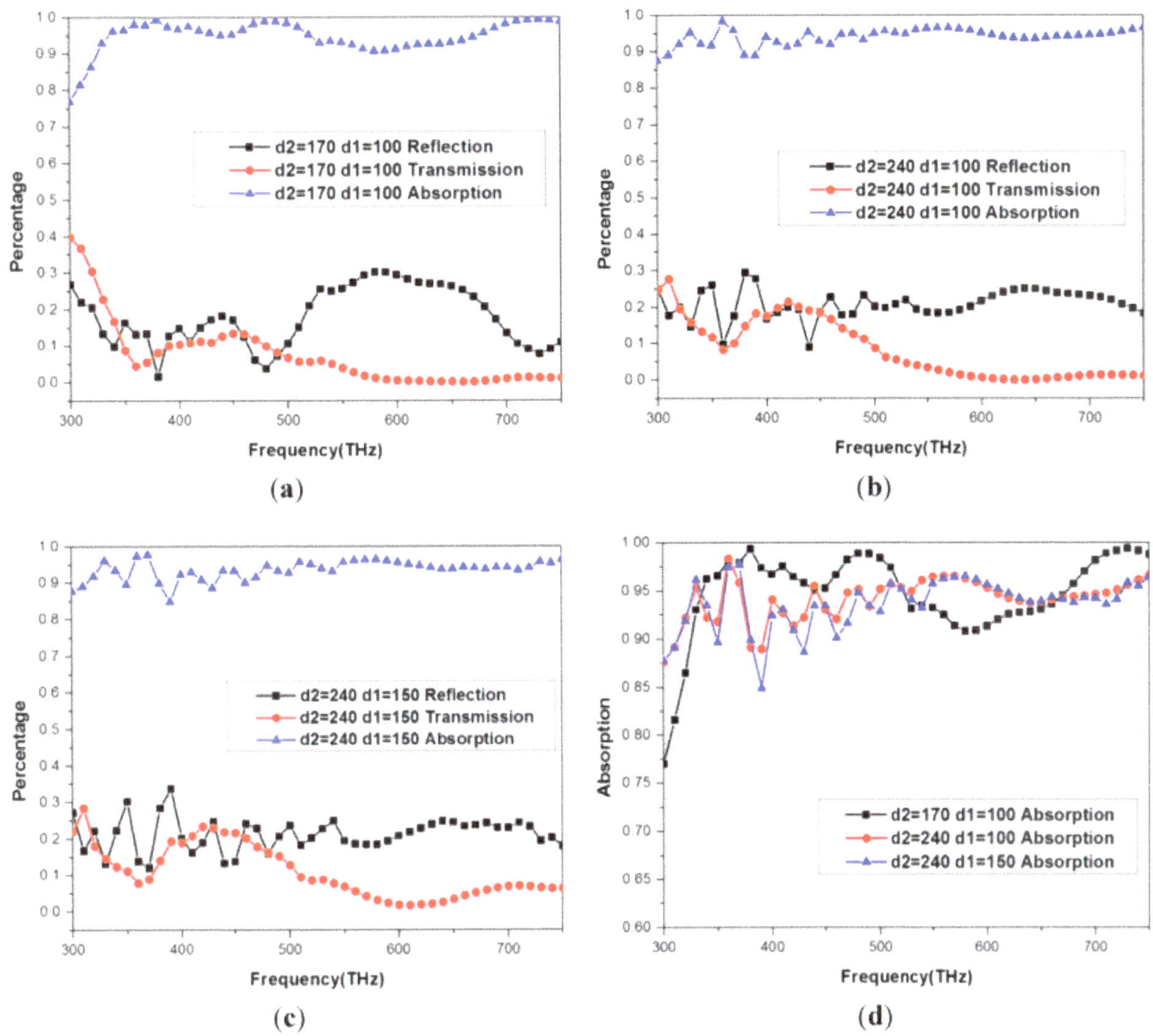


**Figure 9:** FEM results for absorption rates of varies nanowire diameters.

In order to optimize the proposed unit cell, a genetic algorithm (GA) is used because its interactive process can avoid local minima and, therefore, is an ideal optimization method for this particular problem. An optimal photonic structure is created using MATLAB, in combination with HFSSs Visual Basic scripting, where optimization is carried out via a GA on the proposed design. The GA mimics naïve evolution where the objective function, or in this case maximum energy absorption, is used as a measure of fitness for the system. The GA is a three-step optimization process that implements selection, crossover, and mutation over generations (iterations) in order to seek out an optimal structure topology. Through the iterative process geometric parameters, including the unit cell width, *w*, thin film thicknesses $h_1$, $h_2$, $h_3$, and the height, *h*, as well as the diameters, $d_1$, $d_2$, and the unit cell width, *w*, of the core/shell are randomly selected until a global minimum is found (optimal fitness). In this case, the GA is utilized to find the geometric parameters that result in the maximum energy absorption over the entire frequency spectrum. This is accomplished by minimizing the reciprocal of the maximum energy absorption. After structural optimization is carried out, excellent absorptions with an average rate of approximately 90% can be observed over a broad high frequency range from 300 to 750 THz, as shown in Figure 9. It can be found that optimization results are achieved in the optical spectrum, 400–1000 nm (300–750 THz), where most energy comes from and different optimized designs have different maximum absorption frequency ranges. For example, the optimized structure with $d_2$ = 170 $d_i$ = 100 can reach the maximum absorption rate in two frequency ranges of 350–500 THz and 650–750 THz, while in 500–650 THz the absorption rate of this design is the least among all three optimized designs. The obtained optimized structures can be used in solar cells with specifically-emphasized operating frequency ranges.

## CONCLUSIONS

In summary, this paper has demonstrated a broadband, high energy absorption rate of the proposed photonic crystal solar cell that utilizes optical resonance-enhanced light trapping. First, analytic and numerical analysis are carried out on a simple five-layered thin film solar cell in which very good agreement between both models can be found. Next, a photonic crystal design consisting of ZnO/CdS core-shell nanowire arrays is proposed. The energy absorption rate of the proposed photonic nanostructure has been evaluated. An excellent absorption rate is obtained in the high frequency range but ineffective performance is found in the low frequency region. Finally, non-linear optimization is carried out via a genetic algorithm on a photonic crystal solar cell consisting of multilayer thin film structures and periodic core-shell nanowire arrays. The optimized structure has demonstrated excellent performance over a broad frequency range (300 to 750 THz) with an average absorption ratio of 90%. The theoretical and numerical studies on the light trapping and energy absorption in a photonic crystal with multilayer thin film structures and multi-shell nanowire arrays can be very useful to provide guidelines for designing high efficiency solar cell plates. Furthermore, the optimized design of the photonic crystal unit cell can be quite useful in designing solar cells with broadband high energy absorption rates.

## ACKNOWLEDGMENTS

The authors would like to thank Dr. Xiaoming Zhou from Beijing Institute of Technology for his comments and discussions on the theoretical model. This research was supported by the National Science Foundation under award No. EPS-1003970.

## REFERENCES

1. Bermel, P.; Luo, C.; Zeng, L.; Kimerling, L.C.; Joannopoulos, J.D. Improving thin-film crystalline silicon solar cell efficiencies with photonic crystals. *Opt. Express* 2007, *15*, 16986–17000.
2. Zhou, D.; Biswas, R. Photonic crystal enhanced light-trapping in thin film solar cells. *J. Appl. Phys.* 2008, *103*.
3. Chutinan, A.; Li, C.W.W.; Kherani, N.P.; Zukotynski, S. Wave-optical studies of light trapping in submicrometre-textured ultra-thin crystalline silicon solar cells.*J. Phys. D Appl. Phys.* 2011, *44*.
4. Araújo, A.; Barros, R.; Mateus, T.; Gaspar, D.; Neves, N.; Vicente, A.; Martins, R. Role of a disperse carbon interlayer on the performances of tandem a-Si solar cells. *Sci. Technol. Adv. Mater.* 2013, *14*.
5. Zeng, L.; Bermel, P.; Yi, Y.; Alamariu, B.A.; Broderick, K.A.; Liu, J.; Kimerling, L.C. Demonstration of enhanced absorption in thin film Si solar cells with textured photonic crystal back reflector. *Appl. Phys. Lett.* 2008, *93*.
6. Mallick, S.B.; Agrawal, M.; Peumans, P. Optimal light trapping in ultra-thin photonic crystal crystalline silicon solar cells. *Opt. Express* 2010, *18*, 5691–5706.
7. Yi, G.C.; Wang, C.; Park, W.I. ZnO nanorods: Synthesis, characterization and applications. *Semicond. Sci. Technol.* 2005, *20*.
8. Grandidier, J.; Weitekamp, R.A.; Deceglie, M.G.; Callahan, D.M.; Battaglia, C.; Bukowsky, C.R.; Atwater, H.A. Solar cell efficiency enhancement via light trapping in printable resonant dielectric nanosphere arrays. *Phys. Status Solidi*2013, *210*, 255–260.
9. Mendes, M.J.; Morawiec, S.; Mateus, T.; Lyubchyk, A.; Águas, H.; Ferreira, I.; Crupi, I. Broadband light trapping in thin film solar cells with self-organized plasmonic nano-colloids. *Nanotechnology* 2015, *26*.
10. Morawiec, S.; Mendes, M.J.; Filonovich, S.A.; Mateus, T.; Mirabella, S.; Águas, H.; Crupi, I. Broadband photocurrent enhancement in a-Si: H solar cells with plasmonic back reflectors. *Opt. Express* 2014, *22*, A1059–A1070.
11. Hu, L.; Chen, G. Analysis of optical absorption in silicon nanowire arrays for photovoltaic applications. *Nano Lett.* 2007, *7*, 3249–3252.
12. Peng, Q.; Qin, Y. ZnO nanowires and their application for solar cells. In*Nanowires—Implementations and Applications*; Hashim, A., Ed.; INTECH Open Access Publisher: Rijeka, Croatia, 2011; pp. 158–178.
13. Xia, Y.; Yang, P.; Sun, Y.; Wu, Y.; Mayers, B.; Gates, B.; Yin, Y.; Kim, F.; Yan, H. One-dimensional nanostructures: synthesis, characterization, and applications. *Adv. Mater.* 2003, *15*, 353–389.
14. Garnett, E.; Yang, P. Light trapping in silicon nanowire solar cells. *Nano Lett.*2010, *10*, 1082–1087.
15. Myung, Y.; Jang, D.M.; Sung, T.K.; Sohn, Y.J.; Jung, G.B.; Cho, Y.J.; Kim, H.S.; Park, J. Composition-Tuned ZnO-CdSSe Core-Shell Nanowire Arrays. *ACS Nano*2010, *4*, 3789–3800.
16. Seol, M.; Kim, H.; Kim, W.; Yong, K. Highly efficient photoelectrochemical hydrogen generation using a ZnO nanowire array and a CdSe/CdS co-sensitizer. *Electrochem. Commun.* 2010, *12*, 1416–1418.
17. Li, J.; Yu, H.; Wong, S.M.; Zhang, G.; Sun, X.; Lo, P.G.Q.; Kwong, D.L. Si nanopillar array optimization on Si thin films for solar energy harvesting. *Appl. Phys. Lett.* 2009, *95*.
18. Lin, Y.Y.; Lee, Y.Y.; Chang, L.; Wu, J.J.; Chen, C.W. The influence of interface modifier on the performance of nanostructured ZnO/polymer hybrid solar cells.*Appl. Phys. Lett.* 2009, *94*.
19. Wu, J.J.; Chen, Y.R.; Liao, W.P.; Wu, C.T.; Chen, C.Y. Construction of nanocrystalline film on nanowire array via swelling electrospun polyvinylpyrrolidone-hosted nanofibers for use in dye-sensitized solar cells. *ACS Nano* 2010, *4*, 5679–5684.
20. Mokkapati, S.; Beck, F.J.; Polman, A.; Catchpole, K.R. Designing periodic arrays of metal nanoparticles for light-trapping applications in solar cells. *Appl. Phys. Lett.* 2009, *95*.
21. Veselago, V.G. The electrodynamics of substances with simultaneously negative values of $\varepsilon$ and $\mu$. *Sov. Phys. Usp.* 1968, *10*, 509–514.
22. Zhou, X.; Hu, G. Total transmission condition for photon tunnelling in a layered structure with metamaterials. *J. Opt. A Pure Appl. Opt.* 2007, *9*.

CHAPTER 14

# A SMART COLORFUL SUPERCAPACITOR WITH ONE DIMENSIONAL PHOTONIC CRYSTALS

Cihui Liu,[1] Xing Liu,[1] Hongyun Xuan,[1] Jiaoyu Ren,[1] and Liqin Ge[a,1]

State Key Laboratory of Bioelectronics, School of Biological Science and Medical Engineering, Southeast University, Nanjing 210096, China

## ABSTRACT

**To meet the pressing demands for portable and flexible equipment in contemporary society, developing flexible, lightweight, and sustainable supercapacitor systems with large power densities, long cycle life, and ease of strongly required. However, estimating the state-of-charge of existing supercapacitors is difficult, and thus their service life is limited. In this study, we fabricate a flexible color indicative supercapacitor device with mesoporous polyaniline (mPANI)/Poly(N-Isopropyl acrylamide-Graphene Oxide-Acrylic Acid) (P(NiPPAm-GO-AA)) one dimensional photonic crystals (1DPCs) as the electrode material through a low-cost, eco-friendly, and scalable fabrication process. We found that the state-ofcharge could be monitored by the structural color oscillation due to the change in the photonic band gap position of the 1DPCs. The flexible 1DPCs supercapacitor is thin at 3mm and exhibits good specific capacitance of 22.6F g−1 with retention of 91.1% after 3,000 cycles. This study shows the application of the 1DPCs supercapacitor as a visual ultrathin power source. The technology may find many applications in future wearable electronics.**

**Flexible energy storage devices such as electrochemical capacitors are the best candidates for energy storage devices that can be operated in bio-environments because it is easier to find biocompatible components and to exclude the electrochemical reactions for supercapacitors than for batteries or fuel cells[1]. However, the present supercapacitors could not meet the demand of high energy densities and convenient use kind[2,3,4,5,6]. It is difficult for us to estimate the state-of-charge of existing supercapacitors which leads to blind and excessive charge/discharge thus limiting their service lives[7]. So, it is significant to introduce a visual supercapacitor which can indicate the state-of-charge and extend the supercapacitor life.**

**Nature offers a remarkably simple alternative idea that can offer a stacking structure with visible structural color[8]. Recently, it was found that one dimensional photonic crystals (1DPCs) can demonstrate a well stacking structure with vivid structural color, and the colors could be the monitor of supercapacitors' charge/discharge process[9]. Hence, it is useful to construct a similar stacking structure to imitate the natural 1DPCs to study them directly in man-made systems.**

**Polyaniline (PANI) is a well-studied conducting polymer for supercapacitors due to its high specific capacitance from multiple doping/dedoping states although it often suffers from inferior long-term stability and rate capability due to the**

**Citation:** Cihui Liu, Xing Liu, HongyunXuan, Jiaoyu Ren & LiqinGe, A Smart Colorful Supercapacitor with One Dimensional Photonic Crystals, doi: 10.1038/srep18419.

**swelling and shrinking during the rapid charge/discharge process[10]. More interestingly, the mesoporous structure can effectively solve the problem. Compared with pristine PANI, mesoporous PANI (mPANI) electrode exhibits dramatically improved specific capacitance which can enhance the rate performance and the cyclic stability[11].**

**Graphene oxide (GO), with active oxygen functionalities on their planes facilitate surface, has aroused considerable interest as a new electron transport material in recent materials science and condensed-matter studies[12]. The irreversible capacitance which can be considered as an increase of inert components in graphene-based supercapacitors is mainly caused by the two-dimensional stack of graphene and the reduced conductivity of electrons or protons arising therefrom[13,14]. Aggregation and stacking problems are commonly observed in most of the free-standing graphene papers reported to date[15,16,17]. As a result, the unique properties of individual graphene sheets such as high accessible surface area are compromised or not achievable in the macroscopic graphene paper. In order to improve the capacitance, we fabricate P(NiPAAm-GO-AA) hydrogel. It take advantage of GO's superior electrochemical properties and electron transport ability.**

**Furthermore, in previous studies the degradation of specific capacitance was mainly due to severe swelling and shrinking of PANI during the charge/discharge process, which may cause mechanical degradation and breaking of the chains[18,19,20]. To date, most of the covalent connections in the PANI/GO composite materials which were used to make supercapacitors were individually based on the functional groups of GO or PANI[6,21]. Therefore, efforts to enhance the synergistic effect between PANI and GO to improve the overall performance of PANI/GO supercapacitors are still needed[22]. Note that PANI was widely studied to form composites with carbon nanomaterials for supercapacitors in both two-electrode and three-electrode systems, but the specific capacitors when put mesoporous PANI with hydrogel P(NiPPAm-GO-AA) are not seen yet, so the smart 1DPC supercapacitor was novel which the charging and discharging process can be observed by visible structural color change.**

**In this paper, we have developed a flexible mPANI/P(NiPPAm-GO-AA) 1DPC supercapacitor. This supercapacitor can make visible color change during the charge/discharge process to indicate the state-of-charge. These characteristics may extend the life of the supercapacitor. The flexible 1DPC supercapacitor is thin at 3 mm, and exhibits good specific capacitance of 22.6 F $g^{-1}$ with retention of 91.1% after 3,000 cycles.**

## RESULTS

### Fabrication of a Flexible mPANI/P(NiPPAm-GO-AA) 1DPC Supercapacitor

The fabrication of the flexible supercapacitor is schematically shown inFig. 1a. A thin layer of water soluble sacrificial layer Polyvinyl alcohol (PVOH) film was coated onto a silicon wafer. The PVOH layer was then baked at 110 °C for 1 min to evaporate residual water. The 1DPC structural mPANI/P(NiPAAm-GO-AA) were spin-coated onto the PVOH layer[23]. Then, the conducting PDMS film was placed top down onto the 1DPCs. To make conducting PDMS, we modified PDMS with conducting carbon blacks filler. The interfacial contact area between the 1DPC film and PDMS also influences the efficiency of the transfer process. After 20 min bake treatment, the PDMS film with mPANI/P(NiPAAm-GO-AA) Bragg stack structure was peeled off from the silicon wafer substrate when PVOH layer was dissolved. The flexible supercapacitor is finally fabricated by assembling 1DPC with conducting flexible PDMS, Fig. 1b.

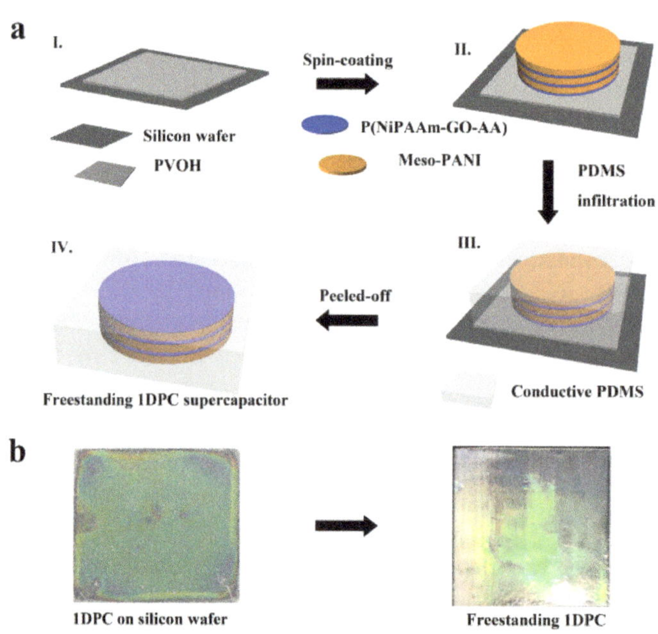


**Figure 1:** (**a**) Fabrication of flexible 1DPCs supercapacitor (I-IV) Schematic illustration showing the freestanding flexible 1DPCs supercapacitor fabricating process. (I) A Polyvinyl alcohol (PVOH) film is placed on a silicon wafer. (II) A P(NiPAAm-GO-AA)/mPANI 1DPC .

## Characterizations of a Flexible mPANI/ P(NiPPAm-GO-AA) 1DPC Supercapacitor

The fabrication of mesoporous polyaniline (mPANI/ P(NiPAAm-GO-AA)) 1DPCs was conducted by alternating mPANI and P(NiPAAm-GO-AA) hydrogel through spin-coating. PANI is one of the most promising conducting polymers for supercapacitor application due to its high conductivity, unique redox properties and easy synthesis approaches. A facile approach has been developed to fabricate mPANI through in situ polymerization using mesoporous silica composite as template. The mesoporous structure can effectively solve the problem that PANI often suffers from inferior long-term stability and rate capability due to swelling and shrinking during the rapid charge/discharge process. Compared with pristine PANI, mPANI electrode exhibits dramatically improved specific capacitance, enhanced rate performance and cyclic stability in 1 mol $L^{-1}$ $H_2SO_4$[24].

P(NiPAAm-GO-AA) is formed through the solution radical copolymerization of NiPAAm, GO and AA with BIS as the initiator in deionized water. The unique hydrogel structure of the GO satisfies several critical requirements for an ideal electrode. The GO in hydrogel network is interconnected and interlocked to prevent it from restacking and to maintain a highly porous monolithic graphene framework. The effective refractive indexes of the P(NiPAAm-GO-AA) and mPANI layers were determined by spectroscopic ellipsometry to be 1.63 and 1.33 respectively. The structural color naturally offers an indication of the lattice structure of the photonic crystal. Color changes can be visually tracked by the naked eye.

The tuning of optical properties of 1DPCs can be achieved by increasing the number of layers, as shown in Fig. 2. An intense and wide Bragg peak (with peak position at 575 nm) is observed; the outcome is a direct consequence of the large refractive index contrast between the mPANI layer and the polymer layer. The photonic band width becomes narrower and the intensity of the peak grows with increasing numbers of layers. The result is in good agreement with the previous theory of the thickness dependence of the optical response of photonic crystal slabs (Fig. 2a).

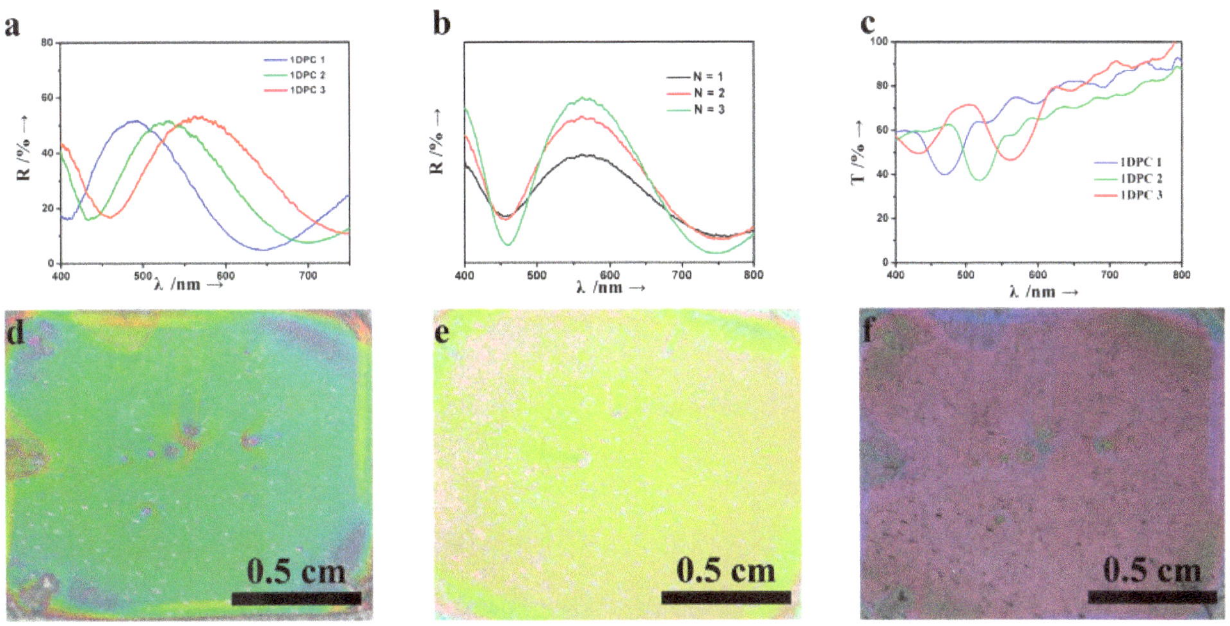


**Figure 2:** Optical properties of flexible 1DPCs supercapacitor.

A change in layer thickness also influences the optical properties of 1DPCs. By increasing the layer thickness, the position of the band gap will make a red shift because of Bragg-Snell's law. According to the significant refractive index difference between two components, the obvious photonic stopband and structural color can be easily obtained in several bilayers. As shown in the reflective spectra (Fig. 2a), the Bragg peak positions are shown to be at 480, 530 and 580 nm, respectively, by fixing the thickness of P(NiPAAm-GO-AA) layer at a proper thickness while changing the thickness of mPANI layer increasingly. The finding agrees with the experiment results. The Bragg peak positions stay unchanged when increasing the numbers of bilayers while keeping the thickness constant (Fig. 2b). The Bragg peak can be manipulated in the full visible range from green (Fig. 2d) to yellow (Fig. 2e) and red (Fig. 2f) by choosing proper periods. Based on reflection spectra characteristics of the mPANI response with electron transfer between emeraldine salt (ES) and emeraldine base (EB) forms, we designed a flexible 1DPC supercapacitor with various stopbands that overlap the whole range of visible reflectance spectra.

The corresponding transmittance spectra are plotted in Fig. 2c, along with the spectra of the multilayers in the PDMS template. The resultant thin Bragg stack sheets were transparent, with optical transmittances at over 90%. As designed, the ordered mPANI/ P(NiPAAm-GO-AA) 1DPCs were stacked to enable high electrical conductivity necessary for the high performance of the resultant supercapacitor (Fig. 2d–f). The film was very uniform over a large area. In our experiments, color uniformity can be achieved over an area of 15 mm × 15 mm at least. The 1DPCs could maintain its color to be invariable for a long time. This ability indicates that the prepared materials and microstructures have perfect stability.

A series of scanning electron microscopy (SEM) and transmission electron microscopy (TEM) images is shown in Fig. 3 to further understand the preparation process of the 1DPC mPANI/ P(NiPAAm-GO-AA) architecture. Figure 3a shows a cross-section view of the 1DPC film. A typical flexible capacitor indicates that the mPANI and P(NiPAAm-GO-AA) hydrogel is in a well-ordered stack arrangement in the PDMS substrate. The PDMS flexible capacitors were peeled off from the silicon wafer substrate that had been synthesized through spin coating. The 1.0 µm-thick Bragg stacks structure could be stably attached onto the flexible PDMS film. It could be peeled off from the substrate traced by the SEM even after many charge/discharge cycles (Fig. 3b). The fracture edge of the film reveals compact layer-by-layer stacking throughout the entire cross-section. The stacking of laminated GO was inhibited by introducing NiPAAm and AA to form a P(NiPAAm-GO-AA) hydrogel microstructure, which shows a clear stratification microstructure of P(NiPAAm-GO-AA) (Fig. 3c). As expected, PANI was well stacked in the 1DPCs to produce high-quality composite films. The mPANI/ P(NiPAAm-GO-AA) 1DPC retained the aligned structure even at a mPANI weight percentage of 35% to 60%.

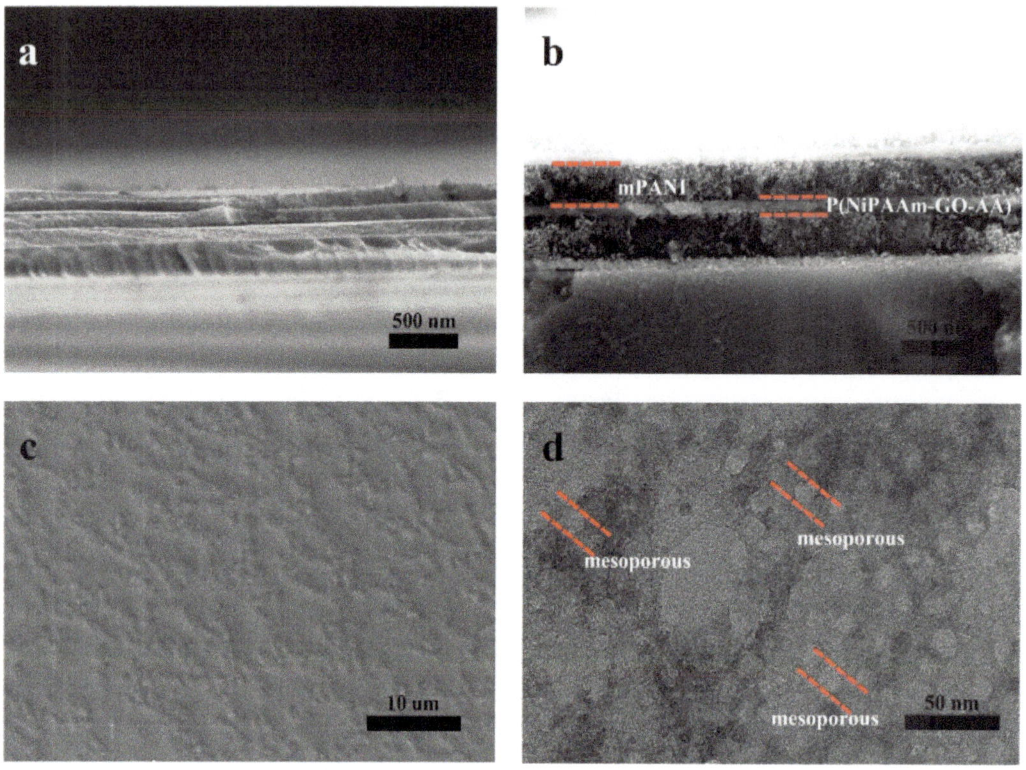


**Figure 3:** SEM and TEM images.

The 1DPC has bright structural colors, which indicate the prominent optical properties of the supercapacitor. For the pure GO template, the top surface of the P(NiPAAm-GO-AA) free-standing film was uniformly covered by a large quantity of small and compact mesoporous PANI. As shown in the cross-section SEM image of the mPANI, no mPANI was observed in the inner layers of the P(NiPAAm-GO-AA) sheets. This phenomenon clearly suggests that for the pure P(NiPAAm-GO-AA) film, the mPANI was stacked in a compact manner on the surface of the GO hydrogel film because of the stacked lamellar structure. Remarkably, a high-resolution TEM image shows that the resultant mPANI sheets had an extremely flat shape and numerous mesoporous structures with size of approximately 10 nm (Fig. 3d). The BET test is in supplementary information Figure S2. The cross-section SEM images clearly demonstrate that the mPANI and P(NiPAAm-GO-AA) have been successfully stacked into a whole flexible PDMS template from the top surface, down to the interior of the architecture.

## Electrochemical evEaluation of the mPANI/ P(NiPAAm-GO-AA) Superceapacitor.

To investigate the electrochemical performance of mPANI/P(NiPAAm-GO-AA) in a real device, a symmetric supercapacitor device was constructed using mPANI/ P(NiPAAm-GO-AA) as the working electrode and Platinum (Pt) as the counter electrode. The good capacitance of mPANI/ P(NiPAAm-GO-AA) 1DPCs films may be attributed to the internal microstructure of the deposited mPANI. This phenomenon can be explained by the decreasing conductivities along with the increasing PANI weight percentage[25]. In consideration of the specific capacitance, the PANI weight percentage of 60% is investigated in the following discussion.

According to a previous study, graphene/PANI supercapacitors cannot provide an effective network for electron transportation during electrochemical reaction if they are stacked loosely[26]. To solve this problem, we made a well-stacked mPANI/P(NiPAAm-GO-AA) 1DPC structure, in which the hydrogel P(NiPPAm-GO-MAA) made the stack dense.

The galvanostatic charge/discharge curves of the obtained mPANI/P(NiPAAm-GO-AA) in different densities were measured to further study their possible application as supercapacitor electrode materials. The typical galvanostatic charge/discharge curves of the mPANI/P(NiPAAm-GO-AA) electrode in 1 M $H_2SO_4$ are presented in Fig. 4. Charge-discharge curves maintain the same shape at a potential range of 0 to 0.7 V in these densities. Therefore, the nanocomposite can experience a wide range of current. Moreover, the discharge/charge efficiency can be obtained according to the ratio of the discharge time to the charge time.

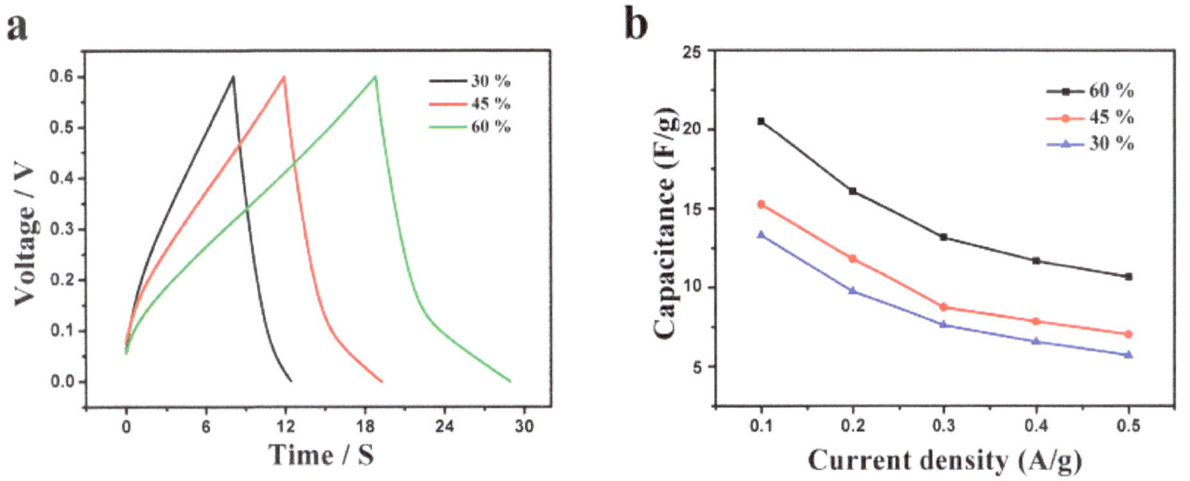


**Figure 4:** (**a**) Galvanostatic charge-discharge profiles of supercapacitors with different mPANI weight percentages; (**b**) Dependence of specific capacitance on mPANI weight percentage and current density.

The resultant supercapacitor exhibited a high electrochemical performance. Figure 4a compares the galvanostatic charge/discharge profiles of the supercapacitors based on different mPANI weight percentages at 0.1 A $g^{-1}$. The maximal specific capacitance was 60%. To further explore the influence of the PANI weight percentage on the specific capacitance, galvanostatic charge/discharge processes were also conducted in increasing current densities from 0.1 to 0.5 A $g^{-1}$.

As shown in Fig. 4b, the specific capacitances were gradually enhanced from 13.8 F $g^{-1}$ to 22.6 F $g^{-1}$, with the increasing mPANI weight percentage due to the improved capacitance from the PANI. With the increasing current densities from 0.1 A $g^{-1}$ to 0.5 A $g^{-1}$, the capacitances were maintained 70.1%, 71.3% and 74.5% at the PANI weight percentages of 30%, 45% and 60%, respectively.

The different mPANI percentage 1DPC film electrodes can be also divided into different thicknesses of the mPANI layer varies from 50 nm to 150 nm. The specific capacitance was increased with the increasing mPANI thickness because of the decreasing resistance along the aligned direction. The influence on the percentage of mPANI in smart supercapacitor is summarized in Fig. 4. A positive electrode was examined to consider the symmetric structure in the supercapacitor. Conducting polymers were shown to change structures under different oxidation states[27].

Figure 5 shows the representative CVs for three different tandem PANI weight percentages (30%, 45%, and 60%) at different scan rates of 2, 5, 20, 25, and 50 mV $s^{-1}$ in 1 M $H_2SO_4$. The rate-dependent cyclic voltammograms of the electrode over a wide range of scan rates are shown in Fig. 4 and they exhibit a typical redox behavior of PANI. The shape of the CV was maintained even up to the high scan rate of 50 mV $s^{-1}$, This result indicates that the mPANI/P(NiPAAm-GO-AA) 1DPCs film has a rapid charge/discharge response within the potential window of 0.6 V in the device[22]. Normally, pure PANI shows two couples of redox peaks: one is due to the transition of PANI from the semiconducting-state to the conductive form, and the other is due to the transition from emeraldine to pernigraniline. The GO electrode also exhibits a pair of peaks that originated from the oxygenous groups on the surface. However, when GO and mPANI are combined into a flexible 1DPC mPANI/P(NiPAAm-GO-AA), their CV curves become different from each component electrode and exhibit the characteristics of both double-layered capacitance and pseudocapacitance. More interestingly, the CV shapes of different PANI percentage change with different mass ratios. As the binder of each working electrode is inactive and occupies the same amount, the change in CV characteristic is due to the nature of the 1DPC materials. The synergetic effect resulting from the interactions of PANI and GO may affect the shape and potential position of the CV curves.

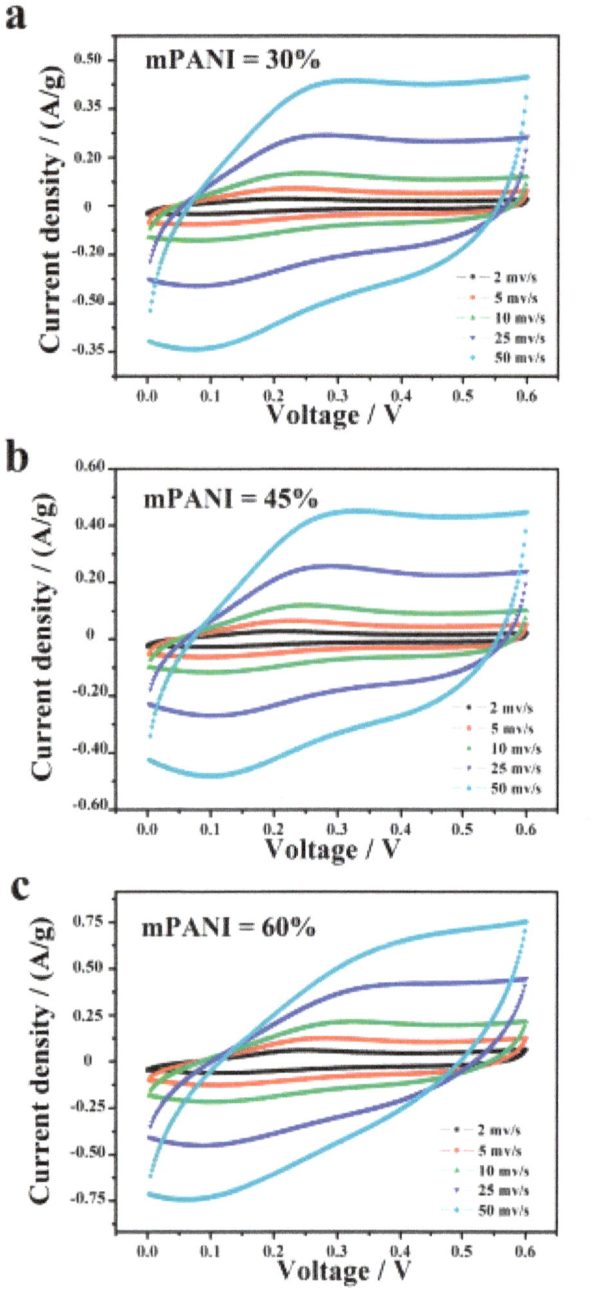


**Figure 5:** Cyclic voltammograms of different mPANI weight percentage with increasing scan rates.

The existence of peak on each CV plot for the samples indicates the existence of the faradic processes. The charge storage in the faradic process is achieved by electron transfer that leads to chemical changes in the electroactive materials. The redox peaks can be ascribed to the changing in mPANI structures and the oxygenated groups attached to the surface of the graphene oxide nanostructures. Besides, the surface chemical functional groups significantly influence the interfacial capacitance by introducing pseudocapacitance as depicted in former works. Thus, graphene oxide sheets with different amounts in the composites may play an important role in determining the CV shapes.

Synthesized materials also exhibit excellent electrochemical behavior in a wide range of scan rates (2 mV $s^{-1}$–50 mV $s^{-1}$) in $H_2SO_4$ electrolyte solution. For example, for mPANI percentage of 30% (Fig. 5a), 45% (Fig. 5b) and 60% (Fig. 5c), the current density of both samples increases with the scan rate and the curve shape is steady, thus indicating the good electrochemical stability of the electrode material. This stability is due to the introduction of the GO hydrogel with a high mechanical property into the 1DPC and the synergetic effect between the two components. The CV shape of 30% mPANI is different from that of 60% mPANI probably because of the different feeding ratios and the different Faradic charge transfer reactions arising from the interaction and synergy among individual components.

The synthesis of PANI nanocomposites and their application in energy storage have aroused great research interests recently. The 1DPC stack structure with mPANI we prepared effectively increased the specific surface area. The increase is one of the reasons for the high specific capacitance. The microstructures of 1DPC films were significantly influenced by thc weight percentage of mPANI, which plays a key role during the fabrication of the capacitor. The further increase in mPANI weight percentage led to the increase in 1DPC capacity. However, as expected, the optical transmittance decreased with the increased thickness of the mPANI sheet. Thus, the 60% weight percentage of the mPANI sheet with thickness of 150 nm was mainly studied in this work. To further understand the electrochemical stability of the mPANI/P(NiPAAm-GO-AA) electrode, consecutive charge–discharge cycles were measured at a current density (Fig. 6a). The mPANI/P(NiPAAm-GO-AA) electrode was shown to maintain 91.1% of its initial capacitance after 3000 cycle tests. The finding suggests that the mPANI/P(NiPAAm-GO-AA) electrode exhibits excellent long-term cycle ability and a high degree of reversibility in consecutive charge/discharge cycles. Therefore, the mPANI/P(NiPAAm-GO-AA) 1DPC is shown to have good cyclic stability as a supercapacitor electrode material[28].

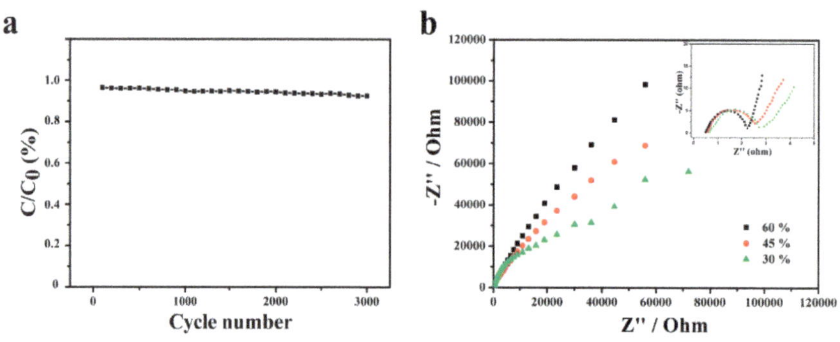


**Figure 6:** (**a**) Dependence of specific capacitance on cycle number at a current density of 0.5 A/g; (**b**) Nyquist plots of 3 different mPANI weight percentage flexible electrodes. Inset is the enlarged plots of the high-frequency region.

The mPANI/P(NiPAAm-GO-AA) electrode exhibits high discharge/charge stability, which indicates excellent discharge/charge rate control capability and electrochemical reversibility. PANI commonly suffers from poor cyclic stability as an electrode material for supercapacitors because of the cyclic mechanical stress caused by material swelling and shrinking during the charge/discharge cycle. The enhanced electrochemical performance can be firstly attributed to the introduction of GO hydrogel into the 1DPC material and mesoporous structure. This Bragg stack synthesis procedure enables the resultant supercapacitors to be less susceptible to fatigue caused by cyclic mechanical stress.

Electrochemical impedance spectroscopy (EIS) is an important tool for the study of the characteristics of conductivity, mechanistic analysis of interfacial processes and structure and charge transport at the material electrolyte interface. Figure 6b shows the Nyquist diagrams for PANI, P(NiPAAm-GO-AA) and the resultant mPANI/ P(NiPAAm-GO-AA) 1DPC. The semicircle in the high-frequency region and the straight line in the low-frequency region correspond to the electron-transfer-limited processes and diffusional-limited electron-transfer processes, respectively. The diameter of the semicircle is one of the limiting factors for the power density of supercapacitors, as it corresponds to the charge transfer resistance at the electrode-electrolyte interface. Compared with that of pure PANI, the charge transfer resistance of the mPANI/ P(NiPAAm-GO-AA) electrode is reduced remarkably because of the formation of the hierarchical nanocomposite due to the use of P(NiPAAm-GO-AA) as a support. In the low frequency, the slope of the 45° portion of the Nyquist plots is called the Warburg resistance ($R_w$) and is a result of the frequency dependence of ion diffusion in the electrolyte on the electrode interface. The $R_w$ value for the 60% mPANI is larger than that of the 30% and 45% mPANI electrode, which demonstrates a better accessibility of the electrolyte ions to the 30% and 45% mPANI electrode. These EIS results support our CV analyses of the enhanced performance of the flexible electrode.

Interestingly, the supercapacitor exhibits different colors during the charge diacharge process caused by the shift of the 1DPC stopbands. The structural color changes because of the variation of the refractive index of the mPANI in different charge/discharge processes (Fig. 7). The 1DPCs film make a red shift from green (Fig. 7a) to yellow (Fig. 7b) after a charging process, as the mPANI changes to the fully conducting form and the refractive index increases (Supporting Information Fig. S3 and Movie [SI]). By contrast, the 1DPC film turns back to green after the discharge process as the refractive index decreases. The effective refractive index of the mPANI is obtained by spectroscopic ellipsometry. Upon a full charge with a voltage of 0.6 V, the electrode turns yellow. When the supercapacitor is discharged to 0.6 V, the electrode changs to green.

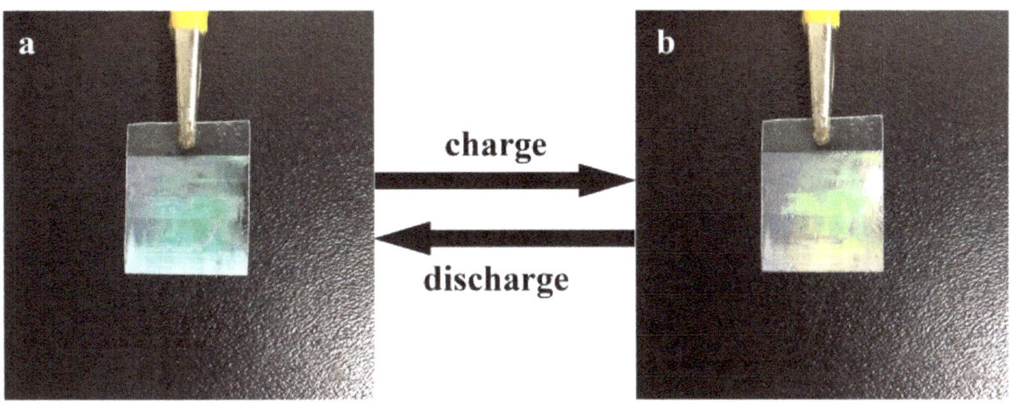


**Figure 7:** Structural color change of flexible 1DPCs supercapacitor during charging-discharging process.

This phenomenon indicates that the supercapacitor under the charge/discharge process can be monitored by the naked eyes. Although PANI has been widely explored in various sensing materials including electrochromatic nanocomposites, to the best of our knowledge, this work represents a better combination of chromatic devices and supercapacitors using PANI.

## DISCUSSION

This attractive supercapacitor is attributed to the Bragg stack with tunable structural color and a good combination of a mesoporous conducting polymer and GO hydrogel. These flexible mPANI/P(NiPAAm-GO-AA) 1DPCs supercapacitors can be an ideal candidate for state-of-charge monitoring application because of their distinct color change under charge/ discharge process.

The good supercapacitive performance of mPANI/ P(NiPAAm-GO-AA) 1DPCs may be largely attributed to their unique structure. First, 1DPCs that consist of mesoporous PANI have high porosity, and thus the electrolyte can easily penetrate through the Bragg stack for efficient redox reactions during the charge storage process. Second, they offer an ordered lamellar structure when GO is combined with the P(NiPAAm-GO-AA) hydrogel. Third, the unique Bragg stack structure can significantly enlarge the active surface area and improve the structural integrity as well. As a result, the stare-of-charge of a flexible 1DPC supercapacitor can be monitored by the structural color change, and the specific capacitance and cycling stability are enhanced.

In summary, a novel smart and colorful supercapacitor with high performance is developed in this study. This supercapacitor can make visible color changes during the charge/discharge process to indicate the state-of-charge. These characteristics may extend the life of the supercapacitor. The flexible 1DPC supercapacitor is thin at 3 mm and exhibits good specific capacitance of 22.6 F $g^{-1}$ with retention of 91.1% after 3,000 cycles. This work can serve as a starting point in designing smart supercapacitors with 1DPCs, which cannot be done with their conventional counterparts. We demonstrate the application of the 1DPC supercapacitor as an ultrathin power source. This technology may find extensive applications in future wearable electronics.

## METHOD

### Synthesis of mPANI and P(NiPPAm-GO-AA).

All the chemicals were directly used after purchase without further purification. In a typical synthesis, the mesoporous PANI was uniformly synthesized on the surface of mesoporous $SiO_2$ ($mSiO_2$) viain situ polymerization. The $mSiO_2$ was achieved from Nanjing Dongjian Biological Technology Co., LTD. 0.47 g of $mSiO_2$ was dispersed in 40 mL of deionized water and stirred for 2 h. Then 20 mL of aniline monomers (solvent: 1 mol $L^{-1}$ HCl) with the concentration of 0.25 mol $L^{-1}$ was added into the above mixed solution. After stirred for another 2 h, 20 mL of ammonium persulfate solution (the same concentration as that of aniline monomers) was slowly added into the above solution and the system temperature was controlled below 4 °C in an ice-water bath. As the reaction proceeded, the color of the solution gradually turned emerald. After the reaction continued for 4 h, the $mSiO_2$/PANI was obtained after filtration and washing. Finally, silica was removed by the treatment with hydrofluoric acid under magnetic stirring for 24 h and the obtained materials was mPANI. P(NiPPAm-GO-AA) polymer hydrogel was prepared by free-radical polymerization. NIPAM (120.0 mg), AA monomer (3.0 mg), GO (3.0 mg), BIS (12.0 mg) and $H_2O$ (2.0 mg) were mixed in a flask. Meanwhile, nitrogen gas was also purged into the solution for 30 min to remove dissolved oxygen. The solution was stirred at 350 rpm for 4 h at 70 °C. As a result, P(NiPPAm-GO-AA) polymer hydrogel was formed as white suspension in aqueous phase.

### Fabrication of Flexible mPANI/P(NiPPAm-GO-AA) 1DPCs Supercapacitor.

A thin layer of water soluble sacrificial layer PVOH film was coated onto a silicon wafer. The PVOH layer was then baked at 110 °C for 1 min to evaporate residual water. The 1DPC structural mPANI/ P(NiPAAm-GO-AA) were spin-coated onto the PVOH layer. Then, the conducting PDMS film was placed top down onto the 1DPCs. To make conducting PDMS, we modified PDMS with conducting carbon blacks filler. The interfacial contact area between the 1DPC film and PDMS also influences the efficiency of the transfer process. After 20 min bake treatment, the PDMS film with mPANI/P(NiPAAm-GO-AA) Bragg stack structure was peeled off from the silicon wafer substrate when PVOH layer was dissolved. The flexible supercapacitor is finally fabricated by assembling 1DPC with conducting flexible PDMS.

### Structural Characterization and Analysis

Field emission scanning electron microscope (FESEM, Zeiss Ultra Plus) was used to see the structure of the flexible 1DPCs supercapacitor. Transmission electron microscope (TEM, JEM2100EX) was used to obtain the image of $mSiO_2$ and mPANI. Reflective spectra were recorded using a fiber optic UV-vis spectrometer (Ocean Optic HR2000CG).

### Electrochemical Measurements

All the electrochemical experiments were carried out using Autolab PGSTAT302N (Metrohm Autolab, Netherlands) at room temperature. The electrical impedance spectroscopy (EIS) measurements were performed at open circuit potential with a sinusoidal signal over a frequency range from 100 kHz to 10 mHz at an amplitude of 10 mV. The cycle life tests were conducted by galvanostatic charge/discharge measurements. Galvanostatic charge–discharge measurements of half cells, symmetric supercapacitors and hybrid devices were carried out galvanostatically at various current densities with voltage windows specific to the materials using a battery test system Chi660D (Chenhua Inc., Shanghai, China).

The gravimetric capacitances ($C_{wt}$) of flexible 1DPCs electrodes derived from galvanostatic discharge curves were calculated based on the following equation 1:

$$C_{wt} = 2(I\Delta t)/(m\Delta V) \quad (1)$$

where $I$ is the constant discharge current, $\Delta t$ is the time for a full discharge, $m$ is the net mass of one electrode and $\Delta V$ represents voltage drop on discharging (excluding the $V_{drop}$)[29,30]

The spectral position of the reflection maximum, $\lambda_B$ of a multilayer made of two types of alternating films of thicknesses $t_1$ and $t_2$ and refractive indices $n_1$ and $n_2$, respectively, depends on the incident angle with respect to the surface normal, $\theta$, as expressed in the equation 2:

$$\lambda_B = 2d\sqrt{n^2-(sin\ \theta)^2} \quad (2)$$

Where d is the unit cell size and n is the average refractive index of the multilayer, which is in turn given by equation 3:

$$d = \frac{t_1 n_1 + t_2 n_2}{n} \quad (3)$$

For more 1DPCs theoretical equation in Supplementary Information.

## ACKNOWLEDGMENTS

The authors thank for the financial support of pre-research foundation for national science foundation and the Fundamental Research Funds for the Central Universities. This work is also supported by the fund for the analysis and testing of the large-scale apparatus from the Southeast University. Prof. Dr. Ge also thanks for the support from Prof. Gu Zhongze.

## REFERENCES

1. Chu, S. & Majumdar, A. Opportunities and challenges for a sustainable energy future. Nature 488, 294–303 (2012).
2. Lee, J. A. et al. Ultrafast charge and discharge biscrolled yarn supercapacitors for textiles and microdevices. Nat. Commun. 4, 1970 (2013).
3. El-Kady, M. F. & Kaner, R. B. Scalable fabrication of high-power graphene micro-supercapacitors for flexible and on-chip energy storage. Nat. Commun.

4. 4, 1475 (2013). 4. Ghidiu, M. et al. Conductive two-dimensional titanium carbide 'clay' with high volumetric capacitance. Nature 516, 78–81 (2014).

5. Hur, J. et al. DNA hydrogel-based supercapacitors operating in physiological fluids. Sci. Rep. 3, 1282 (2013).

6. Yan, C. Y. et al. Highly Stretchable Piezoresistive Graphene-Nanocellulose Nanopaper for Strain Sensors. Adv. Mater. 26, 2022–2027 (2014).

7. Zhang, S. & Pan, N. Supercapacitors Performance Evaluation. Adv. Energ. Mater. doi: 10.1002/aenm.201401401 (2014) .

8. Zhao, Y. et al. Bio-inspired variable structural color materials. Chem. Soc. Rev. 41, 3297–3317 (2012).

9. Puzzo, D. P. et al. Color from colorless nanomaterials: Bragg reflectors made of nanoparticles. J. Mater. Chem. 19, 3500–3506 (2009).

10. Liu, C. et al. Patterned one-dimensional photonic crystals with acidic/alkali vapor responsivity. RSC Adv. 4, 27281–27285 (2014).

11. Anu, Prathap. M. U. et al. Synthesis of mesostructured polyaniline using mixed surfactants, anionic sodium dodecylsulfate and nonionic polymers and their applications in H2O2 and glucose sensing. Colloids Surf. B 89, 108–116 (2012).

12. Xiang, Q., Yu, J. & Jaroniec, M. Graphene-based semiconductor photocatalysts. Chem. Soc. Rev. 41, 782–796 (2012).

13. Kim, M., Lee, C. & Jang, J. Fabrication of Highly Flexible, Scalable, and HighPerformance Supercapacitors Using Polyaniline/Reduced Graphene Oxide Film with Enhanced Electrical Conductivity and Crystallinity. Adv. Funct. Mater. 24, 2489–2499 (2014).

14. Lai, L. et al. Preparation of Supercapacitor Electrodes through Selection of Graphene Surface Functionalities. ACS Nano 6, 5941–5951 (2012).

15. Zhang, E. et al. Robust and thermo-response graphene-PNIPAm hybrid hydrogels reinforced by hectorite clay. Carbon 62, 117–126 (2013).

16. Chi, K. et al. Freestanding Graphene Paper Supported Three-Dimensional Porous Graphene-Polyaniline Nanoconnposite Synthesized by Inkjet Printing and in Flexible All-Solid-State Supercapacitor. ACS Appl. Mater. Interfaces 6, 16312–16319 (2014).

17. Zhang, J. et al. A high-performance asymmetric supercapacitor fabricated with graphene-based electrodes. Energ. Environ. Sci. 4, 4009–4015 (2011).

18. Patil, D. S. et al. Polyaniline based electrodes for electrochemical supercapacitor: Synergistic effect of silver, activated carbon and polyaniline. J. Electroanal. Chem. 724, 21–28 (2014).

19. Hou, J., Liu, Z. & Zhang, P. A new method for fabrication of graphene/polyaniline nanocomplex modified microbial fuel cell anodes. J. Power. Souces 224, 139–144 (2013).

20. Yu, P. et al. Polyaniline Nanowire Arrays Aligned on Nitrogen-Doped Carbon Fabric for High-Performance Flexible Supercapacitors. Langmuir 29, 12051–12058 (2013).

21. Long, C. L. et al. Supercapacitors Based on Graphene-Supported Iron Nanosheets as Negative Electrode Materials. ACS Nano 7, 11325–11332 (2013).

22. Wang, H. et al. Effect of Graphene Oxide on the Properties of Its Composite with Polyaniline. ACS Appl. Mater. Interfaces 2, 821–828 (2010).

23. Howell, I. R. et al. Strain-tunable one dimensional photonic crystals based on zirconium dioxide/slide-ring elastomer nanocomposites for mechanochromic sensing. ACS Appl. Mater. Interfaces 7, 3641–3646 (2015).

24. Wang, Q. et al. Mesoporous polyaniline film on ultra-thin graphene sheets for high performance supercapacitors. J. Power. Souces 247, 197–203 (2014).

25. Xu, D. et al. Fabrication of Free-Standing Hierarchical Carbon Nanofiber/Graphene Oxide/Polyaniline Films for Supercapacitors. ACS Appl. Mater. Interfaces 6, 200–209 (2014).

26. Tong, Z. et al. Layered polyaniline/graphene film from sandwich-structured polyaniline/graphene/polyaniline nanosheets for highperformance pseudosupercapacitors. J. Mater. Chem. A 2, 4642–4651 (2014).

27. Chen, X. et al. Smart, Stretchable Supercapacitors. Adv. Mater. 26, 4444–4449 (2014).

28. Fan, H. et al. Hierarchical nanocomposite of polyaniline nanorods grown on the surface of carbon nanotubes for high-performance supercapacitor electrode. J. Mater. Chem. 22, 2774–2780 (2012).

29. Xu, Y. et al. Holey graphene frameworks for highly efficient capacitive energy storage. Nat. Commun. 5, 4554 (2014).

30. Zhang, F. et al. A high-performance supercapacitor-battery hybrid energy storage device based on graphene-enhanced electrode materials with ultrahigh energy density. Energ. Environ. Sci. 6, 1623–1632 (2013).

CHAPTER 15

# DISSIPATIVE AND DISPERSIVE OPTOMECHANICS IN A NANOCAVITY TORQUE SENSOR

Marcelo Wu[1,2], Aaron C. Hryciw[2], Chris Healey[1,2], David P. Lake[1], Harishankar Jayakumar[1], Mark R. Freeman[2,3], John P. Davis[3], and Paul E. Barclay[1,2]

[1]Institute for Quantum Science and Technology, University of Calgary, Calgary, Alberta T2N 1N4, Canada

[2]National Institute for Nanotechnology, Edmonton, Alberta T6G 2M9, Canada

[3]Department of Physics, University of Alberta, Edmonton, Alberta T6G 2E9, Canada

## ABSTRACT

**Dissipative and dispersive optomechanical couplings are experimentally observed in a photonic crystal split-beam nanocavity optimized for detecting nanoscale sources of torque. Dissipative coupling of up to approximately 500 MHz/nm and dispersive coupling of 2 GHz/nm enable measurements of sub-pg torsional and cantileverlike mechanical resonances with a thermally limited torque detection sensitivity of 1.2×10−20 Nm/Hz−−−√ in ambient conditions and 1.3×10−21 Nm/Hz−−−√ in low vacuum. Interference between optomechanical coupling mechanisms is observed to enhance detection sensitivity and generate a mechanical-mode-dependent optomechanical wavelength response.**

## SUMMARY

In recent years, optical cavities wherein light follows a closed path—have proven to be remarkable tools for nanoscale experiments. Cavity optomechanics uses light to measure and control the motion of optical resonators. By amplifying the interaction between light and mechanical vibrations, cavity-optomechanical sensors have recently enabled unprecedented sensitivity in measuring minute displacements of on-chip devices. These sensors can be used for detecting environmental stimuli such as acceleration, measuring small fields, and sensing particles, which have applications in diverse areas of physics and engineering. We describe the fabrication and optimization of an optomechanical sensor for detecting small amounts of torque using dissipative coupling, a new effect that has been discussed theoretically but has not—before now—been used for torque sensing.

Optomechanical sensors generally rely on dispersive coupling, in which the displacement of the mechanical device changes the wavelength of its optical resonance. This signal is amplified by the cavity before being transmitted to the output. In our case, we use an optical fiber to both input light into the device and collect the signal at the output. We use two nanomechanical resonators, which serve as optical mirrors that can move independently. Our device also allows its displacement to affect the length of time that light stays inside the cavity. Furthermore, the proximity of the optical fiber to the cavity plays a role by influencing how the light is transmitted. Both of these effects, collectively referred to as dissipative coupling, are imprinted on the output as an enhancement in the original signal. Using both dispersive and dissipative coupling effects, we have achieved an order-of-magnitude improvement in torque sensitivity in ambient conditions and in vacuum.

**Citation:** Marcelo Wu, Aaron C. Hryciw, Chris Healey, David P. Lake, Harishankar Jayakumar, Mark R. Freeman, John P. Davis, and Paul E. Barclay, DOI: 10.1103/PhysRevX.4.021052.

The record torque sensitivity exhibited by this device will be useful for studying magnetic materials and sensing magnetic fields. Furthermore, moving to the low-temperature regime will further increase the sensitivity of the nanocavity.

Optical measurement and control of mechanical vibrations are at the heart of many technological and fundamental advances in physics and engineering, from sensitive displacement and force detection [1–9] to proposed observation of gravitational waves [10] and studies of the quantum properties [11] of massive objects [12–14]. Nano photonic implementations of cavity-optomechanical systems [15] localize light to sub wavelength volumes, enhancing optomechanical coupling between photons and phonons of nanomechanical structures [6,16,17]. Harnessing this optomechanical interaction has enabled milestone experiments, including ground-state cooling [13,18], mechanical squeezing of light [19], and optomechanically induced transparency [20,21]. Typically, optomechanical coupling arises in cavity-optomechanical systems from a dispersive dependence of the nano cavity resonance frequency on the nanocavity geometry, which is modulated by mechanical excitations. In this paper, we demonstrate that dissipative optomechanical coupling, where mechanical excitations modulate the nanocavity photon lifetime, can also play a crucial role in the optical transduction of nanomechanical motion. In particular, we demonstrate the dissipative-enhanced optomechanical readout of a cantilever integrated directly within a nanocavity and realize an optomechanical torque detector whose sensitivity of approximately1.3×10−21 Nm/Hz−−−√ promises to significantly advance detection of phenomena in studies of nanomagnetic [22,23] and mesoscopic [24] condensed-matter systems, optical angular momentum [25], and magnetometry [26]. We also observe interference between dissipative and dispersive coupling mechanisms, which reveals details of the nature of the nanomechanical motion and may open new avenues in optomechanical control [27–31].

The nanocavity optomechanical system studied here, an example of which is shown in Fig. 1(a), provides a unique platform for studying dispersive and dissipative optomechanical couplings and their impact on sensing and measurement. These photonic crystal "split-beam" nanocavities support high optical quality factor ( Qo) modes localized between two cantilever nanomechanical resonators, which are patterned to also serve as optical "mirrors." The mirrors can move independently and support mechanical resonances whose properties can be customized through design of their mechanical supporting structure. In the device under study, one of the mirrors is suspended by a single mechanical support. Mechanical resonances of this mirror can be efficiently actuated by coupled sources of torque, as illustrated in Fig. 1(b), potentially allowing sensitive readout of a variety of nanomagnetic and mesoscopic systems [22–26].

Optomechanical coupling in split-beam nanocavities is strongest when mechanical motion of the mirrors modifies the nanocavity mirror gap, effectively changing the nanocavity length, resulting in a dispersive coupling to the nanocavity optical frequency ωo.

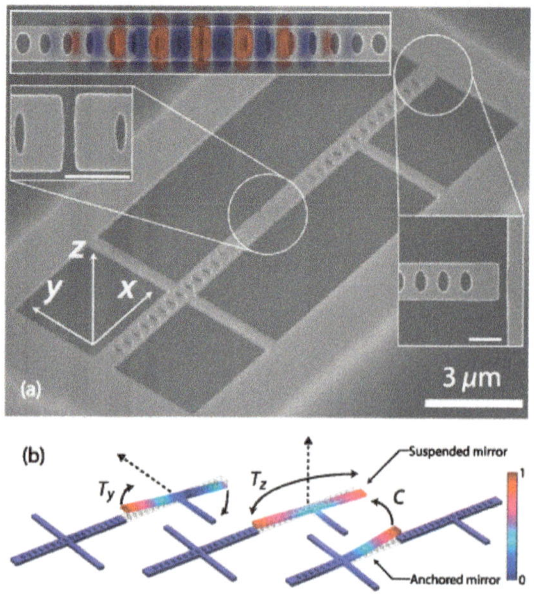


**Figure. 1:** (a) Scanning electron micrograph of a split-beam nanocavity. Top left inset: Top view of the nanocavity overlaid with the field distribution ( Ey) of the optical mode. Left inset: 60-nm-wide nanocavity central gap. Right inset: Gap separating the suspended nanobeam from the device layer. Inset scale bars: 500 nm. (b) Displacement fields of split-beam nanocavitymechanical modes of interest. Dotted arrows indicate the position and direction of torque for efficient actuation.

A more striking property of split-beam nanocavities is the strong dependence of the nanocavity internal photon decay rate γi on the mirror gap, resulting in dissipative optomechanical coupling [27–31]. Additional dissipative coupling arises when the motion of the mirror modulates the nanocavity external photon decay rate γe into an external coupling waveguide. These interactions can be probed by monitoring fluctuations in the transmission T(λ) of a waveguide coupling light into and out of the nanocavity, as illustrated in Fig. 2(a). For cavity-optomechanical systems operating in the sideband-unresolved regime, a shift dx in the mechanical resonator position modifies T by

$$dT = \left( g_{\mathrm{OM}} \frac{\partial T}{\partial \omega_o} + g_i \frac{\partial T}{\partial \gamma_i} + g_e \frac{\partial T}{\partial \gamma_e} \right) dx \tag{1}$$

where $g_{\mathrm{OM}} = d\omega_o/dx$ is the dispersive optomechanical coupling coefficient, and $g_i = d\gamma_i/dx$ and $g_e = d\gamma_e/dx$ are the intrinsic and external dissipative coupling coefficients, respectively. The derivatives in Eq. (1) can be derived from cavity-waveguide coupled-mode theory and are given in the Appendix. A key feature is that$|\partial T/\partial \gamma i,e|$ are unipolar, whereas $|\partial T/\partial \omega_o|$ is bipolar, as illustrated in Fig. 2(b). Interference between these terms can result in an asymmetric optomechanical wavelength response dT(λ)/dx with respect to detuning $\Delta\lambda=\lambda-\lambda_o$. Maximum contributions to dT from dispersive, dissipative intrinsic, and dissipative external optomechanical coupling mechanisms scale with $Q_o/\omega_o\{(1-T_o)g_{\mathrm{OM}}, 4(1-T_o)g_i, 8T_o g_e\}$ and occur when $\Delta\lambda - \{\delta\lambda/2, 0, 0\}$,, respectively, where $T_o$=

$T(\lambda_o)$ and $\delta\lambda = \lambda_o/Q_o$. Notably, transduction via ge does not vanish when $T_o \rightarrow 1$, i.e., for undercoupled nanocavities ( $\gamma_e \ll \gamma_i$). However, $g_e$ itself does vanish as the fiber taper moves further away from the cavity, resulting in degradation of the transduction signal.

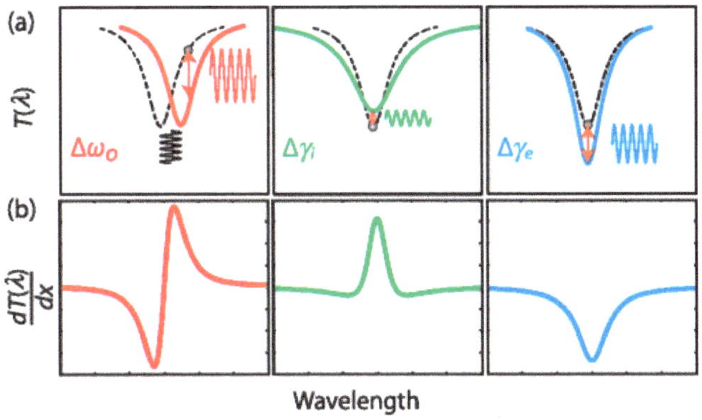


**Figure. 2:** Illustration of the effect of mechanical displacement on the optical response of a cavity with (left) dispersive, (center) dissipative intrinsic, and (right) dissipative external optomechanical coupling. (a) Change in the resonance line shape. (b) Amplitude of the optomechanically actuated signal.

The critical role of optomechanical coupling for a wide class of sensing applications is revealed by the minimum force detectable by a cavity-optomechanical system[26]:

$$F_{\min}(\omega) = \left[\frac{4k_B T_e m\omega_m}{Q_m} + \frac{S_n}{[G(\lambda)|\chi(\omega)|]^2}\right]^{1/2}. \quad (2)$$

$F_{\min}(\omega)$ describes the minimum actuating force required to obtain a unity signal-to-noise ratio in the presence of thermal noise and technical measurement noise. The lower bound on $F_{\min}(\omega)$ is fixed by thermal fluctuations of the mechanical resonator, given by the first term in Eq. (2), and determined by the mechanical resonator frequency $\omega_m$, effective mass m, mechanical quality factor $Q_m$, and temperature $T_e$, where $k_B$ is Boltzmann's constant. Overcoming technical noise $S_n$ requires a combination of large optomechanical gain G(λ) and driving the system at frequency ω, where the mechanical susceptibility $\chi(\omega) = [m(\omega^2 - \omega_m^2 - i\omega\omega_m/Q_m)]^{-1}$ is large. In many cavity-optomechanical systems, including the split-beam nanocavities studied here, $F_{\min}(\omega)$ is thermally limited at room temperature, where the thermal phonon population exceeds 106 for MHz-frequency mechanical resonators. In such systems, further reducing the effects of technical noise is advantageous to allow sensitive off-resonance detection, to improve the measurement resolution, and to enhance the ultimate device sensitivity in the case of low- $T_e$ operation. In the measurements presented below, dominant contributions to $S_n$ are from laser noise and photon shot noise, followed by detector noise. Note that back action noise is not included in Eq. (2), due to its negligible effect in the regime of operation studied here.

The optomechanical gain $G(\lambda) = \eta g_{ti}|dT/dx(\lambda)|P_i$ is determined by Eq. (1), and, for a given waveguide input power $P_i$, waveguide transmission efficiency η, and photodetector gain $g_{ti}$, can be increased through large gOM,i,e, high $Q_o$, and optimally tuning λ within the nanocavity optical mode linewidth. As discussed above, dissipative external coupling can play an important role in maximizing G. However, dissipative coupling is often small compared to dispersive contributions and, to date, has only been reported experimentally in hybrid cavity-nanomechanical systems where $g_e$~10–20 MHz/nm [32,33]. In the split-beam nanocavities studied below, measurements indicate that $g_{OM}$~ 2 GHz/nm,$g_i$~300–500 MHz/nm, and $g_e$~2–3 MHz/nm.

The split-beam nanocavity devices studied here are fabricated from silicon-on-insulator chips consisting of a 220-nm-thick silicon ($S_i$) layer on a 3-μm-thick silicon-dioxide ( $SiO_2$) layer. Using electron-beam lithography, reactive-ion etching, and a hydrofluoric acid undercut, pairs of cantilever photonic crystal mirrors are defined, with a 60-nm gap between them and another at one mirror end. The resulting split-beam photonic crystal nanocavities, whose design is described in Ref. [34], support high- $Q_o$ optical modes with $\omega_o/2\pi$~200 THz ( $\lambda_o$~1550 nm). The mirror pattern consists of a periodic array of holes, whose dimensions are tapered from circles to elliptical shapes with a profile similar to the gap. Crucially, the band edge of the photonic crystal "air mode" associated with the gap unit cell is phase matched with the band edge of the neighboring elliptical hole unit-cell air mode, minimizing radiation loss in the gap region and creating a smooth "optical potential" for localized modes [34–36]. The high- $Q_o$ optical mode supported in the gap region has a field distribution shown in Fig. 1(a) and is characterized by a mode volume $V_o$~0.3(λo/ $nS_i$)3 and radiation loss limited $Q_o$~$10^4$–$10^6$, depending on the minimum realizable feature size [34]. The design utilized here is predicted from finite-element simulations (COMSOL) to support a mode with $Q_o$~$3.5\times10^4$.

The split-beam nanocavity supports several cantileverlike mechanical resonances suitable for torque detection, whose displacement profiles, calculated from simulations and illustrated in Fig. 1(b), are characterized by effective massm~350–800 $f_g$ and frequency $\omega_m/2\pi$~5–8 MHz (see Table I). The two lowest-frequency modes involve pivoting of the suspended mirror about its support. They are torsional in the $\hat{y}$ and $\hat{z}$ directions and are thus labeled $T_y$ and $T_z$, respectively. The third mode C is an out-of-plane cantileverlike mode of the triply anchored mirror.

**Table 1:** Split-beam nanocavity mechanical mode properties

| Mechanical mode | | Ty | Tz | C |
|---|---|---|---|---|
| ωm/2π | (MHz) | 4.9 | 6.4 | 7.7 |
| m | (fg) | 427 | 805 | 348 |
| Qm (ambient) | | 21 | 83 | 42 |
| Qm (vacuum) | | 1800 | 4400 | 2400 |
| \|z,x\|NF (ambient) | (fm/Hz---√) | 6.3 | 6.9 | ··· |
| \|τ\|min (ambient) | (Nm/Hz---√) | 1.2×10−20 | 1.2×10−20 | ··· |
| \|τ\|min (vacuum) | (Nm/Hz---√) | 1.3×10−21 | 1.7×10−21 | ··· |

The optomechanical properties of the split-beam nanocavities are measured using a dimpled optical fiber-taper waveguide to evanescently couple light into and out of the nanocavity. The dimple is fabricated by modifying the process in Ref. [37] to use a ceramic dimple mold. Measurements are performed both in ambient conditions and in vacuum. A tunable laser source is used to measure T(λ), with the taper either hovering approximately 300–500 nm above the nanocavity or touching one of the nanocavity mirrors. The nanocavity studied here supports an optical mode at$\lambda_o$~1530 nm, with unloaded

$Q_o$~12 000 due to fabrication imperfections, resulting in a dip in T(λ) near $\lambda_o$, as shown in Fig. 3(a). Optomechanical coupling between this mode and nanocavity mechanical resonances is studied by measuring the rf voltage noise spectrum $\bar{S}_{VV}(\lambda,\omega)$ of the optical power transmitted through the fiber taper, using a photoreceiver (New Focus 1811, detector noise 2.5 $\mathrm{nW}/\sqrt{\mathrm{Hz}}$) and a real-time spectrum analyzer (Tektronix RSA5106A). A typical measurement of $\bar{S}_{VV}$, with the fiber hovering above the nanocavity and Δλ~−$\delta\lambda_o$/2, is shown on the right of Fig. 3(a). Three distinct resonances are visible, indicative of the optomechanical transduction of the thermal motion of the $T_y$, $T_z$, and C modes. The resonances are identified with mechanical modes through comparison of measured and simulated ωm and by observing the effect of touching the fiber taper on each of the mirrors. As shown in Fig. 3(b), when the fiber contacts the anchored (suspended) mirror, the C ( $T_y$ and $T_z$) resonance is suppressed, as it is a resonance of the anchored (suspended) mirror.

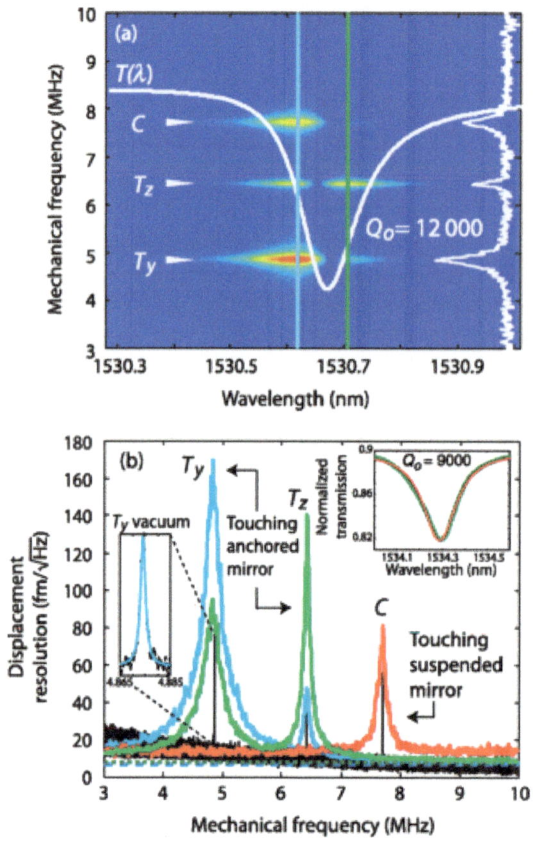


**Figure. 3:** (a) $\bar{S}_{VV}(\lambda,\omega)$ in ambient conditions, with the fiber taper hovering approximately 300 nm above the nanocavity; T(λ) is superimposed in white. Right side in white: $\bar{S}_{VV}(\lambda_b,\omega)$ at $\lambda_b$ indicated by the blue line. (b) Blue (green) data: Calibrated displacement spectrum $S_{xx}^{1/2}$ of $T_y$ $(T_z)$. of $T_y$ ( $T_z$), when the fiber taper is touching the anchored mirror, with λ set at the blue (green) line in (a). Dotted lines indicate the noise floor. Red data: Uncalibrated displacement spectrum ofC with the taper touching the suspended mirror. Black data: Vacuum measurement of the displacement spectrum, uncalibrated. Left inset: Highlight of $T_y$ ( $Q_m$=1800) in vacuum. Top right inset: T(λ) with a fiber-touching device.

The mechanical displacement sensitivity of these measurements can be calibrated from $\bar{S}_{VV}(\omega=\omega_m)$, which is determined by the thermal amplitude of the mechanical resonance [7,8]. From the measured and calculated mechanical mode properties listed in Table I, the noise-floor displacement resolutions |z,x|NF for the $T_y$ and $T_z$modes of approximately 6 and 7 $\mathrm{fm}/\sqrt{\mathrm{Hz}}$, respectively, are measured for$P_i$~25 μW. The minimum detectable torque τmin associated with the angular motion θ of each mechanical mode can be calculated from τ=r×F and Eq. (2). From the mirror length of 7.5 μm and support length of 3 μm, a thermally limited torque sensitivity of the $T_v$ and $T_z$ modes, in ambient conditions, ofτmin~1.2×10$^{-20}$ $\mathrm{Nm}/\sqrt{\mathrm{Hz}}$, and on-resonance technical noise floors of $4\text{–}7\times10^{-22}$ $\mathrm{Nm}/\sqrt{\mathrm{Hz}}$, limited by laser noise and photon shot noise, are extracted. This technical noise floor, corresponding to the second term in Eq. (2), has an effective temperature in the $m_K$ range. Measurements are also performed in low vacuum, where the effect of air damping is reduced. The limit imposed by thermal noise, determined by the first term in Eq. (2), can be reduced by decreasing the mechanical damping of the device. An increase in $Q_m$ of the $T_y$ and $T_z$ modes, from $Q_m^{atm}=21$ and 83 in ambient pressure to $Q_{vacm}$=1800 and 4400 at a relatively low vacuum pressure of 2 $T_{orr}$, is observed, as shown in Fig. 3(b) and summarized in Table I. For a given set of operating conditions, Eq. (2) indicates that this 2-orders-of-magnitude improvement of $Q_m$ will enhance sensitivity by an order of magnitude, resulting in thermally limited τmin~1.3×10−21 and approximately $1.7\times10^{-21}$ $\mathrm{Nm}/\sqrt{\mathrm{Hz}}$ at 2 at 2 $T_{orr}$ for the $T_y$ and $T_z$ modes, respectively. Note that higher $Q_m$ results in a reduced bandwidth of the mechanical response $\chi^{(\omega)}$ and is not always preferred for practical applications.

The observed torque sensitivity in ambient conditions is higher and wider in bandwidth than previously demonstrated optomechanical torque sensors in vacuum[9]. The observed vacuum sensitivity is an order-of-magnitude improvement compared to previous work [9]. Further improvements in detection sensitivity can be realized through improvements in optical and mechanical properties of the devices. Increasing the fiber-cavity coupling efficiency from the relatively weak coupling demonstrated here ( $T_o$~0.92–0.98), using single-sided coupling via an integrated waveguide [38], for example, would increase G by an order of magnitude. Similarly, increasing $Q_o$ to $4\times10^5$ by more accurately fabricating the split-beam nanocavity designs [34] will also enhance G. Combining these improvements, torsional sensitivity could reach $10^{-23}$–$10^{-22}$ $\mathrm{Nm}/\sqrt{\mathrm{Hz}}$.

The role of dissipative optomechanical coupling and its effect on torque-detection sensitivity is studied by measuring the wavelength response of the rf spectrum. Examining $\bar{S}_{VV}(\lambda,\omega)$ and $T(\lambda)$) in Fig. 3(a), it is evident that the optomechanical transduction of each of the three mechanical resonances exhibits a unique λdependence. The particular response stems from differing relative contributions of dissipative and dispersive coupling. $S_{VV}(\lambda,_{\omega m})$ of a purely dispersive cavity-optomechanical system, operating in the unresolved-sideband regime, should follow the slope $|dT/d\lambda|^2$. However, as shown in Fig. 4(a), the $\bar{S}_{VV}(\lambda,\omega_m)$ of $T_y$, $T_z$, and C do not follow $|dT/d\lambda|^2$ and are asymmetric with respect to $\pm\Delta\lambda$. This asymmetry can be characterized by $\zeta^2 = \bar{S}^{+}_{VV}/\bar{S}^{-}_{VV}$, where $\bar{S}^{\pm}_{VV}$ is the maximum rf sideband signal for $\Delta\lambda \gtrless 0$. The slight Fano profile of T(λ), due to nanocavity coupling to higher-order waveguide modes, would result in a ζ~0.8 for purely dispersive optomechanical coupling and does not explain the observed results. In comparison, the $\bar{S}_{VV}(\lambda,\omega_m)$ of the out-of-plane $T_y$ and C modes, shown in Figs. 4(b) and 4(d), are characterized by ζ~0.24 and 0.19, respectively. The in-plane mode $T_z$ is characterized by ζ~1.1, as shown in Fig. 4(c).

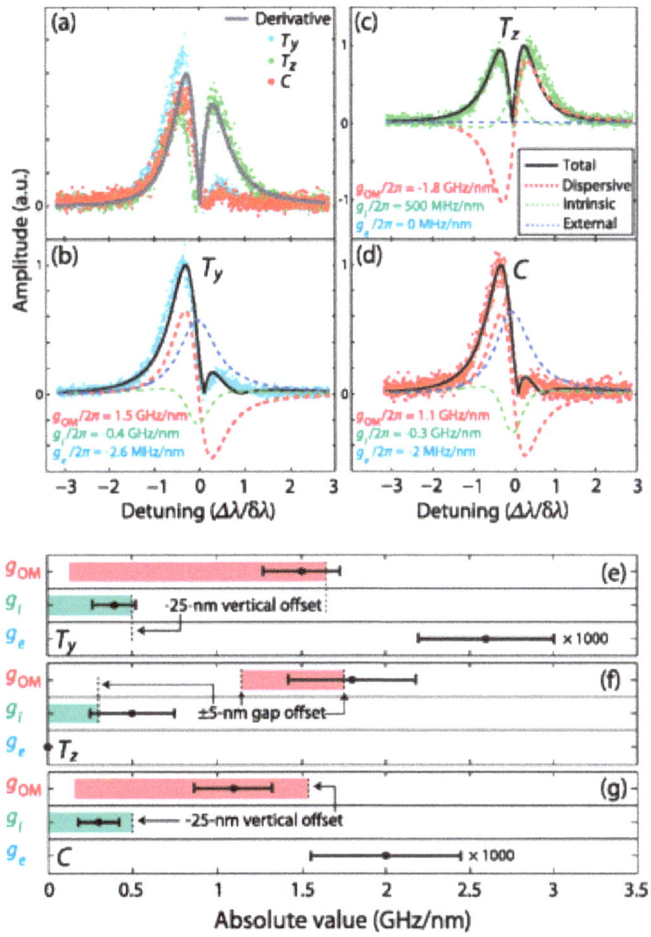


**Figure. 4:** (a) $\bar{S}^{1/2}_{VV}(\lambda,\omega=\omega_m)$ of the C and $T_{y,z}$ modes. The grey line is scaled asdT(λ)/dλ. (b)–(d) Fit (black line) of the optomechanical coupling model to the $\bar{S}^{1/2}_{VV}(\lambda,\omega=\omega_m)$ of the (b) Ty, (c) Tz, and (d) C modes. The dashed colored lines indicate relative contributions from dispersive, external dissipative, and intrinsic dissipative couplings. (e)–(g) Comparison between fit values (black points) and numerically simulated values (shaded regions) of gOM, gi, and ge. The widths of the colored boxes represent numerically simulated ranges due to fabrication imperfections.

The relative contribution of each optomechanical coupling process can be estimated by fitting $\bar{S}^{1/2}_{VV}(\lambda,\omega=\omega_m)$ with a model that includes dispersive, intrinsic dissipative, and external dissipative optomechanical couplings and takes into account the slight Fano shape of T(λ). The resulting fits, estimates for $g_{OM}$, gi, andge, and relative contributions to the optomechanical response are displayed in Figs. 4(b)–4(d). The large asymmetry in $T_y$ and C is attributed primarily to external dissipative coupling, resulting from a variation in the fiber-nanocavity gap caused by the motion of the mirror, and is quantified by $g_e$~−2.6 MHz/nm. The $T_z$ mode is predominantly dispersive, and good agreement with theory is realized with

$g_e$=0. In order to realize best fits in all of the modes, significant intrinsic dissipative coupling must be included, with $g_i$~300–500 MHz/nm. Note that in the case of the out-of-plane C and $T_y$ modes, contributions from $g_e$ effectively double the displacement sensitivity of the optomechanical measurement. Furthermore, even for modest $g_e$, the relative contribution to the optomechanical gain G is significant, despite the weak waveguide-nanocavity coupling used here, owing to $G|_{max} \propto To$ for external dissipative coupling.

The fit values for $g_{OM}$, $g_i$, and $g_e$ are compared with values predicted from numerical simulations, as summarized in Figs. 4(e)–4(g). A range of values for $g_i$ and $g_{OM}$, accounting for uncertainties in device fabrication, is calculated both by directly simulating the optical properties of the nanocavity resonance as a function of mirror displacement and by using perturbation theory (see the Appendix). For $T_z$, the in-plane motion of the suspended mirror contributes to $g_{OM}$ and gi. An uncertainty of±5 nm in the gap size results in the predicted range of $g_{OM}$ and $g_i$ shown in Fig. 4(f). Because of fabrication imperfections unaccounted for in simulations, experimental values can be slightly higher; error bars from the fitting routine, however, fall within the predicted range. For the out-of-plane $T_y$ and C modes, broken vertical symmetry can give rise to significant $g_{OM}$ [3]. Notably, a vertical sagging of the suspended mirror by a plausible offset of 25 nm, as indicated in Figs. 4(e) and 4(g), can give rise to $g_{OM}$ and $g_i$ values comparable to the fit value for$T_z$. Note that renormalization of the nanocavity near field by the waveguide can contribute to $g_{OM}$ but that this effect is not significant for the operating conditions used here. Finally, thc values for $g_e$ extracted from the fits are comparable to the experimentally observed dependence of $\gamma_e$ on the waveguide-nanocavity gap [39].

Several recent studies have explored the potential for exploiting dissipative optomechanical coupling for applications in the quantum regime [27–30]. For many of these proposals, it is desirable to reduce dispersive coupling and maximize dissipative coupling. The former can potentially be achieved in split-beam nanocavities. Simulations indicate (see the Appendix) that internal dissipative coupling of the $T_z$ mode can become dominant if the mirror gap is increased by 50 nm, where $\{g_{OM}, g_i\} \rightarrow \{0, 1.5\ \text{GHz/nm}\}$. In the case of the $T_y$ and C modes,$g_e/g_{OM}$ may be increased by reducing the single mirror "sag" believed to be largely responsible for the appreciable $g_{OM}$ measured for these modes. This may be achieved in devices with symmetrically supported mirrors.

In conclusion, we have observed thermally driven dissipative and dispersive optomechanical couplings in a nanostructure consisting of cantilever nanomechanical resonators integrated directly within an optical nanocavity. This nanocavity is capable of detecting torque with sensitivity of $1.3\times10^{-21}\ \text{Nm}/\sqrt{\text{Hz}}$ in low vacuum and $1.2\times10^{-20}\text{Nm}/\sqrt{\text{Hz}}$ in ambient conditions. This sensitivity surpasses previously demonstrated optomechanical torque detection in vacuum ($4\times10^{-20}\ \text{Nm}/\sqrt{\text{Hz}}$ in Ref. [9]) by an order of magnitude and compares favorably to the performance of magnetic tweezer torque sensors (approximately $10^{-21}$ Nm in Ref. [40]). The low-temperature operation of the existing device would allow sensitivity to be improved by up to 2 orders of magnitude for the technical noise floor measured here. Further optimization of $Q_o$, $Q_m$, m, and the optomechanical coupling strength of our device will serve to further reduce technical and thermal noise [41] and to improve the measurement sensitivity and resolution. For example,$Q_o>10^5$ [34,36], $Q_m>10^4$ [19], and $g_{OM}>10$ GHz/nm [36] are potentially within experimental reach. Realizing a device with this combination of performance would provide a path toward further improving torque sensitivity by orders of magnitude, as well as enhancement of the relative strength of dissipative optomechanical coupling.

## REFERENCES


1. D. Rugar, R. Budakian, H.J. Mamin, and B.W. Chui, Single Spin Detection by Magnetic Resonance Force Microscopy, Nature (London) 430, 329 (2004).
2. O. Arcizet, P.-F. Cohadon, T. Briant, M. Pinard, A. Heidmann, J.-M. Mackowski, C. Michel, L. Pinard, O. Français, and L. Rousseau, High-Sensitivity Optical Monitoring of a Micromechanical Resonator with a Quantum-Limited Optomechanical Sensor, Phys. Rev. Lett. 97, 133601 (2006).
3. M. Li, W.H.P. Pernice, and H.X. Tang, Broadband All-Photonic Transduction of Nanocantilevers, Nat. Nanotechnol. 4, 377 (2009).
4. G. Anetsberger, E. Gavartin, O. Arcizet, Q.P. Unterreithmeier, E.M. Weig, M.L. Gorodetsky, J.P. Kotthaus, and T.J. Kippenberg, Measuring Nanomechanical Motion with an Imprecision below Standard Quantum Limit, Phys. Rev. A 82, 061804 (2010).
5. E. Gavartin, P. Verlot, and T.J. Kippenberg, A Hybrid On-Chip Optomechanical Transducer for Ultrasensitive Force Measurements, Nat. Nanotechnol. 7, 509 (2012).
6. X. Sun, J. Zhang, M. Poot, C.W. Wong, and H.X. Tang, Femtogram Doubly Clamped Nanomechanical Resonators Embedded in a High-q Two-Dimensional Photonic Crystal Nanocavity, Nano Lett. 12, 2299 (2012).
7. A.G. Krause, M. Winger, T.D. Blasius, W. Lin, and O. Painter, A High-Resolution Microchip Optomechanical Accelerometer, Nat. Photonics 6, 768 (2012).
8. Y. Liu, H. Miao, V. Aksyuk, and K. Srinivasan, Wide Cantilever Stiffness Range Cavity Optomechanical Sensors for Atomic Force Microscopy, Opt. Express 20, 18268 (2012).
9. P.H. Kim, C. Doolin, B.D. Hauer, A.J.R. MacDonald, M.R. Freeman, P.E. Barclay, and J.P. Davis, Nanoscale Torsional Optomechanics, Appl. Phys. Lett.102, 053102 (2013).
10. J. Abadie (LIGO Scientific Collaboration), A Gravitational Wave Observatory Operating beyond the Quantum Shot-Noise Limit, Nat. Phys. 7, 962 (2011).
11. D.M. Meekhof, C. Monroe, B.E. King, W.M. Itano, and D.J. Wineland, Generation of Nonclassical Motional States of a Trapped Atom, Phys. Rev. Lett.76, 1796 (1996).
12. W. Marshall, C. Simon, R. Penrose, and D. Bouwmeester, Towards Quantum Superpositions of a Mirror, Phys. Rev. Lett. 91, 130401 (2003).
13. J. Chan, T.P. Mayer Alegre, A.H. Safavi-Naeini, J.T.

Hill, A. Krause, S. Groblacher, M. Aspelmeyer, and O. Painter, Laser Cooling of a Nanomechanical Oscillator into Its Quantum Ground State, Nature (London) 478, 89 (2011).

14. J. D. Teufel, T. Donner, D. Li, J. W. Harlow, M. S. Allman, K. Cicak, A. J. Sirois, J. D. Whittaker, K. W. Lehnert, and R. W. Simmonds, Sideband Cooling of Micromechanical Motion to the Quantum Ground State, Nature (London) 475, 359 (2011).
15. M. Aspelmeyer, T. Kippenberg, and C. Marquardt, Cavity Optomechanics,arXiv:1303.0733.
16. M. Eichenfield, J. Chan, R.M. Camacho, K.J. Vahala, and O. Painter, Optomechanical Crystals, Nature (London) 462, 78 (2009).
17. E. Gavartin, R. Braive, I. Sagnes, O. Arcizet, A. Beveratos, T. J. Kippenberg, and I. Robert-Philip, Optomechanical Coupling in a Two-Dimensional Photonic Crystal Defect Cavity, Phys. Rev. Lett. 106, 203902 (2011).
18. A.H. Safavi-Naeini, J. Chan, J.T. Hill, T.P. Mayer Alegre, A. Krause, and O. Painter, Observation of Quantum Motion of a Nanomechanical Resonator, Phys. Rev. Lett. 108, 033602 (2012).
19. A.H. Safavi-Naeini, S. Groblacher, J.T. Hill, J. Chan, M. Aspelmeyer, and O. Painter, Squeezed Light from a Silicon Micromechanical Resonator, Nature (London) 500, 185 (2013).
20. S. Weis, R. Rivière, S. Deléglise, E. Gavartin, O. Arcizet, A. Schliesser, and T. J. Kippenberg, Optomechanically Induced Transparency, Science 330, 1520 (2010).
21. A.H. Safavi-Naeini, T.P.M. Alegre, J. Chan, M. Eichenfield, M. Winger, Q. Lin, J.T. Hill, D.E. Chang, and O. Painter, Electromagnetically Induced Transparency and Slow Light with Optomechanics, Nature (London) 472, 69 (2011).
22. J.P. Davis, D. Vick, D.C. Fortin, J.A.J. Burgess, W.K. Hiebert, and M.R. Freeman, Nanotorsional Resonator Torque Magnetometry, Appl. Phys. Lett. 96, 072513 (2010).
23. J.A.J. Burgess, A.E. Fraser, F. Fani Sani, D. Vick, B.D. Hauer, J.P. Davis, and M.R. Freeman, Quantitative Magneto-mechanical Detection and Control of the Barkhausen Effect, Science 339, 1051 (2013).
24. A.C. Bleszynski-Jayich, W.E. Shanks, B. Peadecerf, E. Ginossar, F. von Oppen, L. Glazman, and J.G.E. Harris, Persistent Currents in Normal Metal Rings, Science 326, 272 (2009).
25. A.M. Yao and M.J. Padget, Orbital Angular Momentum: Origins, Behavior and Applications, Adv. Opt. Photonics 3, 161 (2011).
26. S. Forstner, S. Prams, J. Knittel, E.D. van Ooijen, J.D. Swaim, G.I. Harris, A. Szorkovszky, W.P. Bowen, and H. Rubinsztein-Dunlop, Cavity Optomechanical Magnetometer, Phys. Rev. Lett. 108, 120801 (2012).
27. F. Elste, S.M. Girvin, and A.A. Clerk, Quantum Noise Interference and Backaction Cooling in Cavity Nanomechanics, Phys. Rev. Lett. 102, 207209 (2009).
28. S. Huang and G.S. Agarwal, Reactive Coupling Can Beat the Motional Quantum Limit of Nanowaveguides Coupled to a Microdisk Resonator, Phys. Rev. A 82, 033811 (2010).
29. Xuereb, R. Schnabel, and K. Hammerer, Dissipative Optomechanics in a Michelson-Sagnac Interferometer, Phys. Rev. Lett. 107, 213604 (2011).
30. T. Weiss and A. Nunnenkamp, Quantum Limit of Laser Cooling in Dispersively and Dissipatively Coupled Optomechanical Systems, Phys. Rev. A 88, 023850 (2013).
31. S.P. Tarabrin, H. Kaufer, F.Y. Khalili, R. Schnabel, and K. Hammerer, Anomalous Dynamic Backaction in Interferomters, Phys. Rev. A 88, 023809 (2013).
32. M. Li, W.H.P. Pernice, and H.X. Tang, Reactive Cavity Optical Force on Microdisk-Coupled Nanomechanical Beam Waveguides, Phys. Rev. Lett. 103, 223901 (2009).
33. G. Anetsberger, E.M. Weig, J.P. Kotthaus, and T.J. Kippenberg, Cavity Optomechanicsand Cooling Nanomechanical Oscillators Using Microresonator Enhanced Evanescent Near-Field Coupling, C.R. Phys. 12, 800 (2011).
34. Hryciw and P.E. Barclay, Optical Design of Split-Beam Photonic Crystal Nanocavities, Opt. Lett. 38, 1612 (2013).
35. P.B. Deotare, M.W. McCutcheon, I.W. Frank, M. Khan, and M. Lončar, High Quality Factor Photonic Crystal Nanobeam Cavities, Appl. Phys. Lett. 94, 121106 (2009).
36. M. Eichenfield, J. Chan, A.H. Safavi-Naeini, K.J. Vahala, and O. Painter, Modeling Dispersive Coupling and Losses of Localized Optical and Mechanical Modes in Optomechanical Crystals, Opt. Express 17, 20078 (2009).
37. C.P. Michael, M. Borselli, T.J. Johnson, C. Chrystala, and O. Painter, An Optical Fiber-Taper Probe for Wafer-Scale Microphotonic Device Characterization, Opt. Express 15, 4745 (2007).
38. S. Groblacher, J. Hill, A. Safavi-Naeini, J. Chan, and O. Painter, Highly Efficient Coupling from an Optical Fiber to a Nanoscale Silicon Optomechanical Cavity, Appl. Phys. Lett. 103, 181104 (2013).
39. K. Srinivasan, P.E. Barclay, M. Borselli, and O. Painter, Optical-Fiber Based Measurement of an Ultra-small Volume High-q Photonic Crystal Microcavity,Phys. Rev. B 70, 081306(R) (2004).
40. J. Lipfert, J.W.J. Kerssemakers, T. Jager, and N.H. Dekker, Magnetic Torque Tweezers: Measuring Torsional Stiffness in DNA and RECA-DNA Filaments, Nat. Methods 7, 977 (2010).
41. S. Forstner, J. Knittel, E. Sheridan, J.D. Swaim, H. Rubinsztein-Dunlop, and W.P. Bowen, Sensitivity and Performance of Cavity Optomechanical Field Sensors, Photonic Sensors 2, 259 (2012).
42. A.N. Cleland and M.L. Roukes, Noise Processes in Nanomechanical Resonators, J. Appl. Phys. 92, 2758 (2002).
43. B.D. Hauer, C. Doolin, K.S.D. Beach, and J.P. Davis, A General Procedure for Thermomechanical Calibration of Nano/Micro-mechanical Resonators, Ann. Phys. (Amsterdam) 339, 181 (2013).

# CHAPTER 16 PHOTONIC CRYSTAL OPTICAL TWEEZERS WITH HIGH EFFICIENCY FOR LIVE BIOLOGICAL SAMPLES AND VIABILITY CHARACTERIZATION

Peifeng Jing[1], JingdaWu[1], GaryW. Liu[2], EthanG. Keeler[1], Suzie H. Pun[2] & LihY. Lin[1]

[1]Department of Electrical Engineering, University of Washington, Seattle WA, 98195, USA

[2]Department of Bioengineering, University of Washington

## ABSTRACT

**We propose and demonstrate a new optical trapping method for single cells that utilizes modulated light fields to trap a wide array of cell types, including mammalian, yeast, and *Escherichia coli* cells, on the surface of a two-dimensional photonic crystal. This method is capable of reducing the required light intensity, and thus minimizing the photothermal damage to living cells, thereby extending cell viability in optical trapping and cell manipulation applications. To this end, a thorough characterization of cell viability in optical trapping environments was performed. This study also demonstrates the technique using spatial light modulation in patterned manipulation of live cell arrays over a broad area.**

## INTRODUCTION

Methods for interacting with biological samples, including single cells, have made great progress in the past couple of decades, revealing high-resolution details on microscopic size-scales. Of particular importance, vibrating micro-electro-mechanical systems (MEMS) resonators possess impressive sensitivity to the mechanical properties of single cells and bio-molecules, resolving properties1·2·3·4·5·6·7 including mass, density, size and stiffness. Detection is accomplished through observation of resonant-frequency shifts induced by mass additions or changes in attached bio-samples. While promising high measurement sensitivity, these devices are hindered by unpredictability in sample placement, since the resonant frequency is affected by both the added mass and its position, thereby causing a loss of measurement precision. To remediate this problem, optical trapping has a unique opportunity for non-invasive sample manipulation and control that could aid these resonant devices8 in obtaining precise measurements. By optical position-fixing, long-term monitoring of single cells and their physical properties is possible. Such measurements can enable critical biological studies, including consideration of cell-growth size dependencies or unregulated growth applicable to understanding cancer mechanisms or those of other diseases9. This relation between mass and cell growth is a fundamental question for biologists, and high-resolution, high-precision measurement can have great potential in medicine and drug discovery.

Optical trapping has become a widely utilized, non-invasive tool for manipulation in biological applications, to place, identify and modify live cells10·11, nano-particles and DNA strands12. However, photodamage to cells limits measurement duration and its application in the life sciences. To address this shortcoming, methods have been developed for increased

**Citation:** Peifeng Jing, JingdaWu, GaryW. Liu, EthanG. Keeler, Suzie H. Pun & LihY. Lin, Photonic Crystal Optical Tweezers with High Efficiency for Live Biological Samples and Viability Characterization, doi:10.1038/srep19924

trapping efficiency – the capability to trap particles measured through trap stiffness and minimum trapping intensity, and thus trapping at a lower-intensity: optoelectronic tweezers (OET)13, plasmonic optical tweezers14·15 and photonic crystal (PhC) waveguides16. Nonetheless, these techniques are not readily compatible with MEMS. OET accomplishes large-scale parallel manipulation with two electrodes to achieve low-intensity optical trapping and avoids the photodamage effect, but integrating OET fabrication methods with current MEMS resonators is not straightforward. Plasmonic optical tweezers use highly localized light intensity to increase the trapping force; although used to trap living cells, such as yeast cells15 and*Escherichia coli* (*E. coli*)17, the large amount of heat generated by the plasmonic surface prohibits long-term manipulation of living cells. PhC waveguides also highly localize light energy and trap particles through evanescent waves; however, the waveguide is not easily coupled with vibrating MEMS structures, and its nano-size trapping cavity is not versatile for various eukaryotic cells, whose sizes can vary from 3–10 μm.

To overcome these compatibility challenges, we propose and demonstrate an approach of low-intensity optical manipulation utilizing the interaction of laser light with a two-dimensional photonic crystal (2D PhC). Photonic crystals have become an important structure for manipulating photons and have demonstrated great success and potential for various applications. Beyond the efforts made on PhC geometry design, they have been utilized in light-trapping and manipulation devices as optical waveguide resonators16, optical fibers18and PhC slabs19·20. Efficient and enhanced optical trapping for small particles has been demonstrated in both the near field16·21 and far field22·23. Our method differentiates itself from other PhC trapping approaches by illuminating the laser light on top of the PhC structure, thus requires no critical light coupling. Furthermore, it enables optical trapping with reduced photodamage by minimizing the required laser intensity, while still maintaining the same trapping capability when manipulating living cells over broad areas (~500 μm$^2$) under the microscope using 20x objective lenses. This technique can be easily integrated with lab-on-chip systems and potentially applied to measuring the mass of single living cells8. In such an application, the proposed method has the added benefit of trapping particles at the PhC substrate surface with the trap pattern determined by the PhC design. By integrating the PhC structure with a MEMS resonant device for mass detection, better vibrational coupling is achieved between the trapped cell and the MEMS resonator which is necessary for mass sensing since the trapped particle needs to move with the vibrating MEMS resonator.

In considering the applicability of optical trapping in long-term cell manipulation and its suitability for the proposed applications, it is essential to understand how optical energy affects the cell over time. Although researchers have used the previously noted methods to manipulate living cells, insufficient viability measurements of living cells have been reported, while existing cell lifetime research for optical traps used only *E. coli* bacteria cells24·25·26·27. These investigations focused more on optimizing wavelength to reduce optical damage, achieving ~10 minute lifetimes at a ~1100 nm wavelength using an objective lens with a high numerical aperture (N.A. = 1.2) and high laser intensity in the specimen plane24. Therefore, experimental research in this area on eukaryotic cells with optical tweezers is not fully explored. Moreover, even though laser wavelength can be optimized, the photodamage is still a severe limit for long-term biological study of living cells.

In our experiments, a simple optical setup was used to guide a loosely focused laser beam onto the surface of a 2D PhC, which then modulates the profile of the laser beam and generates a confined trapping area above the surface of the 2D PhC. We achieved prolonged cell viability (~30 min), confirming that the 2D PhC uses less power while sustaining the same cell trapping force. We also demonstrate higher trap stiffness by trapping polystyrene beads and *E. coli* cells, and these experimental findings were consistent well with finite-difference time-domain (FDTD) simulation results. The highly localized intensity in our method has been a general concern for viability of cells, and this concern extends to plasmonic and PhC waveguide trapping methods. However, we experimentally verify that the viability is largely determined by overall intensity, rather than localized intensity.

## SIMULATION AND EXPERIMENTAL RESULTS

### FDTD SIMULATION

Our proposed method entails designing the PhC structure to modulate the incident optical field and to create efficient optical traps. The PhC comprises a square lattice with a period of 5.8 μm and 3.6 μm-diameter holes. Various hole depths were simulated, and 500 nm was found to be optimum to avoid confining the optical energy inside the holes. This optimized depth also works well with most of the cells in our experiments. After being reflected by the 2D PhC, the phase of light is modulated. The configuration of the optical traps is generated by light modulation and therefore depends on the dimensions of the 2D PhC. If the depth of the holes in the PhC is much greater than the hole diameter, most of the energy is confined inside of the features of the PhC and the particles are trapped by evanescent waves19·20·21. However, if the holes are shallow, most of the modulated light is scattered back into free space and generates an efficient trap positioned above the surface of the 2D PhC. We performed simulations of the modulated light field produced by a PhC with shallow holes using the FDTD method through the open-source software package MEEP28. Figure 1a shows the optical energy density distribution of the modulated light field, presented by both vertical and horizontal cross-sections at different heights. The vertical cross-section shows that the modulated light field generates a focused volume for the optical trap located at about 1.67 μm above the surface of the 2D PhC. The four images on the left side of Fig. 1a are horizontal cross-sections at a height of 0.3, 1.7, 3.0 and 4.3 μm above the PhC surface. Figure 1b shows a three-dimensional shaded surface plot for the horizontal cross-section of the intensity distribution at the trap location. The contour and gradient of the intensity at this height is shown in Fig. 1c. The asymmetry of the gradient along the x- and y-axis is caused by the light polarization. Because the trapping force is proportional to the gradient of the light intensity distribution $(\vec{F} = \frac{\alpha}{2}\nabla\vec{E}^2)$, Fig. 1c also indicates the enhanced trapping potential generated by the PhC-modulated light field. This is further elaborated by Fig. 1d, which shows a comparison between trap potentials (proportional to the intensity distribution) in traditional laser trapping without a PhC and that above the surface of the 2D PhC. The top diagram is the normalized energy density along the *x*-direction of the PhC-modulated light field, and the bottom diagram is the distribution for traditional laser trapping. By increasing the gradient of the intensity distribution, enhanced trapping force is generated above the 2D PhC surface.

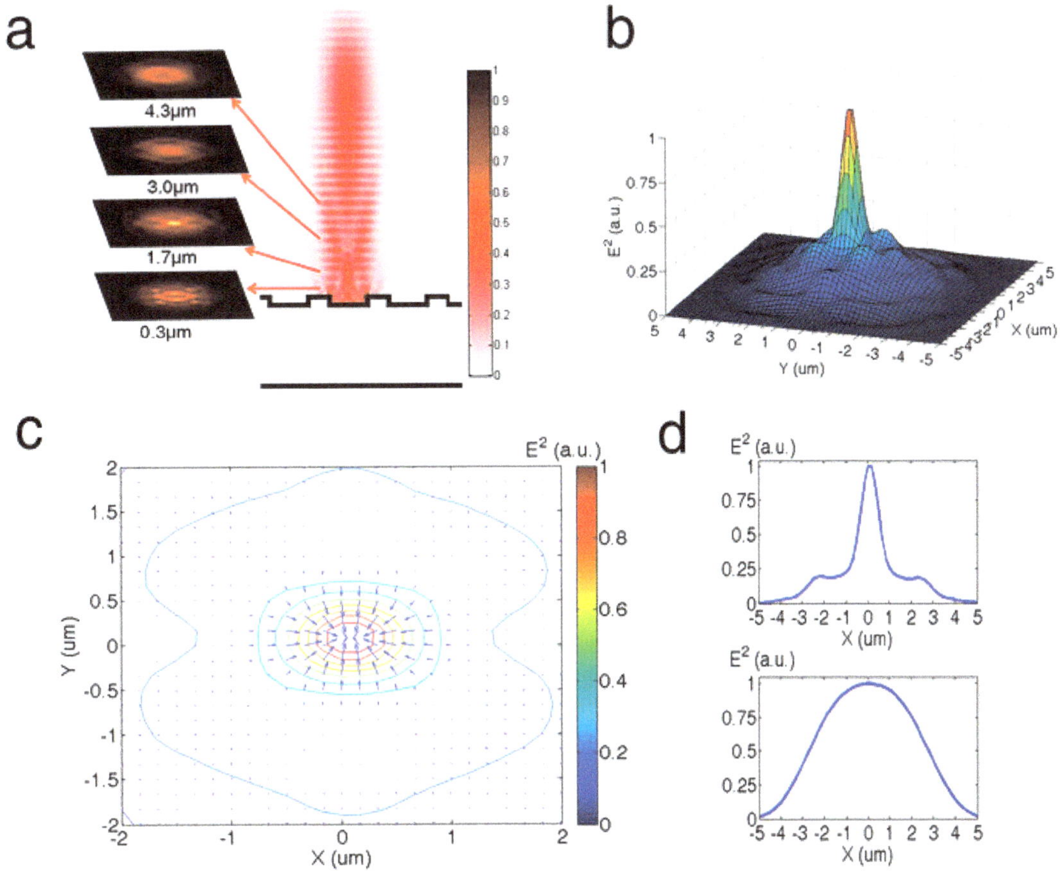


**Figure 1:** FDTD simulation of the laser light field modulated by a 2D PhC, demonstrating enhanced optical trapping through increasing gradient of the intensity distribution.

The PhC is a square lattice with a period of 5.8 μm and 3.6 μm-diameter holes. (**a**) Vertical cross-section of the optical energy density and corresponding horizontal cross-sections at 0.3, 1.7, 3.0 and 4.3 μm above the surface of the 2D PhC. (**b**) 3D surface plot of the optical intensity at the trap location, ~1.67 μm above the surface of the PhC surface. (**c**) Contour and gradient of the optical intensity at the trap plane. (**d**) Comparison of the trap potential between the PhC-modulated light field and the Gaussian profile in traditional laser traps.

Because the incident laser beam is affected by the trapped particle, a comparison of trap enhancement is performed with and without a 2 μm-diameter polystyrene sphere above the PhC. The result insupplementary Fig. 2 shows that the trapped particle above the surface further enhances the optical trap through a slight focusing effect. To investigate the possibility of trap enhancement due to the effect of the standing wave caused by interference of the incident and reflected light, we also used a Gaussian pulsed-laser beam as the light source to eliminate interference. The FDTD simulation result (supplementary Fig. 3) shows a similar light confinement effect, suggests that the enhancement is mainly caused by the reflected light field modulated by the PhC. The shape of the modulated light field is mainly determined by a single hole, while interference from the periodic PhC structure further enhances the intensity of the modulated light. The FDTD simulation result comparing the modulated optical intensity profile from a single hole to that from a PhC structure is shown insupplementary Fig. 4.

## Two-Dimensional Photonic Crystal Optical Trapping

The utility of the proposed method was first verified using eukaryotic yeast cells, which are typically 3-4 μm in diameter. We used a Nd:$YVO_4$ 1064 nm continuous wave (CW) laser to trap cells with a 50X objective lens (N.A. = 0.55). The details of the optical setup are described in the Online Methods section. Figure 2a shows an optical microscope image of the 2D PhC structure, and the snapshots in Fig. 2b show dynamic trapping and dragging of a yeast cell along the 2D PhC surface.

Fabrication of the PhC is described in the Methods section. A 1.2 μm layer of parylene-C was deposited on top of the PhC platform to create a thin, transparent polymer film that protects and better planarizes the PhC surface.

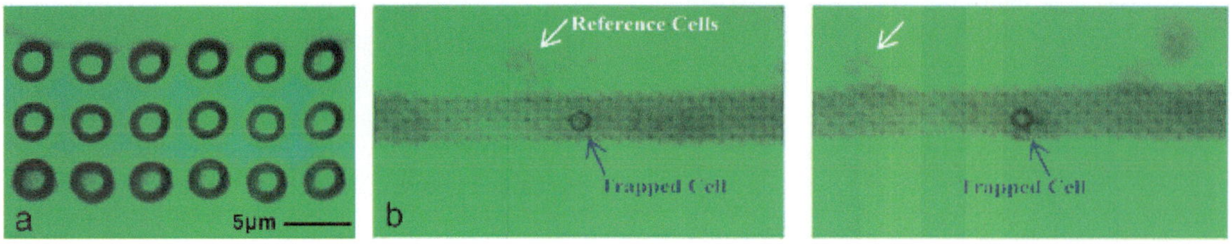


**Figure 2:** (**a**) An optical microscope image of the 2D PhC. (**b**) Images of yeast cells being trapped (left) and dragged (right) on the surface of the 2D PhC.

To characterize the trapping efficiency, the stiffness and minimum trapping intensity were measured by trapping polystyrene beads of sizes ranging from 0.2–1 μm. Stiffness is defined as $k = \frac{k_B T}{\langle x \rangle^2}$, where $k$ is the stiffness, $\langle x \rangle^2$ is the position variance of the trapped particles, $k_B$ is the Boltzmann constant and $T$ is temperature. The minimum trapping intensity is measured under microscope. The intensity is the average intensity determined by dividing the optical power under the microscope objective lens by the area of the Gaussian beam. A single particle is trapped at higher power, and the power is reduced until the particle escapes from the trap. The lowest power and intensity are then recorded as the minimum trapping power and intensity. Figure 3ashows the relative frequency histogram of the Brownian motion of a trapped 750 nm polystyrene bead under 5.8 μW/μm$^2$ laser intensity with (red) and without (blue) the 2D PhC platform. The beads are more firmly trapped above the 2D PhC platform, and their motion distribution is narrower than the distribution without the 2D PhC, Theses findings are consistent with the FDTD simulation results. The histograms follow Gaussian distributions, allowing calculation of the stiffness by the equipartition method16. The trap stiffness is obtained by characterizing the displacement histogram under a specific incident optical intensity. The stiffnesses were 0.156 pN/μm with PhC and 0.030 pN/μm without PhC. The stiffness is theoretically calculated to be 0.25 pN/μm with PhC based on FDTD simulation results. This is higher than the experimental result due to the parylene layer above photonic crystal which has a higher refractive index than water and therefore reduces the intensity of the modulated light field after reflection. We expect the measured trap stiffness to become higher if no covering layer or materials with lower refractive indices are used. Figure 3b shows the minimum trapping intensity required to confine the beads and the stiffness measurement for various sized polystyrene beads. Two sets of data are shown: the measured minimum intensity for trapping polystyrene beads with sizes ranging from 0.2–1 μm; the trap stiffness measured at an intensity of 0.35 mW/μm$^2$ (minimum intensity to trap the 0.2 μm beads) above the surface of the 2D PhC for the polystyrene beads (red curve). The polystyrene beads were initially trapped with higher intensity above the center of a hole in the 2D PhC. The intensity was then decreased, and the minimum trapping intensity value was recorded when the trap released the beads. The minimum trapping intensity inFig. 3b is significantly lower than with conventional optical tweezers, which typically requires ~1 mW/μm$^2$ 13.

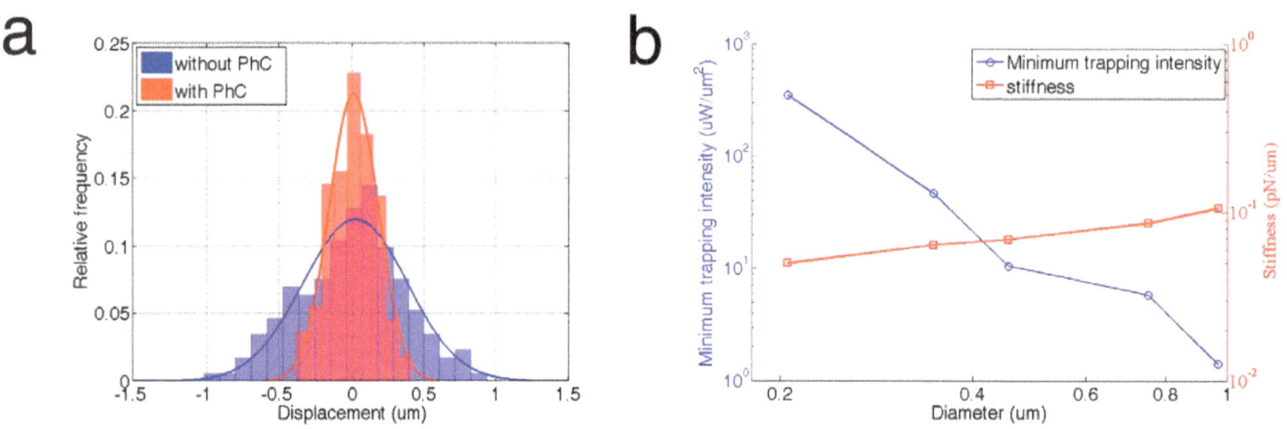


**Figure 3:** Optical trapping efficiency characterizations.

(**a**) Relative frequency histograms of ramdom movement for trapping 750 nm polystyrene beads at a 5.8 μW/μm$^2$ laser intensity with (red) and without (blue) the 2D PhC platform. The stiffness are 0.156 pN/μm and 0.030 pN/μm, respectively. (**b**) Minimum trapping intensity (blue) and stiffness (red) for different sizes of polystyrene beads at the laser intensity of 0.35 mW/μm$^2$.

We then performed similar measurements using *E. coli* bacteria, which are typically 0.5 × 2 μm rod-like cells. Using traditional optical tweezers, the measured stiffness was 0.0095 N/μm, while on the 2D PhC platform the measured stiffness was 0.088 pN/μm, resulting in an enhancement factor ~10.

## Cell Viability Characterization

Although reports have proposed a linear relationship between *E. coli*viability and laser power25, a systematic characterization of eukaryotic cell viability under optical trapping has not yet been performed. Given the promise of our developed platform for optical trapping and manipulation of live cells, we aimed to measure the viability of live NIH/3T3 mammalian fibroblast, yeast, and *E. coli* cells under optical trapping. To measure viability, we suspended cells in media containing propidium iodide (PI) (PBS+1% BSA+2 μg/mL PI). PI is a commonly used cytofluorometric indicator to identify apoptotic and necrotic cells. As PI is cell membrane-impermeable, this dye is excluded from live cells. However, during cellular apoptosis or necrosis the cell membrane loses its selectivity, and PI is able to permeate the cell and intercalate between DNA bases. PI binding induces an emission spectrum shift from 535 nm to 617 nm and enhances PI fluorescence 20- to 30-fold29. Therefore, apoptotic and necrotic cells may be identified by PI (red) staining. We employed a single-blind method to evaluate cell viability. One scientist set the laser power while another scientist, blinded to the power value, observed and recorded the cell lifetime. The 1064 nm near-infrared CW laser was used to trap cells under a 50× objective lens, and a mercury lamp was used to excite the PI dye during the experiments. Figure 4a shows the time-dependent morphology of a trapped NIH/3T3 cell under 36.0 μW/μm$^2$ laser intensity. The cell maintained its morphology for more than 30 min before blebbing, as seen by the protrusion from the main cell body. Blebbing of mammalian cells in damaged cells correlates with cells death from either apoptosis or necrosis30, and was used as a preliminary indication of cell death.Figure 4b shows the measurement results of NIH/3T3 cell lifetime versus PI staining. The time until blebbing (blue curve) and when the cell stained positive for PI (red curve) was observed at each of the laser intensities. As shown in Figure 4b, blebbing occurred before the cell stained positive for PI, indicating that the PI dye may be constrained by diffusion into the cell and bind with the DNA; therefore, in these experiments PI staining was not an immediate indicator of cell necrosis. As evident from the trends in Figure 4b, the time interval between blebbing and PI staining was not constant and was dependent on the optical power.

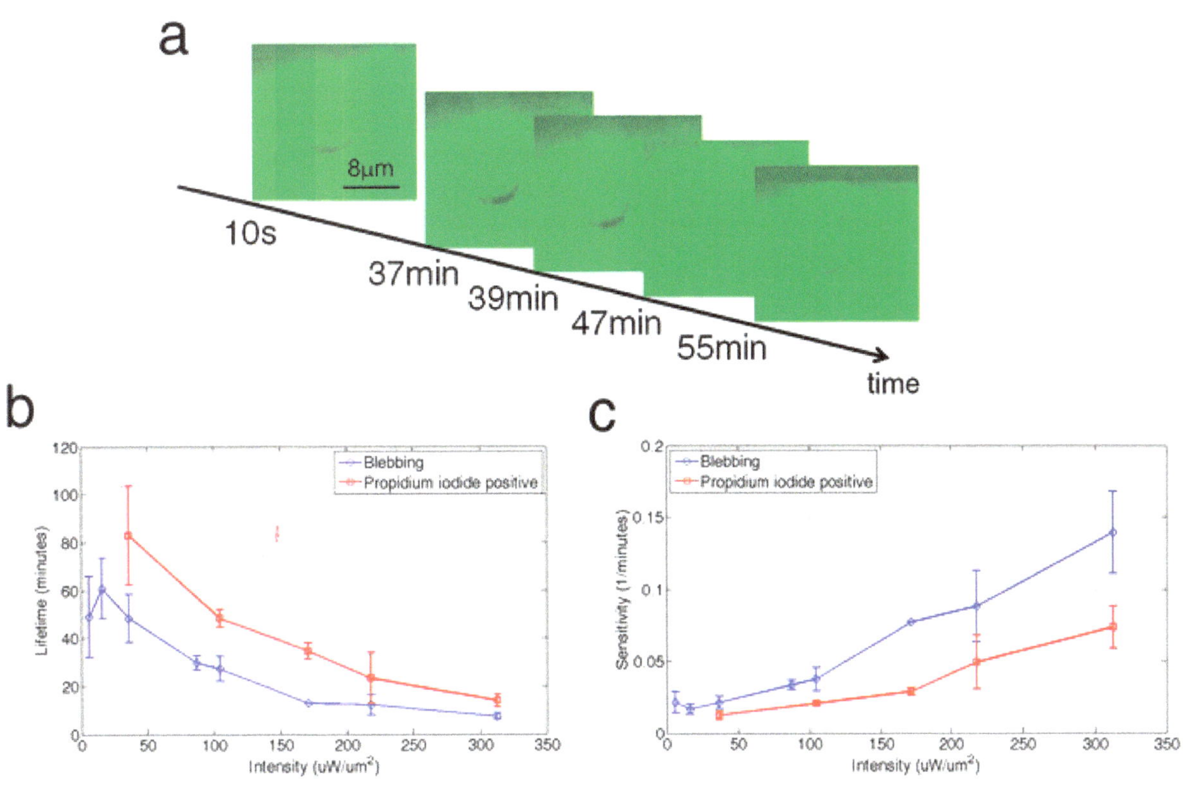


**Figure 4:** Cell viability characterizations.

(**a**) Status of NIH/3T3 mammalian cells under 36.0 μW/μm$^2$ illumination. The cell began blebbing after 30 minutes. (**b**) Lifetime measurements of theNIH/3T3 cells. The blue and red curves represent the times until blebbing and PI staining occurred, respectively. (**c**) Sensitivity of mammalian cells, Defined as 1/min of the lifetimes reported in (**b**).

Sensitivity is used as another parameter to quantify photodamage to optically trapped cells and is defined as the reciprocal of the cell lifetime24. Figure 4c shows the sensitivity of mammalian cells based on the measurement results in Fig. 4b. The sensitivities using both cell blebbing and PI staining lifetimes exhibit an approximately linear relationship with the laser intensity, which is consistent with the results reported by Neuman *et al.* using *E. coli* cells. We also performed the same characterization on *E. coli* cells, and the results are shown insupplementary Fig. 5. The sensitivity flattens at low laser intensity and approaches 20 minutes in lifetime, which matches the replication time of *E. coli* cells.

One question in enhanced optical trapping through highly-localized optical fields is whether the effect of photodamage depends on the localized intensity even if the overall intensity is lower. To answer this question, we performed viability measurements of the trapped cell versus incident optical intensity with and without the PhC platform.Figure 5 compares the lifetime and sensitivity of yeast cells under both trapping methods. The blue curves represent trapping with traditional optical tweezers (without PhC), and the red curves represent trapping on the 2D PhC platform. The two curves are very similar, suggesting that the localized high-intensity light above the surface of the 2D PhC does not compromise cell viability.

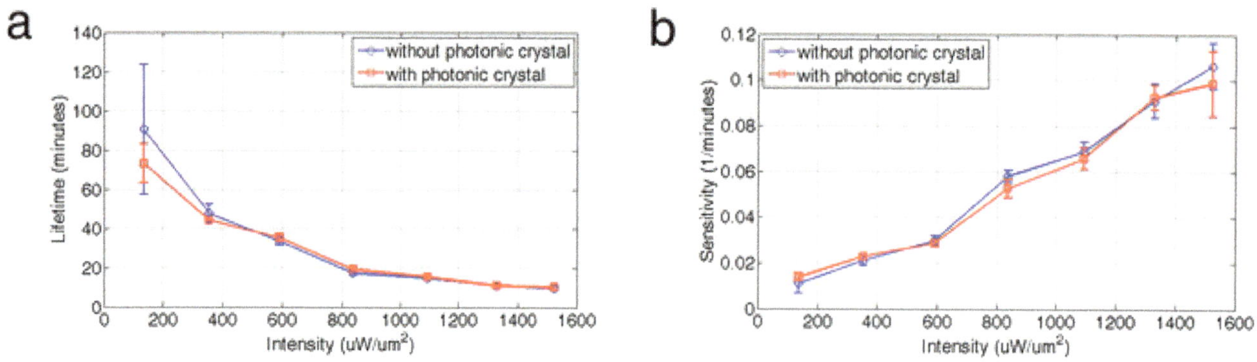


**Figure 5:** Effect of highly localized optical fields on cell viability.

(**a**) Lifetime versus optical intensity for trapping yeast cells with and without the 2D PhC platform. (**b**) The corresponding cell sensitivity to light intensity. The results show that the cell viability is primarily determined by the overall optical intensity, not the localized intensity.

One advantage of the proposed method over plasmonic optical tweezers is that the locations of the optical traps are not at predetermined "hot spot". As the incident beam moves, the trapped particle moves with the beam as shown in Fig. 2. To further demonstrate this versatility and the potential of parallel manipulation over a broad area on the 2D PhC platform, we changed the objective lens to 20X (N.A.=0.22), performed optical trapping and measured the trap stiffness on the 2D PhC. Figure 6a shows the relative frequency histogram of Brownian motion for trapped 964 nm-diameter polystyrene beads, and Fig. 6b shows the relative frequency histogram of motion for trapped *E. coli* bacteria. The same measurements were performed using the 50X objective lens, and the results were compared. The trap stiffness appears to be unaffected by the objective lens magnification, indicating that particles can be trapped with loosely focused light under a larger field of view of the microscope. As an example, Fig. 6c demonstrates parallel manipulation of yeast cells using a spatial light modulator (SLM).

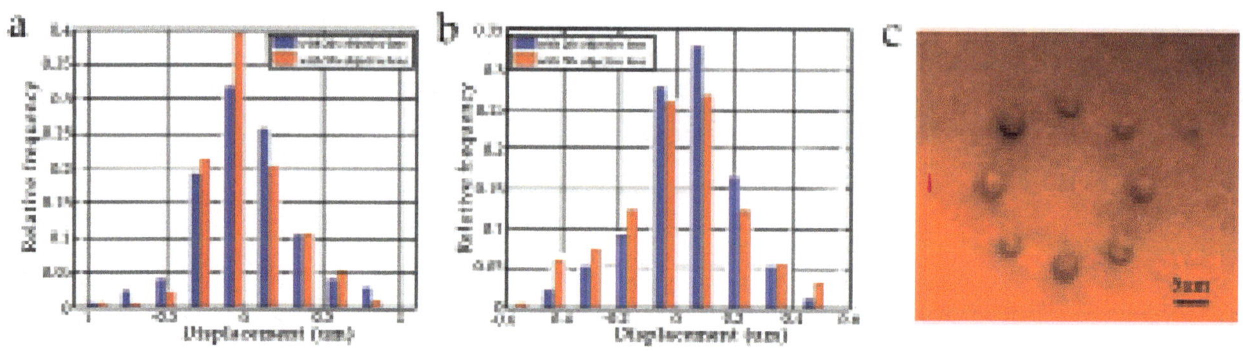

**Figure 6:** (**a**) Histogram of displacement for trapped 964 nm polystyrene beads under 20X and 50X objective lenses with a 28 μW/μm² laser intensity, with stiffnesses of 0.04 pN/μm and 0.06 pN/μm, respectively. (**b**) Histogram of displacement for trapped *E. coli* under 20X and 50X objective lenses with a 28 μW/μm² laser intensity, with stiffnesses of 0.09 pN/μm and 0.15 pN/μm, respectively. (**c**) An example of optical trapping with yeast cells using a pre-programed SLM.

## CONCLUSION

In this work, we demonstrate a novel optical trapping method for live eukaryotic and bacteria cells utilizing a 2D PhC. As shown by both FDTD simulation and experimental results, modulation of the incident light field by the PhC structure produces a trapping volume above the 2D PhC that enhances the trapping force and prolongs the trapping duration of live cells. By thoroughly quantifying the viability and sensitivity of live cells when trapped by traditional optical tweezers and the proposed methods, we experimentally verify that cell viability is mainly determined by the overall optical intensity instead of the highly-localized optical intensity produced by this new method.

We also demonstrate that this straightforward optical system could easily enable large-scale trapping under microscope by using smaller numerical aperture lenses, while still maintaining a similar trapping efficiency with the reduced light intensity. Being compatible with lab-on-chip systems, such as MEMS resonators, this method could have significant impact on cell control and manipulation capabilities thereby enabling numerous applications and improving measurement accuracy for biological studies.

## METHODS

### Two-Dimensional Photonic Crystal Fabrication

The photonic crystal (PhC) was fabricated in silicon due to its high-dielectric properties and its well-established processing techniques. Using photolithography, a PhC feature pattern was transferred to a photoresist etch mask. The pattern had a square lattice orientation consisting of circular holes with diameters of 3.6 µm and a periodicity of 5.8 µm. The holes were etched using a parallel-plate reactive-ion etcher (RIE) to about 500 nm in depth.

The process stripped the photoresist and chemical-vapor deposited a 1.2 µm layer of parylene-C to create a thin, transparent polymer film that protects and better planarizes the PhC surface in addition to giving adequate refractive index contrast with silicon. Parylene-C is also bio-compatible and provides conformal encapsulation of the device.

### NIH/3T3 Mammalian Cell Culture

NIH/3T3 fibroblasts (ATCC) were cultured according to supplier instructions. Cells were cultured in DMEM+10% FBS and passaged every three days or at confluency with $3 \times 10^5$ cells in a T25 flask.

### Yeast Cell Culture

EBY100 cells were grown in SDCAA media at 30 °C for about 20 h to reach a stationary state before passage into a PBS (pH = 7.4). The genotype of the EBY100 cell strain is described by Chao *et al.*31. The sample was centrifuged before experiments, and the fsupematant removed to leave the sediments pellet.

### *Escherichia coli* Culture

Strain ER2738 *Escherichia coli* (*E. coli*) were cultured according to supplier (New England BioLabs) instructions. Briefly, a single colony was inoculated into the LB media and allowed to grow overnight. Finally, the bacteria were diluted for the trapping study.

### Propidium Iodide Staining

For assessing NIH/3T3 cell viability under laser optical trapping, cells were trypsinized, washed and resuspended in a propidium iodide (PI) solution (PBS+1% BSA+2 µg/mL PI) at a density of $1.0 \times 10^6$ cells/mL. For assessing yeast and E. *coli* viability under optical trapping, cells were grown overnight with vigorous shaking and diluted 1:1 with a concentrated PI solution for the final dilution.

## OPTICAL TRAPPING ON THE PHC PLATFORM

### Optical Setup

We used a simple optical setup shown in Supplementary Fig. 1 to guide a loosely focused Nd:$YVO_4$ 1064 nm laser onto the surface of a 2D PhC through a Zeiss Axio Imager fluorescence microscope. The interaction between the incident light and the PhC generates a diffraction pattern above the surface of the 2D PhC to trap the cells. The center of the laser beam is aligned with the center of a hole in the PhC. We then added a 10 µL specimen droplet onto the 2D PhC and covered it with a glass slide with about 0.03 mm spacing. The optical trapping results were recorded by a CCD camera connected to the microscope.

### Stiffness Measurement

We used Tracker (version 4.84), a freeware video analysis and modeling software, to determine the positions of the trapped particles. The stiffness was calculated by the equipartition method using $\frac{1}{2}k\langle x\rangle^2 = \frac{1}{2}k_B T$, where $k$ is the stiffness, $\langle x\rangle^2$ is the position variance of the trapped particles, $k_B$ is the Boltzmann constant and $T$ is temperature.

### Minimum Trapping Intensity Measurement

The polystyrene beads were initially trapped with higher intensity above the center of a hole in the 2D PhC. The intensity was then decreased, and the intensity value was recorded when the trap released the beads.

## REFERENCES


1. Burg, T. et al. Weighing of biomolecules, single cells and single nanoparticles in fluid, Nature 446, 1066–1069 (2007).
2. Gupta, A., Akin, D. & Bashir, R. Detection of bacterial cells and antibodies using surface micromachined thin silicon cantilever resonators. J. Vac. Sci. Tech. B 22, 2785–2791 (2004).
3. Park, K. et al. Living cantilever arrays' for characterization of mass of single live cells in fluids. Lab Chip 8, 1034–1041 (2008).
4. Godin, M., Bryan, A. K. & Burg, T. P. Measuring the mass, density, and size of particles and cells using a suspended microchannel resonator. Appl. Phys. Lett. 91, 123121 (2007).
5. Park, K. et al. Measurement of adherent cell mass and growth. Pnas 107(48), 20691–20696 (2010).
6. Lee, J. et al. Suspended microchannel resonators with piezoresistive sensors. Lab Chip 11, 645–651 (2011).
7. Weng, Y. et al. Mass sensors with mechanical traps for weighing single cells in different fluids. Lab Chip 11, 4174–4180 (2011).
8. Keeler, E. G., Wu, J., Jing, P. & Lin, L. Y. MEMS resonator and photonic Ccrystal integration for enhanced cellular mass sensing, Optics in the Life Sciences BW1A.6, OSA Technical Digest (online), OS A, (2015).
9. Popescu, G., Park, K., Mir, M. & Bashir, R. New technologies for measuring cell mass. Lab Chip 14, 646–652 (2014).
10. Grier, D. A revolution in optical manipulation. Nature 424, 810–816 (2003).
11. Xie, C., Goodman, C., Dinno, M. & Li, Y. Real-time Raman spectroscopy of optically trapped living cells and organelles. Opt. Express 12(25), 6208–6214 (2004).

12. Bustamante, C., Bryant, Z. & Smith, S. B. Ten years of tension: single-molecule DNA mechanics. Nature 421, 423–427 (2003).

13. Chiou, P., Ohta, A. & Wu, M. Massively parallel manipulation of single cells and microparticles using optical images. Nature 436, 370–372 (2005).

14. Juan, M. L., Righini, M. & Quidant, R. Plasmon nano-opitcal tweezers. Nature Photon. 5, 349–356 (2011).

15. Huang, L., Maerkl, S. J. & Martin, O. J. E. Integration of plasmonic trapping in a microfluidic environment. Opt. Express 17(8), 6018–6024 (2009).

16. Mandal, S., Serey, X. & Erickson, D. Nanomanipulation using silicon photonic crystal resonators, Nano Lett. 10, 99–104 (2010).

17. Righini, M. et al. Nano-optical trapping of Rayleigh particles and Escherichia coli bacteria with resonant optical antennas. Nano Lett. 9(10), 3387–3391 (2009).

18. Liberale, C. et al. Miniaturized all-fibre probe for three-dimensional optical trapping and manipulation, Nat. Photonics 1, 723–727 (2007).

19. Jaquay, E., Martinez, L., Mejia, C. & Povinelli, M. Light-assisted, templated self-assembly using a photonic-crystal slab. Nano Lett. 13(5), 2290–2294 (2013).

20. Milord, L. et al. Engineering of slow Bloch modes for optical trapping. Appl. Phys. Lett. 106, 121110 (2015).

21. Chen, Y.-F., Serey, X., Sarkar, R., Chen, P. & Erickson, D. Controlled photonic manipulation of proteins and other nanomaterials. Nano Letters 12, 1633–1637, (2012).

22. Jing, P., Wu, J. & Lin, L. Y. Patterned optical trapping with two-dimensional photonic crystal. ACS Photonics 1(5), 398–402 (2014).

23. Wilson, B. et al. Nanostructure-enhanced laser tweezers for efficient trapping and alignment of particles. Opt. Express 18(15), 16005 (2010).

24. Neuman, K., Chadd, E., Liou, G., Bergman, K. & Block, S. Characterization of photodamage to Escherichia coli in optical traps. Biophys. J. 77, 2856–2863 (1999).

25. Mirsaidov, U. et al. Optimal optical trap for bacterial viability, Phys. Rev. E 78, 021910 (2008).

26. Neuman, K. C. & Block, S. M. Optical trapping, Rev. Sci. Instrum. 75, 2787 (2004).

27. Ericsson, M., Hanstorp, D., Hagbreg, P., Enger, J. & Nystrom, T. Sorting out bacterial viability with optical tweezers. J. Bacteriol 182(19), 5551–5555 (2000).

28. Oskooi, A. et al. MEEP: A flexible free-software package for electromagnetic simulation by the FDTD method. Comput. Phys. Commun. 181, 687–702 (2010).

29. Krishan, A. Rapid flow cytofluorometric analysis of mammalian cell cycle by propidium iodide staining. J. Cell Biol. 66(1), 188–193 (1975).

30. Coleman, M. L. et al. Membrane blebbing during apoptosis results from caspase-mediated activation of ROCK I. Nat. Cell Biol. 3, 339–345 (2001).

31. Chao, G. et al. Isolating and engineering human antibodies using yeast surface display. Nat. Protoc. 1(2), 755–768 (2006).

CHAPTER 17

# RECONFIGURABLE PHOTONIC CRYSTALS ENABLED BY PRESSURE-RESPONSIVE SHAPE-MEMORY POLYMERS

Yin Fang[1] , Yongliang Ni[2], Sin-Yen Leo[1] , Curtis Taylor[2], Vito Basile[3] & Peng Jiang[1]

[1]Department of Chemical Engineering, University of Florida, Museum Road, Gainesville, Florida 32611, USA

[2]Department of Mechanical and Aerospace Engineering, University of Florida, Gainesville, Florida 32611, USA

[3] ITIA-CNR, Industrial Technologies and Automation Institute, National Council of Research, Via Bassini, 15, 20133 Milano, Italy

## ABSTRACT

**Smart shape-memory polymers can memorize and recover their permanent shape in response to an external stimulus (for example, heat). They have been extensively exploited for a wide spectrum of applications ranging from biomedical devices to aerospace morphing structures. However, most of the existing shape-memory polymers are thermoresponsive and their performance is hindered by heat-demanding programming and recovery steps. Although pressure is an easily adjustable process variable such as temperature, pressure-responsive shape-memory polymers are largely unexplored. Here we report a series of shape-memory polymers that enable unusual 'cold' programming and instantaneous shape recovery triggered by applying a contact pressure at ambient conditions. Moreover, the interdisciplinary integration of scientific principles drawn from two disparate fields—the fast-growing photonic crystal and shape-memory polymer technologies—enables fabrication of reconfigurable photonic crystals and simultaneously provides a simple and sensitive optical technique for investigating the intriguing shape-memory effects at nanoscale.**

## INTRODUCTION

Shape-memory (SM) in traditional SM polymers (SMPs) is typically achieved in three steps including programming, storage and recovery[1, 2, 3, 4, 5, 6]. Programming involves deforming a bulk SMP sample from its permanent shape to a temporary configuration. This 'hot' process is usually done above a specific transition temperature ($T_{trans}$), such as the polymer glass transition temperature ($T_g$), to leverage the compliant properties of SMPs at high temperature. Once the sample is deformed, it is cooled below $T_{trans}$ to 'freeze' in the temporary shape. Recovery occurs when the sample is reheated to the vicinity of $T_{trans}$, which increases polymer chain mobility and allows the SMP to return to its permanent shape via entropy elasticity[1]. Unfortunately, heat-demanding 'hot' programming is generally used by almost every class of existing SMPs. By contrast, SMPs that can be 'cold' programmed (that is, deformed to a temporary shape at or below room temperature), which could greatly enhance the processability to accommodate broader application requirements (for example, room-temperature operations for the entire SM cycle), are rare[7, 8]. In addition, most of the current SMP applications focus on leveraging the macroscopic SM effects, where the deformation length scale is large (on the order of centimetres). However, an intriguing potential for all SMPs, largely unexplored, is their ability to memorize and change shape at nanoscale[9, 10, 11, 12, 13, 14].

**Citation:** Yin Fang,Yongliang Ni,Sin-Yen Leo,Curtis Taylor,Vito Basile& Peng Jiang, Reconfigurable photonic crystals enabled by pressure-responsive shape-memory polymers, doi:10.1038/ncomms8416.

Here, we report a new type of SMP that enables 'cold' programming and instantaneous, nanoscopic shape recovery at ambient conditions. No heat is needed for both SM programming and recovery steps. Instead the 'cold' programming is induced by capillary pressure produced by water evaporation from the SMP membranes. Contrary to traditional thermoresponsive SMPs, the SM recovery of the new SMPs can be achieved by applying a small contact pressure or drying the films out of solvents with low surface tension (for example, ethanol). Interestingly, by combining this new type of SMP with macroporous photonic crystals, we demonstrate the fabrication of reconfigurable/rewritable photonic crystal micropatterns by a simple direct print approach. Furthermore, *in-situ* nanoindentation tests reveal that the counterintuitive pressure-induced SM recovery is caused by an adhesive pull-off force between the contact substrate and the SMP membrane.

## RESULTS

### Preparation and Characterization of new SMPs

The new pressure-responsive SMPs were discovered in the fabrication of macroporous polymer photonic crystal membranes[15, 16]. Photonic crystals are periodic dielectric structures with a forbidden photonic band gap (PBG) for electromagnetic waves[17, 18, 19]. They may hold the key to continued progress towards all-optical integrated circuits and high-speed optical computing[19].Figure 1 compares the significant differences in SM effects between traditional thermoresponsive SMPs (Fig. 1a–c) and the new stimuli-responsive SMPs (Fig. 1d–f). The new SMPs are photocured copolymers of ethoxylated (20) trimethylolpropane triacrylate (ETPTA, $T_g$~−40 °C provided by the vendor) and polyethylene glycol (600) diacrylate (PEGDA, $T_g$~−42 °C) oligomers (see molecular structures in Fig. 2a) with varying volumetric ratios from 1:1 to 1:6. As the ETPTA-co-PEGDA copolymer with 1:3 ratio showed the optimal SM behaviours, this recipe was adopted throughout the current work if not explicitly stated otherwise. A single $T_g$ of~−42 °C measured by differential scanning calorimetry (Fig. 2b) of an SMP sample indicates the cross-linked copolymer is a homogeneous mixture of the two components. The Young's moduli of the pure ETPTA and PEGDA polymers and their 1:3 copolymer were characterized by *in-situ* nanoindentation tests (see the typical force-depth indentation curve in Fig. 2c). The results in Fig. 2d show that the average Young's moduli for all samples are about 80 MPa. The tensile strength of the 1:3 copolymer membrane measured by using a conventional tensile tester is ~7.5 MPa and the yielding strain is ~0.19 (Supplementary Fig. 1), indicating the copolymer is quite elastic at room temperature. The bulk Young's modulus is caculated to be ~39 MPa, which is lower than the microscopic moduli measured by nanoindentation.

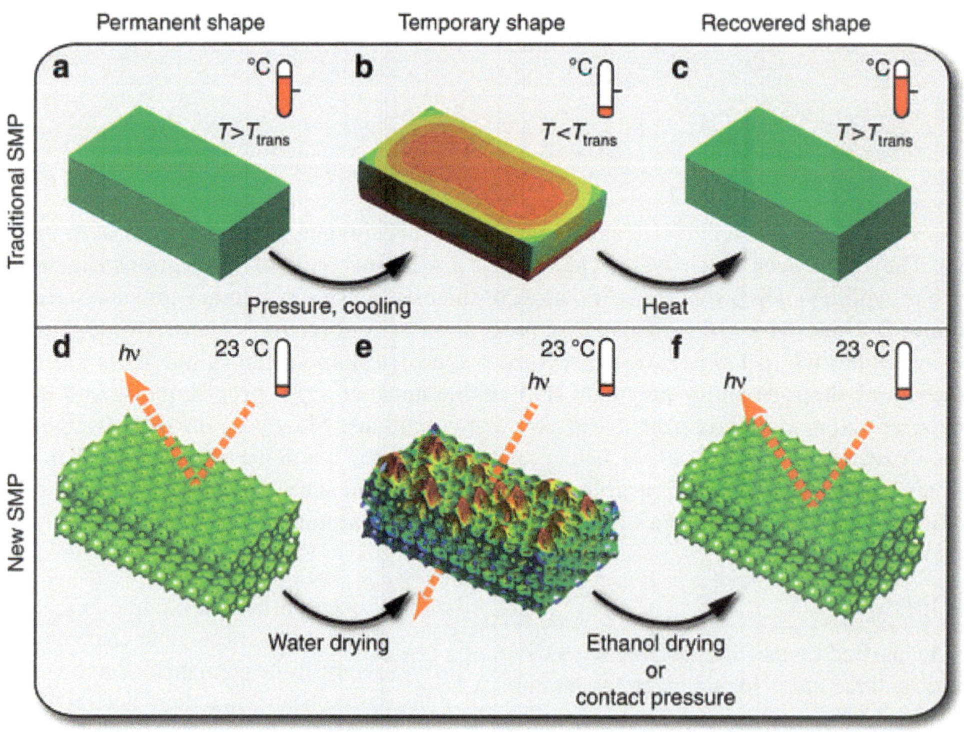


**Figure 1:** Schematic illustration compares different SM effects.

(**a**) A traditional thermoresponsive SMP sample can be deformed into a temporary shape by a 'hot' programming process (at a temperature above a specific $T_{trans}$). (**b**) The temperature is cooled below $T_{trans}$ to 'freeze' in the temporary shape. (**c**) The recovery of the permanent shape can be triggered by applying heat ($T>T_{trans}$) to the strained SMP sample. (**d**) A thin (a few micrometres thick) macroporous SMP photonic crystal with 3D ordered macropores (permanent shape) shows strong Bragg diffraction of visible light. (**e**) The ordered macropores can be deformed to a disordered structure (temporary shape) with no Bragg diffraction induced by an autonomous 'cold' programming process (water drying) at ambient conditions. (**f**) The nanoscopic recovery of the permanent 3D photonic crystal structure can be stimulated by drying the sample out of ethanol or applying an external contact pressure.

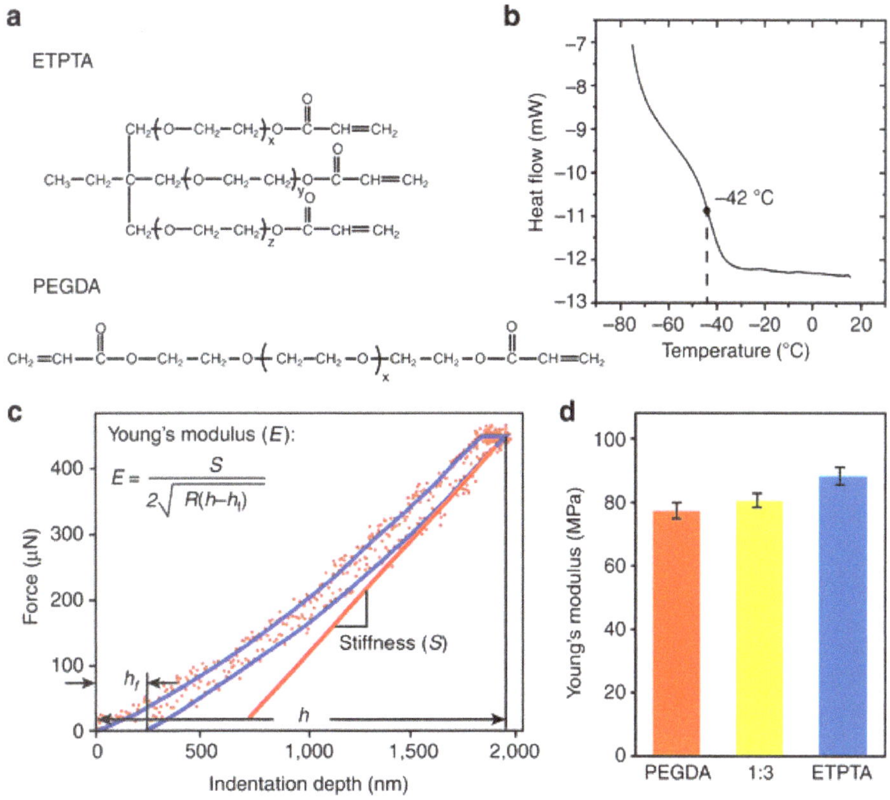


**Figure 2:** Structures and Properties of the new SMPs.

(**a**) Molecular structures of ETPTA ($x+y+z=20$) and PEGDA ($x=12$) oligomers. (**b**) Differential scanning calorimetry plot of a macroporous ETPTA-co-PEGDA copolymer with 1:3 ratio. (**c**) Typical force–depth indentation curve used to measure elastic properties of SMP. (**d**) Average Young's moduli for PEGDA, ETPTA-co-PEGDA copolymer with 1:3 ratio and ETPTA polymers.

## Unusual 'Cold' Programming Caused by Water Evaporation

Macroporous ETPTA-co-PEGDA membranes were fabricated by using self-assembled, three-dimensional (3D) highly ordered silica colloidal crystals as structural templates[15]. After removing the templating silica microspheres by a hydrofluoric acid wash, the resulting macroporous copolymer film immersed in water exhibited iridescent colours caused by Bragg diffraction of visible light from the periodic arrays of polymer macropores. This confirmed the maintenance of the 3D ordered structure of the original silica colloidal crystal. Surprisingly, the shining colours of the macroporous photonic crystal disappeared when the membrane was dried out of water and it became translucent with a pale white appearance (Fig. 3a, also see Supplementary Video 1). This suggests that the 3D periodic structure was lost when water evaporated from the ordered macropores. The cross-sectional scanning electron microscope (SEM) image in Fig. 3b confirms this conjecture, as no ordering of the deformed macropores is shown. Therefore, the new elastic copolymers enable an autonomous 'cold' programming process—the deformation from a 3D ordered permanent structure to a disordered temporary structure can be achieved at ambient conditions by evaporating water from the templated macropores. This is in sharp contrast to traditional SMPs that need to be heated above $T_{trans}$, then deformed to a temporary shape[1].

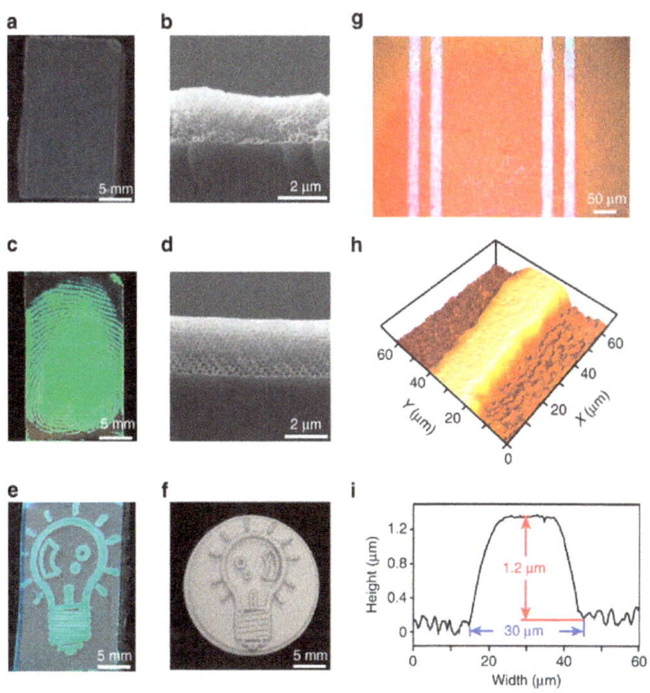


**Figure 3:** Arbitrary photonic crystal patterns printed on the new SMP membranes.

(**a**) Photograph of a translucent macroporous SMP film with disordered macropores. (**b**) Cross-sectional SEM image of the macroporous sample in **a** with deformed macropores. (**c**) Photograph of a green-coloured fingerprint pressed on the sample in **a**. (**d**) Cross-sectional SEM image of an iridescent region in **c** with 3D ordered macropores (300 nm diameter).

(**e**) Photograph of an iridescent 'light bulb' pattern printed on a translucent macroporous SMP membrane templated from 300 nm silica microspheres. (**f**) Photograph of the rubber stamp used in generating the 'light bulb' pattern in **e**. (**g**) Optical microscopic image of micropatterned pairs of double lines on a macroporous SMP membrane. (**h,i**) 3-D AFM image and the height profile of a section of a line in **g**.

## Pressure-Induced SM Recovery

Even more interesting, the recovery of the permanent photonic crystal structure can be triggered at room temperature simply by applying a small contact pressure on the macroporous membranes with collapsed macropores. As illustrated by the fingerprinting process in Supplementary Video 2, an iridescent green-coloured fingerprint (Fig. 3c) immediately appeared on the translucent macroporous copolymer membrane templated from 300 nm silica microspheres. The cross-sectional SEM image in Fig. 3d shows a fingerprinted region with the vivid green colour and the recovered 3D highly ordered macroporous structure is evident. The difference in the surface microstructures between the fingerprint valleys and ridges was characterized by atomic force microscopy (AFM, Supplementary Fig. 2). The surface of the raised fingerprint ridges is apparently much smoother than that of the valleys as confirmed by the magnified AFM images and the surface roughness analysis (Supplementary Table 1). The raising height of the fingerprint ridges above the valleys is estimated to be ~1.5 μm by the AFM depth profile. The gradual transition from a disordered macroporous array in a fingerprint valley to a 3D highly ordered structure in a fingerprint ridge is shown by the SEM image in Supplementary Fig. 3.

To avoid possible body-temperature effects on the macropore recovery in the above fingerprinting process, as well as to verify the feasibility of a new printing-based technology for fabricating arbitrary photonic crystal patterns, we printed a 'light bulb' relief pattern on a rubber stamp (Fig. 3f) onto a translucent SMP copolymer membrane with collapsed 300 nm macropores at room temperature. The final iridescent imprint (Fig. 3e) is a faithful replica of the original relief pattern. Furthermore, standard microfabrication technologies were used in making microscopic patterns on silicon wafers, which were used to imprint the micropatterns on SMP copolymer membranes.Figure 3g shows an optical microscope image of printed pairs of parallel lines with ~30 μm width. The raising up of the smoother line patterns from the rough macroporous surface is clearly presented by the AFM image (Fig. 3h) and the corresponding depth profile (Fig. 3i). Although various technologies for fabricating tunable photonic crystals have been demonstrated using elastic materials (for exmple, elastomers and gels), the temporarily deformed photonic microstructures cannot be memorized[18, 20, 21, 22, 23]. Immediately, they return to the original crystalline lattices once the external stress is released. By contrast, the printed photonic crystal patterns on the new pressure-responsive SMPs are stable over long periods of time. The colourful fingerprints (for example, Fig. 3c) stored at ambient conditions have maintained their vivid colours and clear patterns for more than 2 years. Most importantly, the imprinted photonic crystal patterns can be erased when the SMP membranes are reimmersed in water and then dried out of it (seeSupplementary Video 3). New photonic microstructures can then be printed on the regenerated macroporous SMP membranes. This unique rewriting capability is critical for developing reconfigurable photonic crystals that can adapt various photonic functionalities to accommodate different applications[24, 25]. This reconfigurability can dramatically reduce the complexity and fabrication cost of developing a large number of application-specific devices[26].

## Capillary Pressure-Induced Macropore Collapse

Above we have shown that the new SMP copolymers enable room-temperature operations for the entire SM cycle (from an unusual 'cold' programming process to a contact pressure-induced recovery step). We speculated that the 'cold' programming process was induced by large capillary pressure created by water evaporation from the template macropores, which squeezed the elastic macropores into disordered arrays[27]. Similar macropore collapse was observed for macroporous polymer (for example, polysulfone) reverse osmosis membranes used for water purification[28]. Further insight into the macropore collapse can be gained by considering the capillary pressure ($P_c$) in the Young–Laplace equation, $P_c$=*2 cos /r*, where is the liquid/vapour surface tension, *r* is the radius of the pores and is the contact angle of the liquid on the pore surface[27]. As the measured water contact angles on the copolymers were <20°, *cos* is thus close to 1. One direct evidence supporting this capillary pressure-induced macropore collapse mechanism is that the templated macroporous membrane retained its original 3D ordered structure and iridescent colours when dried out of ethanol (Supplementary Video 4), which has a smaller surface tension than that of water (22.39 versus 72.75 mN m$^{-1}$ at 20 °C). The smaller led to a lower $P_c$ that was not sufficient to squeeze the elastic macropores into disordered arrays. In addition to ethanol, a large variety of solvents with low surface tension (for example, acetone and toluene) can also trigger the same disorder-to-order transition.

Figure 4a–d compare the surface microstructures of a macroporous SMP copolymer membrane dried out of water (Fig. 4a,c) and ethanol (Fig. 4b,d), respectively. The rough, disordered macroporous array was fully recovered to a smooth and ordered structure triggered by ethanol evaporation (Fig. 4e). This disorder-to-order transition and the corresponding translucent-to-iridescent colour change can be characterized by measuring the normal-incidence optical reflection spectra. In Fig. 4f, the sample with disordered macropores (dried out of water) shows no apparent PBG peaks, whereas the ethanol-recovered sample with ordered macropores exhibits a distinct PBG peak with well-defined Fabry–Perot fringes, indicating the high crystalline quality of the solvent-activated macroporous photonic crystal[15]. Importantly, the experimental spectrum of the recovered photonic crystal matches well with the calculated spectrum using a scalar-wave approximation model[29], which assumes a perfect macroporous crystalline lattice. This indicates that the temporarily deformed macropores were fully reopened to their permanent, 3D highly ordered structure triggered by ethanol evaporation.

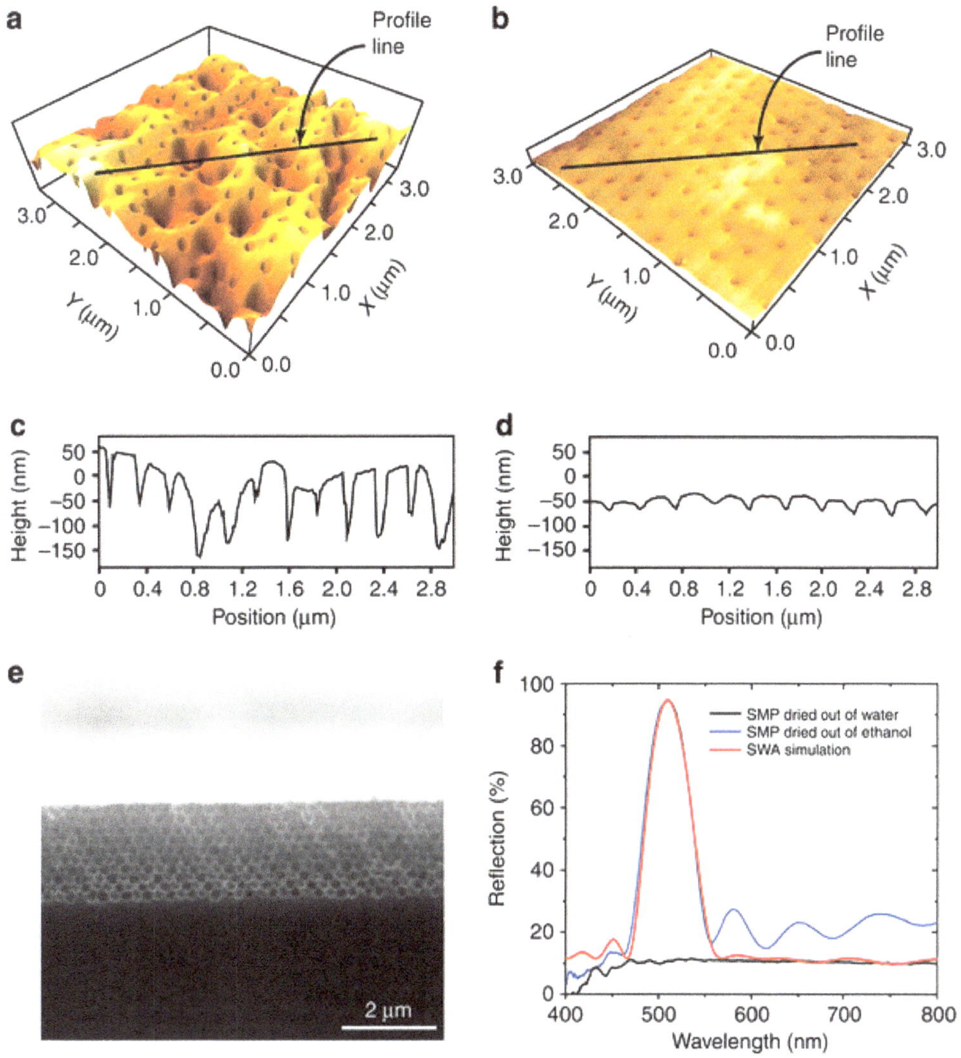


**Figure 4:** Difference in topography and PBGs.

(**a**,**c**) 3D AFM image and the height profile scanned across the profile line for a water-dried SMP sample consisting of 280 nm macropores. (**b**,**d**) 3D AFM image and the height profile for the same sample dried out of ethanol. (**e**) Cross-sectional SEM image of the sample in **b**. (**f**) Normal-incidence optical reflection spectra compares the PBG properties of the macroporous samples dried out of water and ethanol. The scalar-wave approximation-simulated spectrum assuming a perfect macroporous crystalline lattice is also shown to compare with the experimental results.

## Critical Contact Pressure for SM Recovery

By using the ethanol-activated sample as a fully recovered control, we can estimate the critical pressure that is needed to trigger the shape recovery and the nanoscopic strain recovery rate ($R_r$) of the new SMPs through monitoring the PBG properties of the recovered samples under different pressures. Figure 5a shows that different reflection amplitudes of the PBG peaks resulted when various pressures were applied by putting varying weights on a small polydimethylsiloxane piece with a specific area on a macroporous copolymer membrane. Previous work has shown that the PBG optical density of a macroporous photonic crystal is nearly a linear function of its crystalline thickness[15]. We therefore normalized the absolute reflection amplitude (after spectrum baseline correction) of a pressure-recovered SMP sample to that of the ethanol-activated control sample as an indicator of $R_r$ (Fig. 5b). A near-unity $R_r$ was obtained when a 54.4-kPa pressure was applied, whereas a ~50% recovery needed a pressure of 4.21 kPa. The cross-sectional SEM images inFig. 5c,d, which correspond to the samples recovered by applying 7.13 and 27.9 kPa pressure, responsively, reveal that an intermediate pressure only induces the partial recovery of the top layers of the macroporous SMP photonic crystal, leading to the lower reflection amplitude compared with the fully recovered control sample.

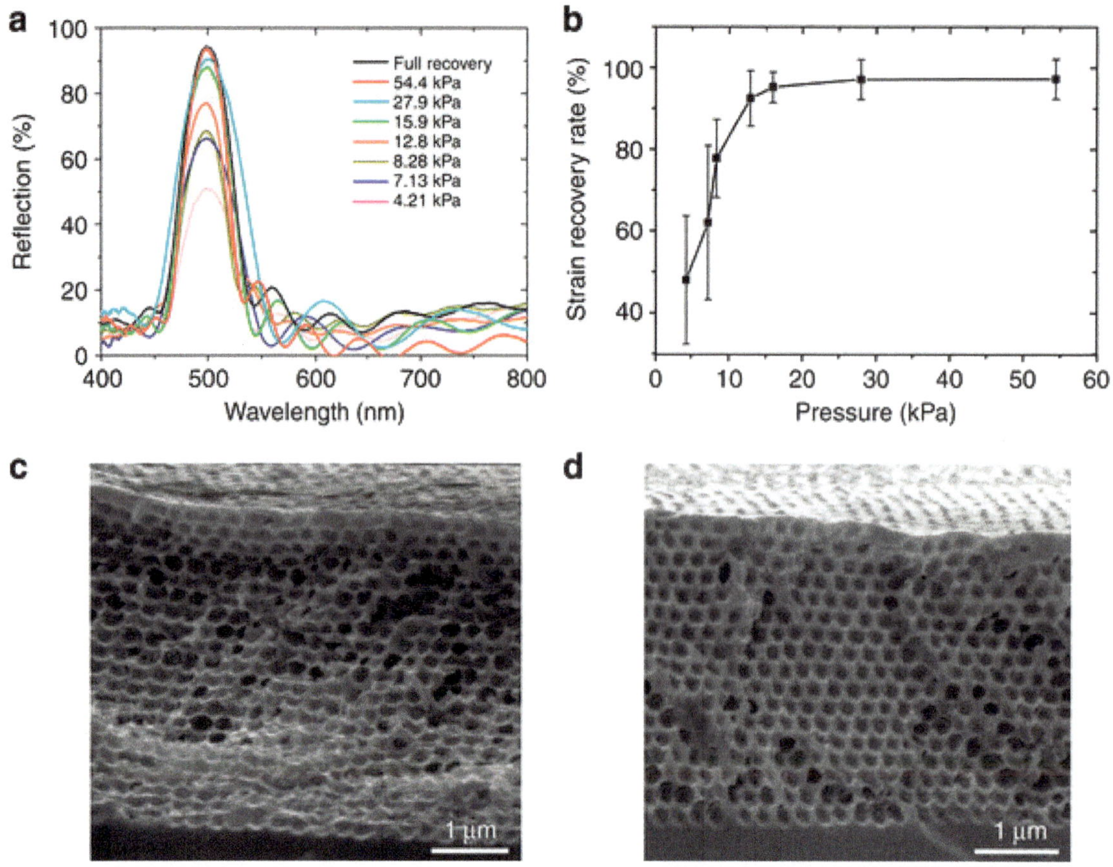


**Figure 5:** Pressure-dependent macropore recovery.

(**a**) Normal-incidence optical reflection spectra obtained from a macroporous SMP membrane consisting of 280 nm macropores under different pressures. The same sample dried out of ethanol was labelled as 'Full Recovery'. (**b**) Normalized absolute reflection amplitude of a recovered SMP sample was used as an indicator of the nanoscopic SMP strain recovery rate under different pressures. (**c**,**d**) Cross-sectional SEM images of a macroporous SMP membrane recovered by 7.13 and 27.9 kPa pressure, respectively.

## Pressure-Induced SM Recovery Mechanisms

Similar to thermoresponsive SMPs, we believe the entropy elasticity[1] is the energetic root cause for the SM effects exhibited by the new pressure-responsive SMPs. When photocured in the interstitial regions of the templating colloidal crystal, the cross-linked polymer chains were primarily in energetically favourable, stress-free configurations. The capillary pressure-induced 'cold' programming squeezed the ordered macropores into disordered arrays with reduced thickness (see Fig. 3b,d), storing excess stresses in the deformed, temporarily configured polymer chains. The strained polymer networks have a strong tendency to recover back to their permanent, stress-free states. However, the observed pressure-induced macropore recovery is counterintuitive as we expect the applied pressure should further deform the collapsed macropores instead of popping them up. To elucidate this unusual shape-recovery mechanism, we conducted *in-situ* nanoindentation tests to characterize the forces in the approaching and retracting processes when a spherical sapphire tip indented the macroporous SMP membrane (Fig. 6). An apparent adhesive pull-off force, caused by the attractive van der Waals interactions and the capillary force induced by the capillary-condensed water meniscus layer between the indenter tip and the SMP membrane[30], is evident in the retraction process. We believe this pull-off force causes the SM recovery of the collapsed macropores. A higher pressure leads to more conformal interactions between the molecules on the tip and the membrane, and thus a larger pull-off force. This could explain the pressure effects on the strain recovery rate (Fig. 5b) and the partial recovery of the top-layer macropores when an intermediate pressure was applied (Fig. 5c). One strong evidence supporting this pull-off mechanism is that the collapsed macropores did not recover back to their ordered structure when a pressure (up to~350 kPa) was applied through compressed air instead of a contacting body or entity (for example, a finger or a rubber stamp), which could exert the pull-off force on the SMP macropores.

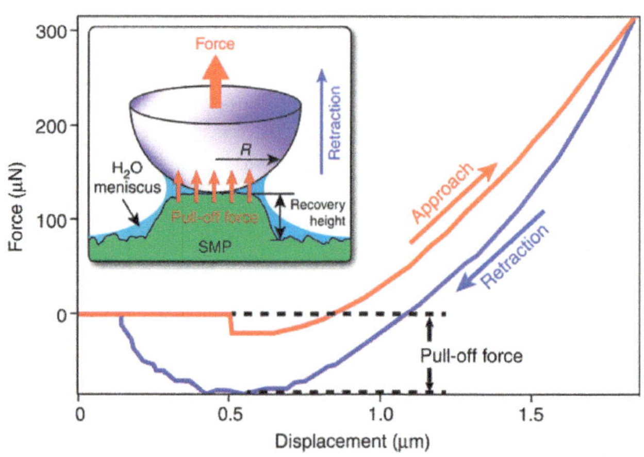


**Figure 6:** Macropore recovery induced by pull-off forces.

A typical indentation force-displacement curve showing approach and retraction segments and pull-off force obtained on a macroporous SMP membrane. Inset shows diagram of macropore recovery during retraction of indenter due to pull-off force caused by van der Waals interactions and the capillary force induced by the capillary-condensed water meniscus layer between the indenter tip and the SMP membrane.

## DISCUSSION

To conclude, we have developed a new type of stimuli-responsive SMP that differs greatly from existing SMPs, as it enables fast response and room-temperature operations for the entire SM cycle. The large capillary pressure generated by water evaporation from the templated macropores induces the unusual 'cold' programming at room temperature. The instantaneous recovery of the temporarily deformed macropores to the permanent, 3D highly ordered photonic crystal structure can be triggered by applying a small contact pressure at ambient conditions. In addition to being pressure responsive, the disorder-to-order transition of the new SMPs can also be stimulated by drying the macroporous SMP membranes out of solvents with low surface tension, such as ethanol and toluene. Importantly, the easily perceived colour change from translucency to iridescence associated with the structural disorder-to-order transition enables a simple and quantitative optical technology for characterizing the intriguing SM effects at nanoscale. Simultaneously, the striking chromogenic effects induced by the recovery of the permanent 3D photonic crystal structure provide opportunities for a wide spectrum of applications ranging from reconfigurable photonic crystal devices to chromogenic pressure and chemical sensors to novel biometric and anti-counterfeiting materials.

## METHODS

### Templated Fabrication of Macroporous SMP Photonic Crystal Membranes

Monodispersed silica microspheres, with diameter ranging from 100 to 600 nm, were synthesized by the standard Stöber method. Silica particles were purified in 200-proof ethanol by multiple centrifugation and redispersion cycles. Next, they were self-assembled on a glass microslide by the convective self-assembly technology to form colloidal crystals[15]. By adjusting the particle volume fraction of the silica/ethanol suspension, the thickness of the colloidal crystal was controlled to 10–50 colloidal monolayers. The microslide with the silica colloidal crystal on its surface was covered by another microslide and a double-sided adhesive tape of ~1 mm thick was used as a spacer in between the microslides. By using capillary force, the interstitial air in between the silica microspheres was replaced by viscous oligomer mixtures consisting of ETPTA (SR415, Sartomer, molecular weight 1,176 kDa, viscosity 225 cps at 25 °C, refractive index 1.470) and PEGDA (SR610, Sartomer, molecular weight 742 kDa, viscosity 90 cps at 25 °C, refractive index 1.468) oligomers with varying volumetric ratios from 1:1 to 1:6. Darocur 1173 (2-hydroxy-2-methyl-1-phenyl-1-propanone, BASF, 1 wt%) was added as the photoinitiator. The sample was transferred to a pulsed ultraviolet curing system (RC 742, Xenon) and the oligomer mixture was rapidly polymerized by exposure to ultraviolet radiation for 4 s. The polymerized film was soaked in a 1 vol% hydrofluoric acid aqueous solution for 4 h and then rinsed with deionized water. The resulting self-standing macroporous polymer membrane showed pale iridescent colours when immersed in water and observed at large viewing angles (>45°) due to the small refractive index contrast between the copolymer (~1.47) and water (1.33).

### Printing Photonic Crystal Patterns on Macroporous SMP Membranes

Using Kimwipes, the free-standing macroporous SMP membrane was dried and the diffractive colours of the film were lost during water evaporation (see Supplementary Video 1). Strikingly iridescent photonic crystal patterns, whose colours are determined by the size of the templating silica microspheres, can be printed on the translucent SMP membranes by using various substances with relief patterns, such as fingers or rubber stamps (see Fig. 3c,e). To generate microscopic photonic crystal patterns, standard photolithography and chlorine reactive ion etching were performed in a class 100 cleanroom, to fabricate micropatterns (for example, parallel lines inFig. 3g) on a silicon wafer. The hard silicon mould was then placed on an SMP membrane and a typical fingerprinting pressure was applied on the mould to transfer the micropatterns. To evaluate the pressure effects on the macroporous SMP strain-recovery rate, we placed different weights (43, 73, 85, 130, 162, 285 and 555 g) on a small polydimethylsiloxane square (1 cm × 1 cm, Sylgard 184, cured at 75 °C for 2 h), to generate various pressures on a macroporous copolymer membrane.

### Sample Characterization

SEM imaging was carried out on a FEI XL-40 FEG-SEM. A thin layer of gold was sputtered onto the samples before imaging. Amplitude-modulation AFM was performed using a MFP-3D AFM (Asylum Research, Inc.) with a Nanosensor PPP-NCHR probe (tip radius <10 nm), to characterize the topography and surface roughness of macroporous SMP membranes. *In-situ* nanoindentation tests were performed with an MFP-3D NanoIndenter (Asylum Research, Inc.) using a spherical sapphire indenter (tip radius ~125 μm). Such configuration of the instrument has a force and displacement resolution <3 μN and 1 nm, respectively. Detailed surface roughness analysis and calculation of Young's modulus by nanoindentation are discussed in Supplementary Information. An Instron model 1122 load frame upgraded with an MTS ReNew system and equipped with a 500-g load cell at a cross-head speed of 0.5 mm $min^{-1}$ was used in testing the tensile strength of the SMP membranes. Differential scanning calorimetric measurements were carried out from −80 to 18 °C at a heating rate of 10 °C $min^{-1}$ using a Seiko DSC 6200 instrument and an empty pan as reference. Normal-incidence optical reflection spectra were obtained using an Ocean Optics HR4000 high-resolution fibre optic visible-near infrared spectrometer with a reflection probe (R-400-7-SR) and a tungsten halogen light source (LS-1). Absolute reflectivity was obtained as the ratio of the sample spectrum and a reference spectrum, which was the optical density obtained from an aluminum-sputtered (1,000 nm thickness) silicon wafer.

## Scalar Wave Approximation Optical Modelling

The scalar wave theory developed for periodic dielectric structures[29] was implemented to calculate the normal-incidence optical reflection spectra from macroporous SMP photonic crystals. In this theory, Maxwell's equations are solved for a periodic dielectric assuming that one may neglect diffraction from all but one set of crystalline planes (for example, the (111) planes in our case). The scalar-wave approximation calculation contains no adjustable parameters, as the size of the macropores and the crystal thickness were independently determined from SEM characterization and the refractive index of the SMP copolymer is known.

## REFERENCES

1. Lendlein, A. Shape Memory Polymers (Springer, 2010).
2. Xie, T. Tunable polymer multi-shape memory effect. Nature 464, 267–270 (2010).
3. Xie, T. Recent advances in polymer shape memory. Polymer (Guildf) 52, 4985–5000 (2011).
4. Mather, P. T., Luo, X. & Rousseau, I. A. Shape memory polymer research. Annu. Rev. Mater. Res. 39, 445–471 (2009).
5. Huang, W. M., Yang, B., Zhao, Y. & Ding, Z. Thermo-moisture responsive polyurethane shape-memory polymer and composites: a review. J. Mater. Chem. 20, 3367–3381 (2010).
6. Lendlein, A., Jiang, H. Y., Junger, O. & Langer, R. Light-induced shape-memory polymers. Nature 434, 879–882 (2005).
7. Yakacki, C. M. et al. Impact of shape-memory programming on mechanicallydriven recovery in polymers. Polymer (Guildf) 52, 4947–4954 (2011).
8. Li, J. & Xie, T. Significant impact of thermo-mechanical conditions on polymer triple-shape memory effect. Macromolecules 44, 175–180 (2011).
9. Wang, Z. et al. Programmable, pattern-memorizing polymer surface. Adv. Mater. 23, 3669–3673 (2011).
10. Xu, H. et al. Deformable, programmable, and shape-memorizing micro-optics. Adv. Funct. Mater. 23, 3299–3306 (2013).
11. Jang, J. -H., Koh, C. Y., Bertoldi, K., Boyce, M. C. & Thomas, E. L. Combining pattern instability and shape-memory hysteresis for phononic switching. Nano Lett. 9, 2113–2119 (2009).
12. Espinha, A., Concepcion Serrano, M., Blanco, A. & Lopez, C. Thermoresponsive shape-memory photonic nanostructures. Adv. Opt. Mater. 2, 516–521 (2014).
13. Schaefer, C. G. et al. Reversible light-, thermo-, and mechano-responsive elastomeric polymer opal films. Chem. Mater. 25, 2309–2318 (2013).
14. Shah, A. A., Schultz, B., Zhang, W. J., Glotzer, S. C. & Solomon, M. J. Actuation of shape-memory colloidal fibres of Janus ellipsoids. Nat. Mater. 14, 117–124 (2015).
15. Bertone, J. F., Jiang, P., Hwang, K. S., Mittleman, D. M. & Colvin, V. L. Thickness dependence of the optical properties of ordered silica-air and airpolymer photonic crystals. Phys. Rev. Lett. 83, 300–303 (1999).
16. Velev, O. D., Jede, T. A., Lobo, R. F. & Lenhoff, A. M. Porous silica via colloidal crystallization. Nature 389, 447–448 (1997).
17. Joannopoulos, J. D., Meade, R. D. & Winn, J. N. Photonic Crystals: Molding the Flow of Light (Princeton University Press, 1995).
18. Arsenault, A. C. et al. From colour fingerprinting to the control of photoluminescence in elastic photonic crystals. Nat. Mater. 5, 179–184 (2006).
19. Rinne, S. A., Garcia-Santamaria, F. & Braun, P. V. Embedded cavities and waveguides in three-dimensional silicon photonic crystals. Nat. Photon. 2, 52–56 (2008).
20. Ge, J., Hu, Y. & Yin, Y. Highly tunable superparamagnetic colloidal photonic crystals. Angew. Chem. Int. Ed. 46, 7428–7431 (2007).
21. Yang, D., Ye, S. & Ge, J. From metastable colloidal crystalline arrays to fast responsive mechanochromic photonic gels: An organic gel for deformationbased display panels. Adv. Funct. Mater. 24, 3197–3205 (2014).
22. Burgess, I. B., Loncar, M. & Aizenberg, J. Structural colour in colourimetric sensors and indicators. J. Mater. Chem. C 1, 6075–6086 (2013).
23. Kang, Y., Walish, J. J., Gorishnyy, T. & Thomas, E. L. Broad-wavelength-range chemically tunable block-copolymer photonic gels. Nat. Mater. 6, 957–960 (2007).
24. Deotare, P. B. et al. All optical reconfiguration of optomechanical filters. Nat. Commun. 3, 846 (2012).
25. Wang, W. S., Xie, N., He, L. & Yin, Y. D. Photocatalytic colour switching of redox dyes for ink-free light-printable rewritable paper. Nat. Commun. 5, 5459 (2014).
26. Grillet, C. et al. Reconfigurable photonic crystal circuits. Laser Photon. Rev. 4, 192–204 (2010).
27. Gregg, S. J. & Sing, K. S. W. Adsorption, Surface Area and Porosity. 2nd edn (Academic Press Inc., 1982).
28. Tsai, J. T. et al. Retainment of pore connectivity in membranes prepared with vapor-induced phase separation. J. Membr. Sci. 362, 360–373 (2010).
29. Satpathy, S., Zhang, Z. & Salehpour, M. R. Theory of photon bands in 3- dimensional periodic dielectric structures. Phys. Rev. Lett. 64, 1239–1242 (1990).
30. Butt, H. J., Cappella, B. & Kappl, M. Force measurements with the atomic force microscope: Technique, interpretation and applications. Surf. Sci. Rep. 59, 1–152 (2005).

CHAPTER 18

# QUANTUM COMPUTATION BASED ON PHOTONS WITH THREE DEGREES OF FREEDOM

Ming-Xing Luo[1], Hui-Ran Li[1], Hong Lai[2] & XiaojunWang[3]

[1] Information Security and National Computing Grid Laboratory, Southwest Jiaotong University, Chengdu 610031, China

[2] School of Computer and Information Science, Southwest University, Chongqing 400715, China

[3] School of Electronic Engineering, Dublin City University, Dublin 9, Ireland

## ABSTRACT

**Quantum systems are important resources for quantum computer. Different from previous encoding forms using quantum systems with one degree of freedom (DoF) or two DoFs, we investigate the possibility of photon systems encoding with three DoFs consisting of the polarization DoF and two spatial DoFs. By exploring the optical circular birefringence induced by an NV center in a diamond embedded in the photonic crystal cavity, we propose several hybrid controlled-NOT (hybrid CNOT) gates operating on the two-photon or one-photon system. These hybrid CNOT gates show that three DoFs may be encoded as independent qubits without auxiliary DoFs. Our result provides a useful way to reduce quantum simulation resources by exploring complex quantum systems for quantum applications requiring large qubit systems.**

## INTRODUCTION

Quantum computer has shown its superiority for solving difficult problems such as the large integer decomposition[1,2,3] and data searching[4,5]. Since its difficult the large integer decomposition is the mathematical foundation of the well-known RSA cryptography which may be used for classical cryptographic applications[6,7,8]. Most of these quantum computation tasks may be completed with evolutions of quantum systems and desired quantum measurements[1,2,3,4,5,9,10]. If the quantum circuit model[11] is applied, these evolutions may be synthesized by series of local quantum gates. Exactly, proper small gates such as the controlled phase-flip (CZ) gate or controlled-not (CNOT) gate combined with single-qubit gates[12,13,14] can be used to implement quantum tasks with multiple qubits. These small gates construct a universal quantum gate set for quantum computing. Up to now, the CNOT gate has been widely implemented using several quantum systems, such as the linear optics[15,16], ion trap[17,18], atom[19,20], and nuclear magnetic resonance[21,22].

The solid-state quantum system has also attracted much attentions in quantum simulations because of its special optical property and scalability[23,24,25,26]. Moreover, electron-spin qubits associated with the nitrogen-vacancy (NV) defect centers are particularly useful. In fact, due to the long room-temperature coherent time[27], the negatively charged NV defect center in the diamond lattice, consisting of a substitutional $^{14}N$ atom and an adjacent vacancy, is an attractive candidate for quantum information processing. It has been used to prepare and detect optical sources[28,29,30,31,32,33], generate hybrid quantum entanglements between the NV center and photon[34],

**Citation:** Ming-Xing Luo, Hui-Ran Li, Hong Lai2 & XiaojunWang, Quantum Computation Based on Photons with Three Degrees of Freedom, doi:10.1038/srep25977.

or electrons[35,36], purify two-photonic hyperentanglement in both the polarization and spatial DoFs[37], or implement the CZ gate between the NV centers assisted by the microsphere cavity[38,39]. The single-electron and nuclear-spin states can be faithfully detected even under ambient conditions[40,41], when the electron spin of the NV defect center couples to nearby $^{13}C$ nuclear spins. Another diamond $NV^-$ center is proposed with six electrons from the nitrogen and three carbons surrounding the vacancy[42], which is confined in a microtoroidal resonator (MTR)[43] with the quantized whispering-gallery mode (WGM). This useful system allows for an ultrahigh-Q and a small mode volume of WGM microresonators[44,45,46], which has been applied to construct quantum gates on electron-spin qubits[47,48] or remote qubits[49,50]. Furthermore, recent experiments have assembled several hybrid systems, where colour centers in diamond nanocrystals or bulk diamond are coupled to the evanescent fields of cavities, which are defined in non-diamond materials coupling to WGMs in a silica microsphere[50,51,52], diamond-GaP micro-disk[53], GaP micro-ring cavities[54], or SiN photonic crystal[55].

Most of previous quantum simulations focused on systems with single DoF[15,16,17,18,19,20,21,22] or hybrid systems[56,57,58,59,60]. A few schemes have considered photons with two DoFs[61,62]. Our recent result[63] presents the independence of two DoFs (polarization DoF and spatial DoF) of photonic system, and then is used to construct the ququart (four-dimensional) quantum logic[64]. Thus quantum simulation resources may be saved one half. In this paper, we further reduce quantum resources by considering photonic systems with three DoFs. Motivated by recent schemes[65,66], each photon may be encoded with two circularly polarized states and four modes, i.e., $|l, I\rangle$ ,$|r, I\rangle$ and $|l, E\rangle$ ,$|r, E\rangle$ for two crystal emissions. Here, l(r) refers to the left (right) side of each cone and I(E) denotes the internal (external) cone, as shown in Fig. 1. A general state is given by the product of one polarization state and two longitudinal momentum states. In the follow, we will investigate these photonic DoFs for the quantum simulation, without using auxiliary DoFs. From the quantum circuit model, CNOT gate will be schematically implemented on these DoFs of photonic states assisted by NV centers. For the symmetry of two spatial modes in each photonic system, fifteen CNOT gates are required to operate on the polarization DoFs and spatial-DoFs of the two-photon or one-photon system. Each gate is completed by interacting photons to auxiliary NV centers, disentangling NV centers, and correcting the emitting photons. These schemes are beyond to previous CNOT gates on the same DoF of two-photon state[14,15,56,57], hybrid CNOT gates on the photon and stationary electron spins in quantum dots[58,59,60,61], or different DoFs of two photons[62,63]. Our theoretical result shows that three DoFs of each photonic system can be used as independent qubits in one quantum task. Hence, two thirds of quantum resources may be saved for quantum simulations with large qubit systems, such as the Shor's algorithm.

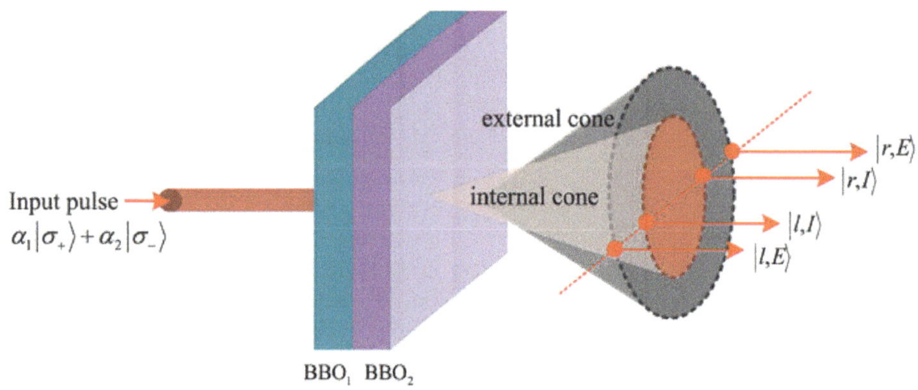


**Figure 1:** Schematic photon generation from the internal (I) and external (E) cone assisted by a two-crystal system.

Polarized state as an input pulse passes through two 0.5 mm thick Type I β-barium-borate (BBO) crystal slabs. The input photon is created at a degenerate wavelength $\lambda=2\lambda_p$ by each BBO crystal along two correlated directions belonging to the lateral surfaces of two SPDC cones, with full aperture angles $\theta_I$ and $\theta_E$, respectively. The internal (I) and external (E) cone correspond to the first and the second crystal, respectively. The annular sections of each emission cone, with approximate diameters $d_I$ and $d_E$ are intercepted by a single eight-hole screen. The dichotomy existing between the I cone and E cone is identified as an independent DoF. The corresponding mode emission as l(r) by referring to the left (right) side of each cone.

## RESULTS

To show the encoding independence of the polarization DoF and two spatial DoFs of each photon, it is necessary to prove that all quantum transformations in $SU(2^n)$ may be implemented on these DoFs. Based on the theory of the universal logic gates[12,13,14], it is sufficient to consider the CNOT gate on any two DoFs of the photonic system. It means that fifteen CNOT gates should be performed on photonic systems with three DoFs, where nine CNOT gates are on the two-photon system (all combinations of three DoFs) and six CNOT gates are on the one photon system. By exploring optical selection rules of the NV center in the crystal cavity, these CNOT gates may be realized without altering DoFs and auxiliary DoFs during implementations. In this case, each photonic DoF can be encoded as an independent qubit in quantum applications.

## Photon with three DOFS

Circularly polarized photon in the state $\alpha_1|L\rangle+\alpha_2|R\rangle$ (left circularly polarized state $|L\rangle$ and right circularly polarized state $|R\rangle$ ) is created at a degenerate wavelength $\lambda=2\lambda_p$ by each BBO crystal along two correlated directions belonging to the lateral surfaces of two SPDC cones, with full aperture angles $\theta_I$ and $\theta_E$, respectively[64,65], as shown inFig. 1. The output state is dependent of these angles. I refers to the internal cone whereas E denotes the external cone, corresponding to the first and the second crystal, respectively. The dichotomy existing between the I cone and E cone is thus identified as an independent DoF, i.e., the corresponding mode emission as l(r) by referring to the left (right) side of each cone[64,65]. If the pump coherence length exceeds more than one order of magnitude the total crystal length, the coherence and indistinguishability between two crystal emissions may be guaranteed[64,65]. Two conical emissions are then transformed into two cylindrical ones by a positive lens with focal length f, located at a distance f from the intermediate point of the second crystal device. By selecting four pairs of correlated modes with an eight-hole screen, $|l, I\rangle$ and $|r, I\rangle$ for the first crystal and $|l, E\rangle$ and $|r, E\rangle$ for the second crystal emission, a general photonic state is prepared as the product of one polarization state and two longitudinal momentum states (or, equivalently, a ququart state) and is expressed as a 3-qubit state:

$$|\phi\rangle = (\alpha_1|L\rangle + \alpha_2|R\rangle) \otimes (\beta_1|l, I\rangle + \beta_2|r, I\rangle + \beta_3|l, E\rangle + \beta_4|r, E\rangle) \tag{1}$$

where $|\alpha_1|^2+|\alpha_2|^2=1$ and $|\beta_1|2+|\beta_2|2+|\beta_3|2+|\beta_4|2=1$. Here, $\beta_j$ are dependent of aperture angles $\theta_I$ and $\theta_E$, and focal length f, which are not goals in this paper[64,65].

## A Diamond NV center Coupled to an MTR with a WGM

Schematic NV center in a diamond embedded in a photonic crystal cavity is shown in Fig. 2. The negatively charged NV center is consisted of a substitutional nitrogen atom and an adjacent vacancy with six electrons. The Λ-type three-level system is realized using specific excited state $|A_2\rangle = (|E_-\rangle|m^+\rangle + |E_+\rangle|m^-\rangle)$ as an ancillary state[66,34]. Here, $|E_\pm\rangle$ are orbital states with angular momentum projection along the NV axis. The ground state is an electronic spin triplet with a splitting of 2.88 GHz between the magnetic sublevels $|0\rangle$ ($m_s=0$) and $|m^\pm\rangle$ ($m_s=\pm1$)[34]. $|A_2\rangle$ may decay into two ground states $|m^-\rangle$ and $|m^+\rangle$ by exciting the NV center with a polarized 2-ns p-pulse that is shorter than the emission timescale, and the reflection may be separated from fluorescence photons using detection timing[34]. The normal boundary condition $\hat{b}_{out} = \hat{b}_{in} + \sqrt{\kappa}\hat{a}$ is used to derive the optical selection rule with the input field $\hat{b}_{in}$, output field $\hat{b}_{out}$ and cavity field operator $\hat{a}$. If spins stay in the ground states most of the time[67], the optical reflection coefficient may be approximately defined in the follow (see Methods)

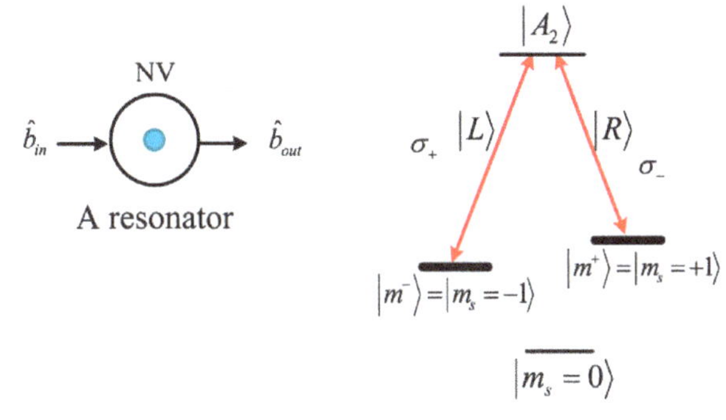


**Figure 2:** Schematic NV center coupling to the resonator and possible Λ-type optical transitions in the NV center.

$\hat{a}_{in}$ and $\hat{a}_{out}$ are the input and output field operators of a waveguide, respectively. The bold levels encode qubits, i.e., $|m^\pm\rangle = |m_s=\pm\rangle$. The transition $|m^-\rangle \rightarrow |A_2\rangle$ is derived by a left circularly polarized photon $\sigma_+$ ($|L\rangle$), and $|m^+\rangle \rightarrow |A_2\rangle$ is derived by a right circularly polarized photon $\sigma_-$ ($|R\rangle$).

$$r(\omega) = \frac{\left(i\delta\omega_c - \frac{\kappa}{2} + \frac{\kappa_s}{2}\right)\left(i\delta\omega_e + \frac{\gamma}{2}\right) + g^2}{\left(i\delta\omega_c + \frac{\kappa}{2} + \frac{\kappa_s}{2}\right)\left(i\delta\omega_e + \frac{\gamma}{2}\right) + g^2} := |r(\omega)|e^{i\theta} \tag{2}$$

where $\delta\omega_c$ and $\delta\omega_e$ are frequency detunings satisfying $\delta\omega_c=\omega_c-\omega$ and $\delta\omega_e=\omega_e-\omega$. $\omega_c$, $\omega$ and $\omega_e$ are the frequencies of the cavity mode, input photon pulse, and NV center, respectively. g is the coupling strength between the cavity and the NV center. $\kappa$, $\kappa_s$ and $\gamma$ are the damping rate of the cavity, cavity side leakage mode, and spontaneous decay rate of the NV center, respectively. If define the cooperativity $C=2g2/(\gamma\kappa)$, the photonic reflection probability[68] is determined by the cooperativity Cand the cavity tuning as follow

$$P = 1 - \frac{1 + 4C + \left(\frac{\delta\omega_c}{\gamma}\right)^2}{1 + 4C + 4C^2 + \left(\frac{\delta\omega_c}{\gamma}\right)^2} \tag{3}$$

Considering the coupling strength $g=0$, an NV center is uncoupled from the cavity (the cold cavity), and the reflection coefficient $r(\omega)$ becomes

$$r_0(\omega) = \frac{i\delta\omega_c - \frac{\kappa}{2} + \frac{\kappa_s}{2}}{i\delta\omega_c + \frac{\kappa}{2} + \frac{\kappa_s}{2}} =: |r_0(\omega)|e^{i\theta_0} \tag{4}$$

Thus the input pulse in the polarized state $|L\rangle$ gains a phase shift $\theta$after reflecting from the hot cavity ($g>0$) with the NV center $|m^-\rangle$, or a phase shift $\theta_0$ after reflecting from the cold cavity ($g=0$) with the NV center $|m^+\rangle$. The input pulse in the state $|R\rangle$ gains a phase shift $\theta_0$after reflecting from the cold cavity with the NV center $|m^-\rangle$, or a phase shift $\theta$ after reflecting from the hot cavity with the NV center $|m^+\rangle$. By choosing a proper frequency detuning $\delta\omega_e=0$[66] and the cooperativity $C \gg 1$, the reflection coefficients may satisfy $|r(\omega)|\approx1$ and $|r_0(\omega)|\approx1$ when the cavity side leakage $\kappa_s$ is negligible. By adjusting the frequencies $\omega$ and $\omega_c$ such that $\delta\omega_c/\kappa\rightarrow0$ and $C \gg 1$, the phase shifts may be realized as $\theta=0$ and $\theta_0=\pi$. Hence, the following optical transition may be obtained as

$$\begin{aligned}&|R\rangle|m^-\rangle \rightarrow -|R\rangle|m^-\rangle,\ |R\rangle|m^+\rangle \rightarrow |R\rangle|m^+\rangle,\\ &|L\rangle|m^-\rangle \rightarrow |L\rangle|m^-\rangle,\ |L\rangle|m^+\rangle \rightarrow -|L\rangle|m^+\rangle.\end{aligned} \tag{5}$$

From this optical transition, an NV center requires a polarization-degenerate cavity mode, which is also suitable in H1 photonic crystals[69,70] and fiber-based cavities[71].

## CNOT Gate on the same DoF of the two-Photon System

Schematic CNOT gate on the same DoF of the two-photon system is shown in Fig. 3. NV centers $e_i$ trapped in the photonic crystal $NV_i$ are initially prepared in the superposition states $|+\rangle_{e_i} := (|m^-\rangle + |m^+\rangle)/\sqrt{2}$. Two input photons $A_1$ and $A_2$ are in the states $|\phi\rangle_{A_i} = (\alpha_{i1}|R\rangle + \alpha_{i2}|L\rangle) \otimes (\beta_{i1}|l_i I_i\rangle + \beta_{i2}|l_i E_i\rangle + \beta_{i3}|r_i I_i\rangle + \beta_{i4}|r_i E_i\rangle)$, i=1, 2.Figure 3(a) is used to complete the CNOT gate on the polarization DoFs of two photons, i.e.,

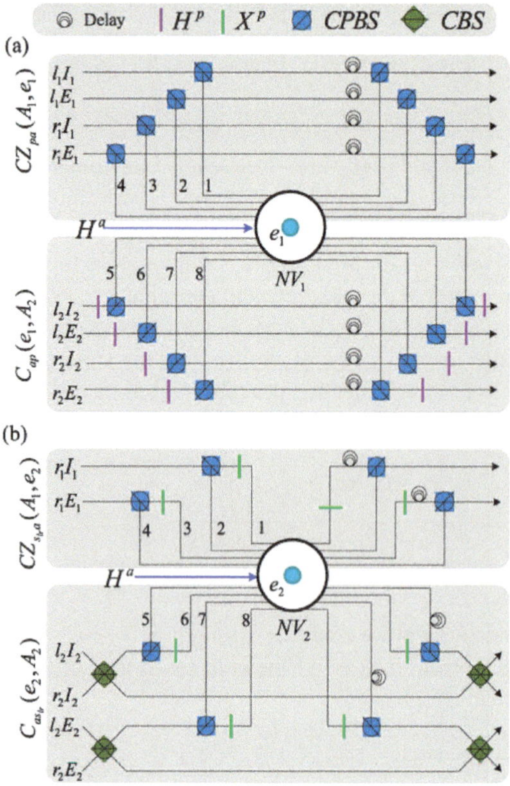


**Figure 3:** Schematic CNOT gate on the same DoF of two photons.

(**a**) Schematic CNOT gate on the polarization DoFs of two photons. CPBS represents a polarizing beam splitter in the circular basis, which transmits $|R\rangle$ and reflects $|L\rangle$. CBS represents a 50% 50 beam splitter to perform the Hadamard operation on the spatial DoF of a photon. X represents a waveplate to implement the bit-flip operation $X^p=|R\rangle\langle L|+|L\rangle\langle R|$. $H^a$ represents the Hadamard operation on the NV center in a cavity. (**b**) Schematic CNOT gate on the spatial DoFs of two photons. $H^p$ represents a half-wave plate (HWP) to perform the Hadamard operation on the polarization DoF of a photon. The numbers 1, 2, …, 8 denote the orders for an input pulse to interact with an NV center.

$$C_{pp}(A_1, A_2) = |R\rangle\langle R| \otimes (|R\rangle\langle R| + |L\rangle\langle L|) + |L\rangle\langle L| \otimes (|R\rangle\langle L| + |L\rangle\langle R|) \tag{6}$$

In detail, the photon $A_1$ from each spatial mode ($l_1I_1$, $l_1E_1$, $r_1I_1$ or $r_1E_1$) evolves as CPBS→$NV_1$→CPBS to complete the following controlled phase gate

$$CZ_{pa}(A_1, e_1) = |R\rangle\langle R| \otimes (|m^-\rangle\langle m^-| + |m^+\rangle\langle m^+|) + |L\rangle\langle L| \otimes (|m^-\rangle\langle m^-| - |m^+\rangle\langle m^+|) \tag{7}$$

on the polarization DoF and the NV center $e_1$ (see Appendix A ofSupplementary Information for details). And then, after one Hadamard operation $H^a$ on the NV center $e_1$ in the $NV_1$, the photon $A_2$ from each spatial mode evolves as $H^p$→CPBS→$NV_1$→CPBS→$H^p$ to complete the following hybrid CNOT gate

$$C_{ap}(e_1, A_2) = |m^-\rangle\langle m^-| \otimes (|R\rangle\langle R| + |L\rangle\langle L|) + |m^+\rangle\langle m^+| \otimes (|R\rangle\langle L| + |L\rangle\langle R|) \tag{8}$$

on the NV center $e_1$ and the polarization DoF of the photon $A_2$ (see Appendix A of Supplementary Information for details). Now, after disentangling the NV center $e_1$ using the measurement under the basis $\{|\pm\rangle := (|m^-\rangle \pm |m^+\rangle)/\sqrt{2}\}$, the phase flip $Z^p=|R\rangle\langle R|-|L\rangle\langle L|$ is performed on the photon $A_1$ from each mode for the measurement outcome $|-\rangle_{e1}$. Thus the CNOT gate $C_{pp}(A_1, A_2)$ has been realized on the photons $A_1$ and $A_2$.

Figure 3(b) is a schematic circuit to complete the CNOT gate on the spatial DoF {l, r}s of two photons, i.e.,

$$C_{ss_{l,r}}(A_1, A_2) = |l\rangle\langle l| \otimes (|l\rangle\langle l| + |r\rangle\langle r|) + |r\rangle\langle r| \otimes (|l\rangle\langle r| + |r\rangle\langle l|) \tag{9}$$

Here, the photon $A_1$ from each spatial mode $r_1I_1$ or $r_1E_1$ evolves as CPBS→$NV_2$→(X→$NV_2$→X)→CPBS to complete the following controlled phase gate

$$CZ_{s_{l,r}a}(A_1, e_2) = |l\rangle\langle l| \otimes (|l\rangle\langle l| + |r\rangle\langle r|) + |r\rangle\langle r| \otimes (|l\rangle\langle l| - |r\rangle\langle r|) \tag{10}$$

on the spatial DoF {l, r} and the NV center $e_2$ in the state $|+\rangle$ (see Appendix B of Supplementary Information for details). Now, after a Hadamard gate $H^a$ performed on the NV center $e_2$ in the $NV_2$, the followed circuit CBS→CPBS→$NV_2$→(X→$NV_2$→X)→CPBS→CBS for each mode pair ($l_2I_2$, $r_2I_2$) or ($l_2E_2$, $r_2E_2$) is used to complete the hybrid CNOT gate on the NV center $e_2$ and the spatial DoF {l, r} of the photon$A_2$ (see Supplementary Information for details), i.e.,

$$C_{as_{l,r}}(e_2, A_1) = |m^-\rangle\langle m^-| \otimes (|l\rangle\langle l| + |r\rangle\langle r|) + |m^+\rangle\langle m^+| \otimes (|l\rangle\langle r| + |r\rangle\langle l|) \tag{11}$$

Now, the CNOT gate $C_{ss_{l,r}}(A_1, A_2)$ may be realized by disentangling the NV center $e_2$ using the measurement under the basis $\{|\pm\rangle\}$, where $Z^p$ is performed on the photon $A_1$ from each spatial mode $r_1I_1$ and $r_1E_1$ for the measurement outcome $|-\rangle_{e2}$.

A similar CNOT gate

$$C_{ss_{I,E}}(A_1, A_2) = |I\rangle\langle I| \otimes (|I\rangle\langle I| + |E\rangle\langle E|) + |E\rangle\langle E| \otimes (|I\rangle\langle E| + |E\rangle\langle I|) \tag{12}$$

holds for the spatial DoF {I, E}s of two photons using an NV center $e_3$trapped in the optical cavity $NV_3$ (see Appendix C of Supplementary Information for details).

## Hybrid CNOT Gate on the different DoFs of the two-Photon System

Figure 4(a) is a schematic circuit to implement the CNOT gate on the polarization DoF of the photon $A_1$ and the spatial DoF {l, r} of the photon$A_2$, i.e.,

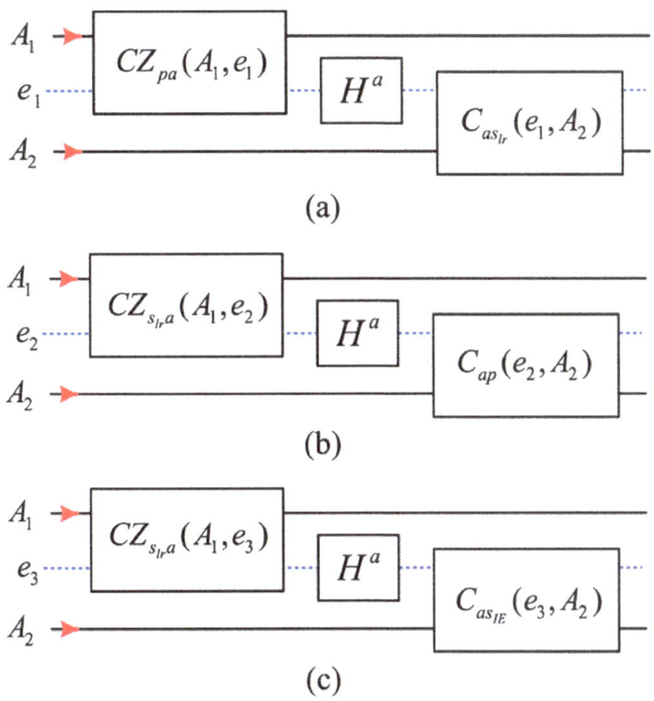


**Figure 4:** Schematic CNOT gate on different DoFs of the two-photon system.

(**a**) Schematic CNOT gate on the polarization DoF of one photon and the spatial DoF {l, r} of the other. (**b**) Schematic CNOT gate on the spatial DoF {l, r} of one photon and the polarization DoF of the other. (**c**) Schematic CNOT gate on the spatial DoF {l, r} of one photon and the spatial DoF {I, E} of the other photon. $e_i$ denote auxiliary NV centers in the NV-cavity $NV_i$, i = 1, 2, 3. The subcircuits $H^a$, $CZ_{pa}$, $C_{ap}$, $CZ_{sa}$ and $C_{as}$ are shown in the Fig. 3.

$$C_{ps_{l,r}}(A_1, A_2) = |R\rangle\langle R| \otimes (|l\rangle\langle l| + |r\rangle\langle r|) + |L\rangle\langle L| \otimes (|l\rangle\langle r| + |r\rangle\langle l|) \quad (13)$$

In fact, similar to the Fig. 3(a), the first controlled phase flip $CZ_{pa}(A_1, e_1)$ in the equation (7) is used to change the photon $A_1$ and the NV center $e_1$ from $|\phi\rangle_{A_1}|+\rangle_{e_1}$ to $CZ_{pa}(A_1, e_1)(|\phi\rangle_{A_1}|+\rangle_{e_1})$. And then, after one Hadamard operation $H^a$ performed on the NV center $e_1$ in the $NV_1$, the followed circuit CBS → CPBS → $NV_1$ → (X → $NV_1$ → X) → CPBS → CBS for each spatial mode pair ($l_2I_2$, $r_2I_2$) or ($l_2E_2$, $r_2E_2$) is used to complete the CNOT gate $C_{as}(e_2, A_1)$ in the equation (11) on the NV center $e_1$ and the spatial mode {l, r} of the photon $A_2$ (similar to the Fig. 3(b)). After disentangling the NV center $e_1$ using the measurement under the basis {|±⟩ }, the hybrid CNOT gate $C_{ps_{l,r}}(A_1, A_2)$ is realized on the photons $A_1$ and $A_2$, where $Z^p$ is performed on the photon $A_1$ from each spatial mode $r_1I_1$ and$r_1E_1$ for the measurement outcome $|-\rangle_{e1}$, see Appendix D ofSupplementary Information for details.

Similarly, after the controlled-phase flip $CZ_{pa}(A_1, e'_1)$ on the photon $A_1$and the NV center $e_{1'}$ in the state $|+\rangle$ , a schematic circuit is applied to the photon $A_2$ from two spatial mode pairs ($l_2I_2$, $l_2E_2$) and ($r_2I_2$, $r_2E_2$) to complete the CNOT gate on the NV center $e_{1'}$ and the spatial DoF {I, E} of the photon $A_2$ (see Appendix E of Supplementary Information for details). The hybrid CNOT gate

$$C_{ps_{I,E}}(A_1, A_2) = |R\rangle\langle R| \otimes (|I\rangle\langle I| + |E\rangle\langle E|) + |L\rangle\langle L| \otimes (|I\rangle\langle E| + |E\rangle\langle I|) \quad (14)$$

is implemented on the polarization DoF of the photon $A_1$ and the spatial DoF {I, E} of the photon $A_2$ after disentangling the NV center $e_{1'}$.

Figure 4(b) is used to implement the CNOT gate on the spatial DoF {l, r} of the photon $A_1$ and the polarization DoF of the photon $A_2$, i.e.,

$$C_{s_{l,r}p}(A_1, A_2) = |l\rangle\langle l| \otimes (|R\rangle\langle R| + |L\rangle\langle L|) + |r\rangle\langle r| \otimes (|R\rangle\langle L| + |L\rangle\langle R|) \quad (15)$$

In fact, similar to the evolutions as shown in the Fig. 3(b), the controlled phase gate $CZ_{s_{l,r}a}(A_1, e_2)$ in the equation (10) is performed on the photonA and the NV center $e_2$ in the state $|+\rangle$ to get $CS_{sa}(A_1, e_2)(|\phi\rangle_{A_1}|+\rangle_{e_2})$. And then, after one Hadamard operation $H^a$ on the NV center $e_2$ in the $NV_2$, the followed circuit for the photon $A_2$ from each spatial mode is used to complete the CNOT gate $C_{ap}(e_2, A_2)$ on the NV center $e_2$ and the polarization DoF of the photon $A_2$ (see the Fig. 3(a)). The final joint state is $C_{ap}(e_2, A_2)CS_{s_{l,r}a}(A_1, e_2)(|\phi\rangle_{A_1}|\phi\rangle_{A_2})$. Finally, by disentangling the NV center $e_2$ using the measurement under the basis {|±⟩ }, the hybrid CNOT gate $C_{s_{l,r}p}(A_1, A_2)$ is realized, where $-I^p$ will be performed on the photon $A_1$ from each spatial mode $r_1I_1$ and $r_1E_1$ for the measurement outcome $|-\rangle_{e2}$, see Appendix F of Supplementary Information for details. Moreover, if the second part of the present circuit above is applied to the photon $A_1$ from two spatial modes $l_1I_1$ and $l_1E_1$, the CNOT gate is implemented on the spatial DoF {I, E} of the photon $A_1$ and the polarization DoF of the photon $A_2$, see Appendix G of Supplementary Information for details.

Figure 4(c) is used to implement the CNOT gate on the spatial DoF {l, r} of the photon $A_1$ and the spatial DoF {I, E} of the photon $A_2$, i.e,

$$C_{s_{l,r}s_{I,E}}(A_1, A_2) = |l\rangle\langle l| \otimes (|I\rangle\langle I| + |E\rangle\langle E|) + |r\rangle\langle r| \otimes (|I\rangle\langle E| + |E\rangle\langle I|) \quad (16)$$

In detail, similar to the evolutions as shown in the Fig. 3(b), the controlled phase gate $CZ_{s_{l,r}a}$ in the equation (10) is performed for the photon A and the NV center $e_3$ in the state $|+\rangle$ to get $CZ_{s_{l,r}a}(A_1, e_3)(|\phi\rangle_{A_1}|+\rangle_{e_3})$. And then, after one Hadamard operation $H^a$ on the NV center $e_3$, the followed circuit for the photon $A_2$ from each spatial mode is used to realized the CNOT gate $C_{as_{I,E}}(e_3, A_2) = |m^-\rangle\langle m^-| \otimes (|I\rangle\langle I| + |E\rangle\langle E|) + |m^+\rangle\langle m^+| \otimes (|I\rangle\langle E| + |E\rangle\langle I|)$ on the NV center $e_3$ and the spatial DoF {I, E} of the photon $A_2$. The final joint state is $C_{as_{I,E}}(e_3, A_2) CZ_{s_{l,r}a}(A_1, e_3)(|\phi\rangle_{A_1}|+\rangle_{e_3}|\phi\rangle_{A_2})$. Now, by disentangling the NV center $e_3$ using the measurement under the basis $\{|\pm\rangle\}$, $C_{s_{l,r}s_{I,E}}(A_1, A_2)$ may be deterministically realized, where $-I^p$ will be performed on the photon $A_1$ from each spatial mode $l_1E_1$ and $r_1E_1$ for the measurement outcome $|-\rangle_{e3}$, see Appendix H of Supplementary Information for details. Similarly, the CNOT gate may be implemented on the spatial DoF {I, E} of the photon $A_1$ and the spatial DoF {l, r} of the photon $A_2$, see Appendix I of Supplementary Information for details.

## Hybrid CNOT Gate on different DoFs of one Photon

Figure 5 is a schematic circuit to implement the CNOT gate $C_{ps_{l,r}}$ in the equation (13) on the polarization DoF and the spatial DoF {l, r} of the photon $A_1$. In detail, similar to the Fig. 3(a), the controlled-phase flip$CZ_{pa}$ in the equation (7) is used to change the photon $A_1$ and the NV center $e_1$ from $|\phi\rangle_{A_1}|+\rangle_{e_1}$ to $CZ_{pa}(A_1, e_1)(|\phi\rangle_{A_1}|+\rangle_{e_1})$. And then, after one Hadamard operation $H^a$ performed on the NV center $e_1$, the followed CNOT gate $C_{asl}$ in the equation (11) is performed on the NV center $e_1$ and the spatial DoF {l, r} of the photon $A_1$ (similar to the Fig. 3(b)). After disentangling the NV center $e_1$ using the measurement under the basis $\{|\pm\rangle\}$, the hybrid CNOT gate $C_{ps_{l,r}}$ is realized on the photon $A_1$, where $Z^p$will be performed for the photon $A_1$ from each mode for the measurement outcome $|-\rangle_{e1}$, see Appendix J of Supplementary Information for details. Moreover, if the CNOT gate $C_{as_{IE}}(e_1, A_1)$ is performed on the NV center $e_1$ and the photon $A_1$ after $CZ_{pa}(A_1, e_1)$, the hybrid CNOT gate $C_{ps_{I,E}}$ in the equation (14) is realized on the photon $A_1$after properly disentangling the auxiliary NV center, see Appendix K ofSupplementary Information for details.

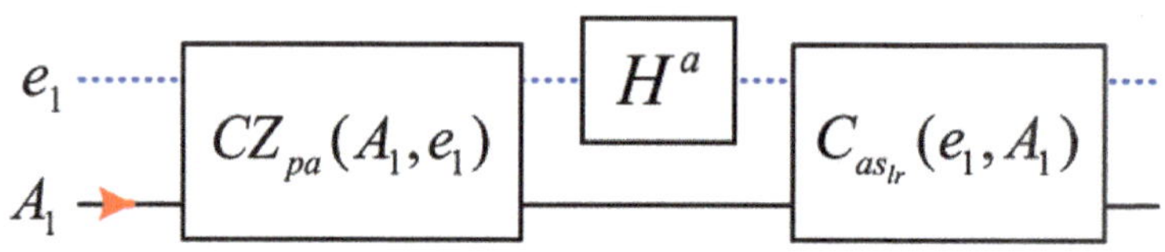


**Figure 5:** Hybrid CNOT gate on different DoFs of one photon.

$e_1$ denotes an auxiliary NV center in the NV-cavity $NV_1$. The subcircuits $CZ_{pa}$ and $C_{as}$ are shown in the Fig. 3.

For the hybrid CNOT gate on the spatial DoF {l, r} and the spatial DoF {I, E} of the photon $A_1$, the photon $A_1$ from the spatial modes $r_1I_1$ and $r_1E_1$passes through CBS, $-I$, CBS, sequentially. The photon $A_1$ evolves as follows

$$\begin{aligned}
|\phi_1\rangle_{A_1} \xrightarrow[\text{mode pairs } (r_1I_1,\, r_1E_1)]{CBS} & \alpha_0|R\rangle|l_1I_1\rangle + \alpha_1|R\rangle|l_1E_1\rangle + \alpha_4|L\rangle|l_1I_1\rangle + \alpha_5|L\rangle|l_1E_1\rangle \\
& + \alpha_2'|R\rangle|r_1I_1\rangle + \alpha_3'|R\rangle|r_1E_1\rangle + \alpha_6'|L\rangle|r_1I_1\rangle + \alpha_7'|L\rangle|r_1E_1\rangle \\
\xrightarrow[\text{mode } r_1E_1]{-I} & \alpha_0|R\rangle|l_1I_1\rangle + \alpha_1|R\rangle|l_1E_1\rangle + \alpha_4|L\rangle|l_1I_1\rangle + \alpha_5|L\rangle|l_1E_1\rangle \\
& + \alpha_2'|R\rangle|r_1I_1\rangle - \alpha_3'|R\rangle|r_1E_1\rangle + \alpha_6'|L\rangle|r_1I_1\rangle - \alpha_7'|L\rangle|r_1E_1\rangle \\
\xrightarrow[\text{mode pairs } (r_1I_1,\, r_1E_1)]{CBS} & \alpha_0|R\rangle|l_1I_1\rangle + \alpha_1|R\rangle|l_1E_1\rangle + \alpha_4|L\rangle|l_1I_1\rangle + \alpha_5|L\rangle|l_1E_1\rangle \\
& + \alpha_3|R\rangle|r_1I_1\rangle + \alpha_2|R\rangle|r_1E_1\rangle + \alpha_7|L\rangle|r_1I_1\rangle + \alpha_6|L\rangle|r_1E_1\rangle
\end{aligned} \quad (17)$$

where $\alpha_2' = (\alpha_2' + \alpha_3')/\sqrt{2}$, $\alpha_3' = (\alpha_2' - \alpha_3')/\sqrt{2}$, $\alpha_6' = (\alpha_6' + \alpha_7')/\sqrt{2}$, and $\alpha_7' = (\alpha_6' - \alpha_7')/\sqrt{2}$. Similar circuit may be used to realize the hybrid CNOT gate on the spatial DoF {I, E} and the spatial DoF {l, r} of the photon $A_1$, see Appendix K of Supplementary Information for details. Moreover, the CNOT gates on the spatial mode DoF and the polarization DoF of the photon $A_1$ are easily realized by two flip waveplates on two spatial modes $r_1I_1$ and $r_1E_1$, or $l_1E_1$ and $r_1E_1$, respectively.

## DISCUSSIONS

In ideal conditions, one may neglect the cavity side leakage, and the reflection coefficients satisfy $|r_0(\omega)| \approx 1$ and $|r(\omega)| \approx 1$. The corresponding fidelities of the present CNOT gates are nearly 100%. In experiment, the general fidelity is defined by $F = \int |\langle\Phi|\rho_f|\Phi\rangle|$, where $|\Phi\rangle$ is the ideal final state without side leakage while $\rho_f$ is the final state under a real situation with side leakage. In the resonant condition $\delta\omega_e = 0$, if the cavity side leakage is considered, the optical selection rule for the NV-cavity system given by the equation (5) becomes

$$\begin{aligned}
&|R\rangle|m^-\rangle \to -|r_0||R\rangle|m^-\rangle,\ |R\rangle|m^+\rangle \to |r||R\rangle|m^+\rangle, \\
&|L\rangle|m^-\rangle \to |r||L\rangle|m^-\rangle,\ |L\rangle|m^+\rangle \to -|r_0||L\rangle|m^+\rangle.
\end{aligned} \quad (18)$$

Due to the exchangeability of two spatial DoFs of one photon with respect to random initial photons, the fidelities and efficiencies are evaluated for four CNOT gates: CNOT gate on two polarization DoFs, CNOT gate on two spatial DoFs, CNOT gate on the polarization DoF of one photon and the spatial DoF of the other photon, and CNOT gate on the polarization and spatial DoFs of one photon system, as shown inFigs 6 and 7, respectively. Generally, large cooperativity C and low relative detuning $\delta\omega_c/\kappa$ are required for high fidelities and efficiencies. For the diamond NV centers, the photoluminescence is partially unpolarized, and the emission with ZPL is only 4% of the total emission. ZPL with zero phonon line is only 4% of $\gamma = 2\times 15$ MHz31. For the diamond NV center in a MTR with WGM mode system, $|r(\omega)| \approx 0.95$ when $C \geq 18$34 with small detuning $\delta\omega_c/\kappa \approx 0$; $|r(\omega)| \approx 1$ when $C \geq 50$ with small detuning $\delta\omega_c/\kappa \approx 0$ for $\kappa \approx 1$ GHz or $\kappa \approx 10$ GHz. For our CNOT gates, if $C \geq 18$ and $\delta\omega_c/\kappa \approx 0.1$, their fidelities are greater than 82.6% and their efficiencies are greater than 75.4%. If $C \geq 50$ and $\delta\omega_c/\kappa \approx 0.1$, their fidelities are greater than 98.4% and their efficiencies are greater than 94.7%.

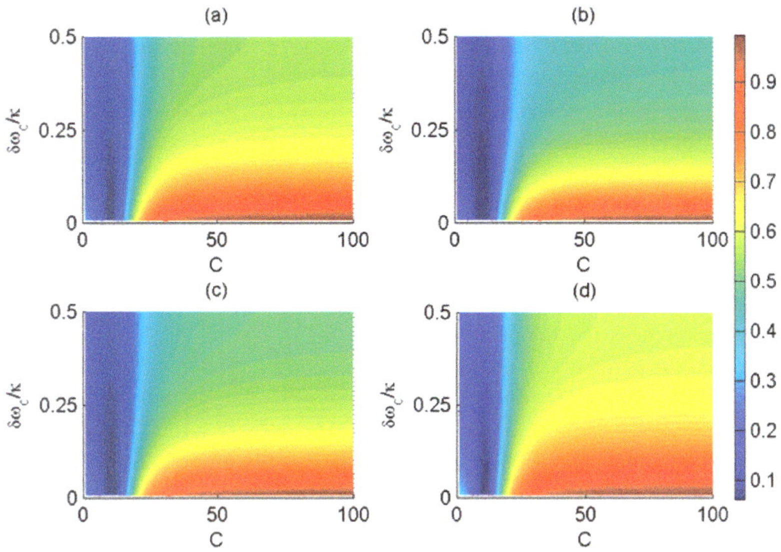


**Figure 6:** Average fidelities of the present CNOT gates vias the cooperativity C **and relative detuning $\delta\omega_c/\kappa$**.

(**a**) The average fidelity of the CNOT gate on the polarization DoFs of two photons; (**b**) The average fidelity of the CNOT gate on the spatial DoFs of two photons; (**c**) The average fidelity of the hybrid CNOT gate on the polarization and spatial DoFs of the two-photon system; (**d**) The average fidelity of the hybrid CNOT gate on the polarization and spatial DoFs of the one photon system. The average fidelity is computed as the expectation of random input photons.

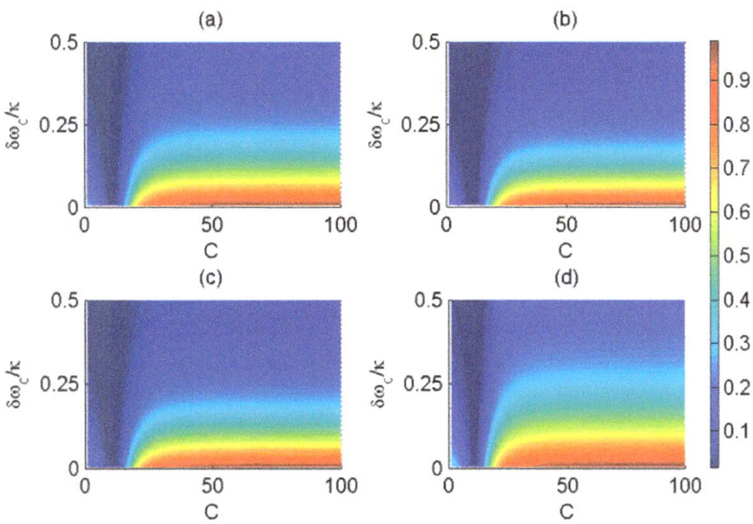


**Figure 7:** Average efficiencies of the present CNOT gates vias the cooperativity C **and relative detuning $\delta\omega_c/\kappa$**.

(**a**) The average fidelity of the CNOT gate on the polarization DoFs of two photons; (**b**) The average fidelity of the CNOT gate on the spatial DoFs of two photons; (**c**) The average fidelity of the hybrid CNOT gate on the polarization and spatial DoFs of the two-photon system; (**d**) The average fidelity of the hybrid CNOT gate on the polarization and spatial DoFs of the one photon system. The average fidelity is computed as the expectation of random input photons.

In conclusion, we have investigated the possibility of quantum simulations using photon systems with three DoFs. We have constructed fifteen schematic CNOT gates operating on the spatial and polarization DoFs of the two-photon system or one-photon system. Different from previous CNOT gate on the same DoF of the two-photon system[14,15,56,57], our schemes are based on different DoFs of two photons or one photon. Compared with hybrid implementations on the photon and stationary electron spins in quantum dots[58,59,60,61], the present CNOT circuits are ultimately realized on the photon system, and the electron spins in NV center are auxiliary resources to build the correlation between photons. The present schemes have shown that two different spatial DoFs may be viewed as independent qubits simultaneously, which has beyond previous independence of the polarization and spatial DoFs[62,63]. Although different DoFs may be easily exchanged in terms of encoding, the schematic operations are inconvenient for photon systems with two different spatial DoFs. The main reason is that the hybrid CNOT gates are not realized in one-shot manners. Thus, it is difficult to exchange these DoFs during applications, where different DoFs may be used as different encoding types such as the quantum Shor algorithm or the quantum search algorithm. Hence, our results are distinct from all previous quantum logic gates on different photons[14,15,56,57]. Our theoretical schemes have shown that three DoFs of photon systems may be independent in quantum information processing. Two thirds of the quantum resources may be saved in quantum simulations. With the recent experiments of the NV-cavity system[33,34,35], our schemes are expected to be applicable for the entanglement distribution or large-scale quantum computation.

## METHODS

### A diamond NV Center Coupled to an MTR with a WGM

The master equation of the whole system may be expressed by a Lindblad form as follows

$$\dot{\rho} = i\hbar[\rho, H] + \mathcal{H}_1\rho + \mathcal{H}_2\rho, \tag{19}$$

where $H=H_1+H_2+H_3+H_4$. $H_1 = \omega\hat{b}^\dagger\hat{b}$ is the Hamiltonian of an input photon pulse. $H_2 = g(\hat{a}\sigma_+ + \hat{a}^\dagger\sigma_-)$ is the standard Jaynes-Cummings Hamiltonian for a two-level system interacting with a single electromagnetic mode by applying the rotating wave approximation and dropping the energy nonconserving terms. $\sigma_-$ and $\sigma_+$ are the Pauli raising and lowering operators, respectively. g is the coupling strength between the cavity and $X^-$. $H_3 = \frac{\omega_c}{2}\sigma_z$ is the Hamiltonian of the dipole. $\sigma_z$ is the Pauli operator for the population inversion. $H_4 = \sqrt{\kappa}(\hat{a}^\dagger\hat{b}_{in} - \hat{a}\hat{b}_{in}^\dagger)$ is the interaction between the excitation field and system. $\mathcal{H}_1\rho = -\frac{\kappa+\kappa_s}{2}(\hat{b}^\dagger\hat{b}\rho + \rho\hat{b}^\dagger\hat{b} - 2\hat{b}\rho\hat{b}^\dagger)$ accounts for the damping of the input photon pulse. $\mathcal{H}_2\rho = \frac{\gamma}{2}(2\sigma_-\rho\sigma_+ - \sigma_+\sigma_-\rho - \rho\sigma_+\sigma_-)$ accounts for spontaneous emission of the dipole. The input-output optical relation of the NV center system may be calculated from the Heisenberg equations[67] in terms of the cavity field operator $\hat{a}$, input pulse field $\hat{b}$ and dipole operator $\sigma_-$,

$$\frac{d\hat{a}}{dt} = -\left(i\delta\omega_c + \frac{\kappa}{2} + \frac{\kappa_s}{2}\right)\hat{a} - g\hat{\sigma}_- - \sqrt{\kappa}\hat{b}_{in},$$
$$\frac{d\sigma_-}{dt} = -\left(i\delta\omega_e + \frac{\gamma}{2}\right)\hat{\sigma}_- - g\sigma_z\hat{a} + \sqrt{\gamma}\hat{b}_{in}, \tag{20}$$

If spins stay in the ground states most of the time [ $\langle\sigma_-\rangle = -1$], the cavity output $\hat{b}_{out}$ is connected with the input field by the standard input-output relation by a reflection coefficient r(ω).

### Measurement of the NV center e in Cavity

To measure the NV center e of an entangled system $\alpha|m^-\rangle_e|\Omega_1\rangle + \beta|m^+\rangle_e|\Omega_2\rangle$, an auxiliary photon c in the state $\frac{1}{\sqrt{2}}(|R\rangle + |L\rangle)$ may be used as follows. Let the photon c pass through one CPBS to split the circular polarizations $|R\rangle$ and $|L\rangle$, and the right-circular polarization $|R\rangle$ interact with the cavity system, and its output combine with $|L\rangle$ of the photon c using the other CPBS. Thus, the joint system evolves

$$\frac{\alpha}{\sqrt{2}}(|R\rangle + |L\rangle)_c|m^-\rangle_e|\Omega_1\rangle + \frac{\beta}{\sqrt{2}}(|R\rangle + |L\rangle)_c|m^+\rangle_e|\Omega_2\rangle$$
$$\mapsto \frac{\alpha}{\sqrt{2}}(|R\rangle - |L\rangle)_c|m^-\rangle_e|\Omega_1\rangle + \frac{\beta}{\sqrt{2}}(|R\rangle + |L\rangle)_c|m^+\rangle_e|\Omega_2\rangle. \tag{21}$$

Hence, the NV center e can be determined by measuring the photon in the orthogonal basis $\left\{\frac{1}{\sqrt{2}}(|R\rangle \pm |L\rangle)\right\}$. The NV center is $|m^-\rangle$ or $|m^+\rangle$ for the measurement outcome $\frac{1}{\sqrt{2}}(|R\rangle + |L\rangle)$ or $\frac{1}{\sqrt{2}}(|R\rangle - |L\rangle)$, respectively.

## REFERENCES

1. Nielsen, M. A. & Chuang, I. L. (ed.) [Quantum Computation and Quantum Information] [216–271] (Cambridge University Press, Cambridge, 2000).
2. Shor, P. W. Polynomial-time algorithms for prime factorization and discrete logarithms on a quantum computer. SIAM J. Comput. 26, 1484–1509 (1997).
3. Li, J., Peng, X., Du, J. F. & Suter, D. An efficient exact quantum algorithm for the integer square-free decomposition problem. Sci. Rep. 2, 260 (2012).
4. Grover, L. K. Quantum mechanics helps in searching for a needle in a haystack. Phys. Rev. Lett. 79, 325–328 (1997).
5. Farhi, E. et al. A quantum adiabatic evolution algorithm applied to random instances of an NP-Complete problem. Science 292, 472–475 (2001).
6. Li, B., Yu, Z.-H. & Fei, S.-M. Geometry of quantum computation with qutrits. Sci. Rep. 3, 2594 (2013).

7. Guo, P., Wang, J., Geng, X. H., Kim, C. S. & Kim, J. U. A variable threshold-value authentication architecture for wireless mesh networks. J Internet Tech. 15, 929–936 (2014).
8. Fu, Z., Sun, X., Liu, Q., Zhou, L. & Shu, J. Achieving efficient cloud search services: multi-keyword ranked search over encrypted cloud data supporting parallel computing. IEICE Trans. Commun. 98, 190–200 (2015).
9. Li, J., Li, X., Yang, B. & Sun, X. Segmentation-based image copy-move forgery detection scheme. IEEE Trans. Inf. Forensics Security 10, 507–518 (2015).
10. Luo, M.-X., Chen, X.-B., Yang Y.-X. & Wang, X. Geometry of quantum computation with qudits. Sci. Rep. 4, 4044 (2014).
11. Deutsch, D. Quantum computational networks. Proc. R. Soc. Lond. A 425, 73–90 (1989).
12. Barenco, A. et al. Elementary gates for quantum computation. Phys. Rev. A 52, 3457–4467 (1995).
13. Sleator, T. & Weinfurter, H. Realizable universal quantum logic Gates. Phys. Rev. Lett. 74, 4087–4090 (1995).
14. Shende, V., Bullock, S. S. & Markov, I. L. Synthesis of quantum-logic circuits. IEEE Tran. Comput. AID Design 26, 1000–1010 (2006).
15. Knill, E., Laflamme, R. & Milburn, G. J. A scheme for efficient quantum computation with linear optics. Nature 409, 46–52 (2001).
16. Nemoto, K. & Munro, W. J. Nearly deterministic linear optical controlled-NOT gate. Phys. Rev. Lett. 93, 250502 (2004).
17. Cirac, J. I. & Zoller, P. Quantum computations with cold Trapped Ions. Phys. Rev. Lett. 74, 4091 (1995).
18. Riebe, M. et al. Process tomography of Ion Trap quantum gates. Phys. Rev. Lett. 97, 220407 (2006).
19. Monroe, C., Meekhof, D. M., King, B. E., Itano, W. M. & Wineland, D. J. Demonstration of a fundamental quantum logic gate, Phys. Rev. Lett. 75, 4714 (1995).
20. Isenhower, L. et al. Demonstration of a neutral atom controlled-NOT quantum gate. Phys. Rev. Lett. 104, 010503 (2010).
21. Vandersypen, L. M. et al. Experimental realization of Shor's quantum factoring algorithm using nuclear magnetic resonance. Nature 414, 883–887 (2001).
22. Jones, J. A., Vedral, V., Ekert, A. & Castagnoli, G. Geometric quantum computation using nuclear magnetic resonance. Nature 403, 869–871 (2000).
23. Li, X. et al. An all-optical quantum gate in a semiconductor quantum dot. Science 301, 809–811 (2003).
24. Romero, G., Ballester, D., Wang, Y. M., Scarani, V. & Solano, E. Ultrafast quantum gates in circuit QED. Phys. Rev. Lett. 108, 120501 (2012).
25. Stojanović, V. M., Fedorov, A., Wallraff, A. & Bruder, C. Quantum-control approach to realizing a Toffoli gate in circuit QED. Phys. Rev. B 85, 054504 (2012).
26. Yamamoto, T., Pashkin, Y. A., Astafiev, O., Nakamura, Y. & Tsai, J. S. Demonstration of conditional gate operation using superconducting charge qubits. Nature 425, 941–944 (2003).
27. Balasubramanian, G. et al. Ultralong spin coherence time in isotopically engineered diamond. Nature Mater. 8, 383–387 (2009).
28. Charnock, F. T. & Kennedy, T. A. Combined optical and microwave approach for performing quantum spin operations on the nitrogen-vacancy center in diamond. Phys. Rev. B 64, 041201 (2001).
29. Jelezko, F., Gaebel, T., Popa, I., Gruber, A. & Wrachtrup, J. Observation of coherent oscillations in a single electron spin. Phys. Rev. Lett. 92, 076401 (2004).
30. Epstein, R. J., Mendoza, F. M., Kato, Y. K. & Awschalom, D. D. Anisotropic interactions of a single spin and dark-spin spectroscopy in diamond. Nature Phys. 1, 94–98 (2005).
31. Gaebel, T. et al. Room-temperature coherent coupling of single spins in diamond. Nature Phys. 2, 408–413 (2006).
32. Hanson, R., Mendoza, F. M., Epstein, R. J. & Awschalom, D. D. Polarization and readout of coupled single spins in diamond. Phys. Rev. Lett. 97, 087601 (2006).
33. Fuchs, G. D., Dobrovitski, V. V., Toyli, D. M., Heremans, F. J. & Awschalom, D. D. Gigahertz dynamics of a strongly driven single quantum spin. Science 326, 1520–1522 (2009).
34. Togan, E. et al. Quantum entanglement between an optical photon and a solid-state spin qubit. Nature 466, 730–735 (2010).
35. Neumann, P. et al. Quantum register based on coupled electron spins in a room-temperature solid. Nature Phys. 6, 249–253 (2010).
36. Zheng, A. S., Li, J. H., Yu, R., Lü, X. Y. & Wu, Y. Generation of Greenberger-Horne-Zeilinger state of distant diamond nitrogenvacancy centers via nanocavity input-output process. Opt. Express 20, 16902–16912 (2012).
37. Ren, B. C. & Deng, F. G. Hyperentanglement purification and concentration assisted by diamond NV centers inside photonic crystal cavities. Laser Phys. Lett. 10, 115201 (2013).
38. Jelezko, F. et al. Observation of coherent oscillation of a single nuclear spin and realization of a two-qubit conditional quantum gate. Phys. Rev. Lett. 93, 130501 (2004).
39. Yang, W. L., Yin, Z. Q., Xu, Z. Y., Feng, M. & Du, J. F. One-step implementation of multiqubit conditional phase gating with nitrogenvacancy centers coupled to a high-Q silica microsphere cavity. Appl. Phys. Lett. 96, 241113 (2010).
40. Yang, W. L., Xu, Z. Y., Feng, M. & Du, J. F. Entanglement of separate nitrogen-vacancy centers coupled to a whispering-gallery mode cavity. New J. Phys. 12, 113039 (2010).
41. Dayan, B. et al. A photon turnstile dynamically regulated by one atom. Science 319, 1062–1065 (2008).
42. Spillane, S. M. et al. Ultrahigh-Q toroidal microresonators for cavity quantum electrodynamics. Phys. Rev. A 71, 013817 (2005).

43. Louyer, Y., Meschede, D. & Rauschenbeutel, A. Tunable whispering-gallery-mode resonators for cavity quantum electrodynamics. Phys. Rev. A 72, 031801 (2005).

44. Vahala, K. J. Optical microcavities. Nature 424, 839–846 (2003).

45. Wei, H.-R. & Deng, F.-G. Compact quantum gates on electron-spin qubits assisted by diamond nitrogen-vacancy centers inside cavities. Phys. Rev. A 88, 042323 (2013).

46. Wang, T.-J. & Wang, C. Universal hybrid three-qubit quantum gates assisted by a nitrogen-vacancy center coupled with a whispering-gallery-mode microresonator. Phys. Rev. A 90, 052310 (2014).

47. Luo, M.-X., Li, H.-R. & Wang, X. Teleportation of a controlled-NOT gate for photon and electron-spin qubits assisted by the nitrogen-vacancy center. Quantum Infor. & Comput. 15, 1397–1419 (2015).

48. Luo, M.-X. & Wang, X. Universal remote quantum computation assisted by the Cavity input-output process. Proc. R. Soc. Lond. A 471, 20150274 (2015).

49. Park, Y.-S., Cook, A. K. & Wang, H. Cavity QED with diamond nanocrystals and silica microspheres. Nano Lett. 6, 2075–2079 (2006).

50. Schietinger, S., Schröder, T. & Benson, O. One-by-One coupling of single defect centers in nanodiamonds to high-Q modes of an optical microresonator. Nano Lett. 8, 3911–3915 (2008).

51. Barbour, R. J., Dinyari, K. N. & Wang, H. A composite microcavity of diamond nanopillar and deformed silica microsphere with enhanced evanescent decay length. Opt. Express 18, 18968–18974 (2010).

52. Barclay, P. E., Fu, K. M. C., Santori, C. & Beausoleil, R. G. Chip-based microcavities coupled to nitrogen-vacancy centers in single crystal diamond. Appl. Phys. Lett. 95, 191115 (2009).

53. Fu, K.-M., Barclay, P., Santori, C., Faraon, A. & Beausoleil, R. Low temperature tapered-fiber probing of diamond NV ensembles coupled to GaP microcavities. New J. Phys. 13, 055023 (2011).

54. McCutcheon, M. W. & Loncar, M. Design of a silicon nitride photonic crystal nanocavity with a Quality factor of one million for coupling to a diamond nanocrystal. Opt. Express 16, 19136–19145 (2008).

55. Duan, L.-M. & Kimble, H. J. Scalable photonic quantum computation through cavity-assisted interactions. Phys. Rev. Lett. 92, 127902 (2004).

56. Ren, B. C. & Deng, F. G. Hyper-parallel photonic quantum computation with coupled quantum dots. Sci. Rep. 4, 4623 (2014).

57. Hua, M., Tao, M. J. & Deng, F. G. Universal quantum gates on microwave photons assisted by circuit quantum electrodynamics. Phys. Rev. A 90, 012328 (2014).

58. Hu, C. Y., Young, A., O'Brien, J. L., Munro, W. J. & Rarity, J. G. Giant optical Faraday rotation induced by a single-electron spin in a quantum dot: applications to entangling remote spins via a single photon. Phys. Rev. B 78, 085307 (2008).

59. Bonato, C. et al. CNOT and Bell-state analysis in the weak-coupling cavity QED regime. Phys. Rev. Lett. 104, 160503 (2010).

60. Reiserer, A., Kalb, N., Rempe, G. & Ritter, S. A quantum gate between a flying optical photon and a single trapped atom. Nature 508, 237–240 (2014).

61. Wei, H. R. & Long, G. L. Hybrid quantum gates between flying photon and diamond nitrogen-vacancy centers assisted by optical microcavities. Sci. Rep. 5, 12918 (2015).

62. Luo, M.-X. & Wang, X. Parallel photonic quantum computation assisted by quantum dots in one-side optical microcavities. Sci. Rep. 4, 5732 (2014).

63. Luo, M.-X., Deng, Y., Li, H.-R. & Ma, S.-Y. Photonic ququart logic assisted by the cavity-QED system. Sci. Rep. 5, 13255 (2015).

64. Vallone, G., Ceccarelli, R., De Martini, F. & Mataloni, P. Hyperentanglement of two photons in three degrees of freedom. Phys. Rev. A 79, 030301 (2009).

65. Vallone, G., Donati, G., Ceccarelli, R. & Mataloni, P. Six-qubit two-photon hyperentangled cluster states: Characterization and application to quantum computation. Phys. Rev. A 81, 052301 (2010).

66. Manson, N. B., Harrison, J. P. & Sellars, M. J. Nitrogen-vacancy center in diamond: Model of the electronic structure and associated dynamics. Phys. Rev. B 74, 104303 (2006).

67. Walls, D. F. & Milburn, G. J. (ed.) [Quantum Optics] [250–350] (Springer-Verlag, Berlin, 1994).

68. Kimble, H. J. Strong interactions of single atoms and photons in cavity QED. Phys. Scr. 76, 127 (1998).

69. Chen, Q., Yang, W. L., Feng, M. & Du, J. F. Entangling separate nitrogen-vacancy centers in a scalable fashion via coupling to microtoroidal resonators. Phys. Rev. A 83, 054305 (2011).

70. Xiao, Y. F. et al. Quantum electrodynamics in a whispering-gallery microcavity coated with a polymer nanolayer. Phys. Rev. A 81, 053807 (2010).

71. Luxmoore, I. J. et al. Restoring mode degeneracy in H1 photonic crystal cavities by uniaxial strain tuning. Appl. Phys. Lett. 100, 121116 (2012).

# INDEX